Madalyn Aslan, author of *What's Your Sign?,* is the only astrologer whose readings have been auctioned at Christie's and who appears in *Who's Who 2004*. She has published numerous columns and predicts regularly on television and radio. She holds a BA from Cornell University, an MFA from Sarah Lawrence College, and a lectureship at the College of Psychic Studies. Madalyn's monthly horoscopes can be found on **www.MadalynAslan.com**.

MADALYN ASLAN'S
JUPITER SIGNS

How to Improve Your Luck,
Career, Health, Finances,
Appearance, and Relationships
Through the New Astrology

MADALYN ASLAN

A PLUME BOOK

PLUME
Published by the Penguin Group
Penguin Group (USA) Inc., 375 Hudson Street, New York, New York 10014, U.S.A.
Penguin Group (Canada), 10 Alcorn Avenue,
Toronto, Ontario, Canada M4V 3B2 (a division of Pearson Penguin Canada Inc.)
Penguin Books Ltd, 80 Strand, London WC2R 0RL, England
Penguin Ireland, 25 St Stephen's Green, Dublin 2, Ireland (a division of Penguin Books Ltd)
Penguin Group (Australia), 250 Camberwell Road,
Camberwell, Victoria 3124, Australia (a division of Pearson Australia Group Pty Ltd)
Penguin Books India Pvt Ltd, 11 Community Centre,
Panchsheel Park, New Delhi – 110 017, India
Penguin Books (NZ), Cnr Airborne and Rosedale Roads,
Albany, Auckland, New Zealand (a division of Pearson New Zealand Ltd)
Penguin Books (South Africa) (Pty) Ltd, 24 Sturdee Avenue,
Rosebank, Johannesburg 2196, South Africa

Penguin Books Ltd, Registered Offices: 80 Strand, London WC2R 0RL, England

Published by Plume, a member of Penguin Group (USA) Inc.
Previously published in a Viking Studio edition.

First Plume Printing, October, 2004
10 9 8 7 6 5 4 3 2 1

 REGISTERED TRADEMARK—MARCA REGISTRADA

The Library of Congress has catalogued the Viking Studio edition as follows:

Aslan, Madalyn.
Madalyn Aslan's Jupiter signs : how to improve your luck, career, health, finances, appearance,
and relationships through the new astrology / Madalyn Aslan.
p.   cm.
ISBN 0-670-03149-6 (hc.)
ISBN 0-452-28590-9 (pbk.)
1. Astrology. 2. Jupiter (Planet)—Miscellanea. I. Title.

BF1724.2.J87A85 2003
133.5'36—dc21
2003047961

Printed in the United States of America
Original hardcover design by Jaye Zimet

For Norman and Beverly Mirman,
who have given true Jupiter blessings of love, learning,
and hope to thousands of children

CONTENTS

Jupiter Table IX

INTRODUCTION: How I Know Jupiter Works XI

ARIES JUPITER: The Desire to Soar I

TAURUS JUPITER: The Long Way Impels Us 28

GEMINI JUPITER: Spreading the Word 56

CANCER JUPITER: The Harmonious Human Multitude 83

LEO JUPITER: If You've Got It, Flaunt It! 110

VIRGO JUPITER: Do the Right Thing 139

LIBRA JUPITER: Shining Bright 167

SCORPIO JUPITER: Dig Within 193

SAGITTARIUS JUPITER: Believe That You Can 220

CAPRICORN JUPITER: Going That Extra Mile 247

AQUARIUS JUPITER: The Birth of Cool 276

PISCES JUPITER: We Reach Divinity 306

ACKNOWLEDGMENTS 334

Where Is Your Jupiter?

To use this table, simply locate the year of your birth. You will see at a glance exactly what signs Jupiter was in during that year.

♃♃♃♃♃♃♃♃♃♃♃♃♃♃♃♃♃♃♃♃♃♃♃♃♃♃♃♃♃♃♃♃♃♃♃♃♃

1905	January 1 to March 6	ARI
	March 7 to July 19	TAU
	July 20 to December 3	GEM
	December 4 to December 31	TAU
1906	January 1 to March 8	TAU
	March 9 to July 29	GEM
	July 30 to December 31	CAN
1907	January 1 to August 17	CAN
	August 18 to December 31	LEO
1908	January 1 to September 11	LEO
	September 12 to December 31	VIR
1909	January 1 to October 10	VIR
	October 11 to December 31	LIB
1910	January 1 to November 10	LIB
	November 11 to December 31	SCO
1911	January 1 to December 9	SCO
	December 10 to December 31	SAG
1912	January 1 to December 31	SAG
1913	January 1	SAG
	January 2 to December 31	CAP
1914	January 1 to January 20	CAP
	January 21 to December 31	AQU
1915	January 1 to February 2	AQU
	February 3 to December 31	PIS
1916	January 1 to February 11	PIS
	February 12 to June 24	ARI
	June 25 to October 25	TAU
	October 26 to December 31	ARI
1917	January 1 to February 11	ARI
	February 12 to June 28	TAU
	June 29 to December 31	GEM
1918	January 1 to July 12	GEM
	July 13 to December 31	CAN
1919	January 1 to August 1	CAN
	August 2 to December 31	LEO
1920	January 1 to August 26	LEO
	August 27 to December 31	VIR
1921	January 1 to September 24	VIR
	September 25 to December 31	LIB

1922	January 1 to October 25	LIB
	October 26 to December 31	SCO
1923	January 1 to November 23	SCO
	November 24 to December 31	SAG
1924	January 1 to December 17	SAG
	December 18 to December 31	CAP
1925	January 1 to December 31	CAP
1926	January 1 to January 4	CAP
	January 5 to December 31	AQU
1927	January 1 to January 17	AQU
	January 18 to June 5	PIS
	June 6 to December 31	ARI
1928	January 1 to June 2	ARI
	June 3 to December 31	TAU
1929	January 1 to June 11	TAU
	June 12 to December 31	GEM
1930	January 1 to June 25	GEM
	June 26 to December 31	CAN
1931	January 1 to July 16	CAN
	July 17 to December 31	LEO
1932	January 1 to August 10	LEO
	August 11 to December 31	VIR
1933	January 1 to September 9	VIR
	September 10 to December 31	LIB
1934	January 1 to October 9	LIB
	October 10 to December 31	SCO
1935	January 1 to November 7	SCO
	November 8 to December 31	SAG
1936	January 1 to December 1	SAG
	December 2 to December 31	CAP
1937	January 1 to December 18	CAP
	Decemebr 19 to December 31	AQU
1938	January 1 to May 13	AQU
	May 14 to July 28	PIS
	July 29 to December 28	AQU
	December 29 to December 31	PIS
1939	January 1 to May 10	PIS
	May 11 to October 28	ARI
	October 29 to December 19	PIS
	December 20 to December 31	ARI
1940	January 1 to May 15	ARI
	May 16 to December 31	TAU
1941	January 1 to May 25	TAU
	May 26 to December 31	GEM
1942	January 1 to June 9	GEM
	June 10 to December 31	CAN
1943	January 1 to June 29	CAN

1943	June 30 to December 31	LEO
1944	January 1 to July 24	LEO
	July 25 to December 31	VIR
1945	January 1 to August 24	VIR
	August 25 to December 31	LIB
1946	January 1 to September 24	LIB
	September 25 to December 31	SCO
1947	January 1 to October 22	SCO
	October 23 to December 31	SAG
1948	January 1 to November 14	SAG
	November 15 to December 31	CAP
1949	January 1 to April 11	CAP
	April 12 to June 26	AQU
	June 27 to November 29	CAP
	November 30 to December 31	AQU
1950	January 1 to April 14	AQU
	April 15 to September 13	PIS
	September 14 to November 30	AQU
	December 1 to December 31	PIS
1951	January 1 to April 20	PIS
	April 21 to December 31	ARI
1952	January 1 to April 27	ARI
	April 28 to December 31	TAU
1953	January 1 to May 8	TAU
	May 9 to December 31	GEM
1954	January 1 to May 22	GEM
	May 23 to December 31	CAN
1955	January 1 to June 11	CAN
	June 12 to November 15	LEO
	November 16 to December 31	VIR
1956	January 1 to January 16	VIR
	January 17 to July 6	LEO
	July 7 to December 11	VIR
	December 12 to December 31	LIB
1957	January 1 to February 18	LIB
	February 19 to August 5	VIR
	August 6 to December 31	LIB
1958	January 1 to January 12	LIB
	January 13 to March 19	SCO
	March 20 to September 6	LIB
	September 7 to December 31	SCO
1959	January 1 to February 9	SCO
	February 10 to April 23	SAG
	April 24 to October 4	SCO
	October 5 to December 31	SAG
1960	January 1 to February 28	SAG
	March 1 to June 8	CAP
	June 9 to October 24	SAG

1960 October 25 to December 31 — CAP	**1976** August 23 to October 15 — GEM	**1996** January 3 to December 31 — CAP
1961 January 1 to March 14 — CAP	October 16 to December 31 — TAU	**1997** January 1 to January 20 — CAP
March 15 to August 11 — AQU	**1977** January 1 to April 2 — TAU	January 21 to December 31 — AQU
August 12 to November 2 — CAP	April 3 to August 19 — GEM	**1998** January 1 to February 3 — AQU
November 3 to December 31 — AQU	August 20 to December 29 — CAN	February 4 to December 31 — PIS
1962 January 1 to March 24 — AQU	December 30 to December 31 — GEM	**1999** January 1 to February 11 — PIS
March 25 to December 31 — PIS	**1978** January 1 to April 10 — GEM	February 12 to June 27 — ARI
1963 January 1 to April 2 — PIS	April 11 to September 4 — CAN	June 28 to October 22 — TAU
April 3 to December 31 — ARI	September 5 to December 31 — LEO	October 23 to December 31 — ARI
1964 January 1 to April 11 — ARI	**1979** January 1 to February 27 — LEO	**2000** January 1 to February 13 — ARI
April 12 to December 31 — TAU	February 28 to April 19 — CAN	February 14 to June 29 — TAU
1965 January 1 to April 21 — TAU	April 20 to September 28 — LEO	June 30 to December 31 — GEM
April 22 to September 19 — GEM	September 29 to December 31 — VIR	**2001** January 1 to July 11 — GEM
September 20 to November 15 — CAN	**1980** January 1 to October 26 — VIR	July 12 to December 31 — CAN
November 16 to December 31 — GEM	October 27 to December 31 — LIB	**2002** January 1 to July 31 — CAN
1966 January 1 to May 4 — GEM	**1981** January 1 to November 25 — LIB	August 1 to December 31 — LEO
May 5 to September 26 — CAN	November 26 to December 31 — SCO	**2003** January 1 to August 26 — LEO
September 27 to December 31 — LEO	**1982** January 1 to December 24 — SCO	August 27 to December 31 — VIR
1967 January 1 to January 14 — LEO	December 25 to December 31 — SAG	**2004** January 1 to September 23 — VIR
January 15 to May 22 — CAN	**1983** January 1 to December 31 — SAG	September 24 to December 31 — LIB
May 23 to October 18 — LEO	**1984** January 1 to January 18 — SAG	**2005** January 1 to October 24 — LIB
October 19 to December 31 — VIR	January 19 to December 31 — CAP	October 25 to December 31 — SCO
1968 January 1 to February 25 — VIR	**1985** January 1 to February 5 — CAP	**2006** January 1 to November 22 — SCO
February 26 to June 14 — LEO	February 6 to December 31 — AQU	November 23 to December 31 — SAG
June 15 to November 14 — VIR	**1986** January 1 to February 19 — AQU	**2007** January 1 to December 17 — SAG
November 15 to December 31 — LIB	February 20 to December 31 — PIS	December 18 to December 31 — CAP
1969 January 1 to March 29 — LIB	**1987** January 1 to March 1 — PIS	**2008** January 1 to December 31 — CAP
March 30 to July 14 — VIR	March 2 to December 31 — ARI	**2009** January 1 to January 4 — CAP
July 15 to December 15 — LIB	**1988** January 1 to March 7 — ARI	January 5 to December 31 — AQU
December 16 to December 31 — SCO	March 8 to July 20 — TAU	**2010** January 1 to January 16 — AQU
1970 January 1 to April 29 — SCO	July 21 to November 29 — GEM	January 17 to June 5 — PIS
April 30 to August 14 — LIB	November 30 to December 31 — TAU	June 6 to September 7 — ARI
August 15 to December 31 — SCO	**1989** January 1 to March 9 — TAU	September 8 to December 31 — PIS
1971 January 1 to January 13 — SCO	March 10 to July 29 — GEM	**2011** January 1 to January 21 — PIS
January 14 to June 3 — SAG	July 30 to December 31 — CAN	January 22 to June 3 — ARI
June 4 to September 10 — SCO	**1990** January 1 to August 17 — CAN	June 4 to December 31 — TAU
September 11 to December 31 — SAG	August 18 to December 31 — LEO	**2012** January 1 to June 10 — TAU
1972 January 1 to February 5 — SAG	**1991** January 1 to September 11 — LEO	June 11 to December 31 — GEM
February 6 to July 23 — CAP	September 12 to December 31 — VIR	**2013** January 1 to June 24 — GEM
July 24 to September 24 — SAG	**1992** January 1 to October 9 — VIR	June 25 to December 31 — CAN
September 25 to December 31 — CAP	October 10 to December 31 — LIB	**2014** January 1 to July 15 — CAN
1973 January 1 to February 22 — CAP	**1993** January 1 to November 9 — LIB	July 16 to December 31 — LEO
February 23 to December 31 — AQU	November 10 to December 31 — SCO	**2015** January 1 to August 10 — LEO
1974 January 1 to March 7 — AQU	**1994** January 1 to December 8 — SCO	August 11 to December 31 — VIR
March 8 to December 31 — PIS	December 9 to December 31 — SAG	
1975 January 1 to March 17 — PIS	**1995** January 1 to December 31 — SAG	
March 18 to December 31 — ARI	**1996** January 1 to January 2 — SAG	
1976 January 1 to March 25 — ARI		
March 26 to August 22 — TAU		

How I Know Jupiter Works

Your greatest self. You have one. Everyone has one. It's there inside of each of us, keeping us going, even in the darkest of times. Sometimes we're so close to it that we can feel and taste it. That's living. That's the real thing. All you need to know is *how to get there.*

It is incredible how we trip over ourselves, getting in our own ways. Our destiny might be there, sitting right beside us, and we can't even see it. Research shows that the brain cannot function when under the effects of stress and worry. Panic alters our brain chemistry. In other words, we cannot think when we are anxious. So, examine where the alarms go off in your life. You might call them the asteroids in your world. Which of them might cost you your job, your soul mate, your happiness? Do you want to be lucky and happy? As the Good Witch Glinda said, look no farther than yourself. *Your power was with you all along.*

Want a miracle? You can change your life. You can make miracles happen . . . you just have to know where to look.

The furniture in your planetary house is being moved around every day. You are always changing, always making choices. You can't change past mistakes (although you may heal the damage they've caused), but as far as the future goes, the world is your oyster. That's why Jupiter signs are so important. Astrology, used correctly, isn't just about discovering your character and personality. It's also about making the right decisions, about change, about creating a life that uniquely satisfies you. *It's about participating in your own fate.* Putting your Jupiter sign into practice will help you find luck and success in life. I wrote this book so you can be amazing. No matter how hard the road has been thus far.

So what does it mean, technically, to be a certain Jupiter sign? It means to be born when Jupiter was in one of the zodiac's twelve constellations or signs—Aries, Taurus, Gemini, Cancer, Leo, Virgo, Libra, Scorpio, Sagittarius, Capricorn, Aquarius, or Pisces. There are twelve Jupiter signs, same as the Sun signs. Both work together to help determine your fate. A hot tip for lovers: *Lovers who have the same Jupiter sign can stay together forever!*

Sun signs (and love!) are a great place to start, but there's much more to astrology! You have *ten* different planets in your astrological chart. And they all interact with each other in countless different ways to affect who you are. For a reminder, here are the bodies (discovered so far) affecting us:

The Sun: *around which we revolve, gives us life, motivation, energy, heat, and light.*

Mercury: *delivers and translates thought, messages, speech, and movement.*

The Moon: *guides emotions, the tides, intuition, and subconscious needs.*

Venus: *idealizes our stay on Earth. Enhances love, desire, and beauty.*

Mars: *charges our energy, ambition, and sex drive. Connects us to our mates.*

Saturn: *teaches us the lessons we must learn, giving great wisdom and strength.*

Neptune: *creates our dreams, images, and psychic abilities.*

Pluto: *ignites transformations and drastic changes, including natural disasters on Earth.*

Uranus: *shocks us with revolutions, revelations, and physical electricity.*

Jupiter: *protects us, inspires us, and gives us lucky break after lucky break!*

The question is how are *you* affected? You probably know your Sun sign and how it affects your life force and impulses. You may be familiar with other Sun sign traits and recognize them in people you know—*Oh, I'm Leo, he's Pisces, she's Gemini, and so on.* You may know your different planetary placements, such as which sign your Moon is in. Your Moon and Ascendant (or Rising sign) are considered the most important positions after your Sun sign. But I say your Jupiter is the most important! Think about it: Jupiter gives off nine times the radiation it receives from the Sun. Following the traits of your Jupiter sign increases your luck *ninefold*.

The news from Jupiter is good. Success is yours for the asking. All you have to do is act like your *Jupiter* sign—following the instructions laid out in this book—and you will find luck, success, and magic drawn to you, the gifts of Jupiter's bounty. Your motivation comes from the Sun (your Sun sign), but how you succeed in bringing about success comes from Jupiter (your Jupiter sign). Sun is personality; Jupiter is potential. Sun is descriptive; Jupiter is prescriptive. Jupiter is all about inspiring action and results.

Take some time and ask yourself what has made you lucky in your life. What has made you unlucky? Look back over your life and recount some important events. When you act like your Jupiter sign, you can create miracles in your life. No matter what your Sun sign is, you have a Jupiter sign, which is, ultimately, more important for deciding your destiny. *Sun is character, Jupiter is destiny.*

Why is Jupiter so great and so powerful? From the macrocosm of the universe to the microcosm of your life, Jupiter's effect is awesome. Jupiter emits frequencies radiating more energy than even the *Sun's* frequencies. Jupiter is, in fact, its own sun—scientists believe Jupiter formed from the same mass of material *as* the Sun. Its mass is thirteen hundred times greater than that of our Earth. Midas-like, it expands everything it touches. It acts big and generously, too. Ancient Greeks and Romans called Jupiter the planet of benevolence. Modern-day astrologers call Jupiter the planet of beneficence. Jupiter is, above all else, *helpful*. It saves our lives every day. Scientists argue that Jupiter has protected Earth from mass destruction for *millennia*. Its gravitational field is so powerful that it physically pulls asteroids aimed right at us off their course and deflects them into outer space. It does the same energetically in your life. What might you consider to be destructive asteroids threatening your life? Which obstacles are in your path? Jupiter protects, saves, and makes you lucky.

We are made of molecules that were once stars. It is impossible to not be affected by such a powerful environment as this great universe in which we live. Put yourself into the picture. How can Jupiter's force field affect millions of miles, glacial asteroids, the Earth—*and not you at all?* Without the Sun, there would be no life. The Sun gives us life; Jupiter gives us lucky breaks. It's as simple, and gracious, as that.

What is *your* Jupiter sign in? When are you *most* lucky? Think about what in your life has made you lucky. Now look up your Jupiter sign in the Jupiter table and compare it to your Sun sign. Perhaps you're a modest, self-sacrificing Virgo, and your Jupiter sign is in royal, confident Leo. Ask yourself: Which sign's behavior has brought more luck into your life? If you ever really let yourself act like a Leo, what would happen? Feed its influence and see your luck expand. Watch your former bad luck, unfed, wither and die away.

How does your Jupiter sign make you lucky? One of my clients, Joanna, learned the hard way. She was a Libra and knew what that meant: the Scales, weighing everything, striving to make the perfect decision, wanting everyone around her to be in harmony—a lot of wishful thinking, but not much in the results department; a few early stabs of ambition in her life, to be eventually crushed by self-doubt and conflicting Libra dilemmas and urges.

She'd moved from Ohio to New York at the age of nineteen to be a model, a huge and very frightening step for her to take. When she came to see me, she was twenty-seven, a fashion accessories editor at an upscale magazine based in Manhattan, frustrated and discouraged to dreary distraction. With three helpings of Libra (Sun, Ascendant, and Moon), she suffered with a misplaced idealism that had caused her to experience several tormented years. She was very concerned about other people liking her, about which parties she got invited to, all the materially based hullabaloo of the early nineties. All this began to change when she discovered that her Jupiter, the planet of luck, was in her opposite sign of Aries. "How totally weird," she exclaimed when I first revealed this to her. "But exactly *what* does that *mean?* I'm a Libra . . . aren't I supposed to *act* like a Libra?"

Not exactly. But it took two more daunting years. By then Joanna was twenty-nine, had split up with her boyfriend, was sick of being broke, hanging on in the same sterile studio apartment, stuck in the same boring and humiliating job, which, in spite of the free samples of perfume and scarves, just wasn't "doing it for her."

Finally she came to trust and to try her Jupiter luck and to not be scared of it.

At first, her Jupiter sign flew in the face of all she thought about herself. She saw herself as cool, observant, with a funny, slightly sarcastic (but not mean) bent. She had friends who were going places but never seemed to want to take her with them. She felt she gave and gave, and got nothing in return. The last thing she wanted to be was aggressive and selfishly oblivious, like one of those "trendy fashion idiots." It also didn't help that her older sister was an Aries, with Jupiter in Aries, too, and the *last thing* Joanna wanted to be was like her sister. I pointed out that we often resist what is good for us because it reminds us of something we'd rather keep in the past!

We looked back on her life together, and I helped Joanna to see that every "lucky" thing that had ever happened to her was when she acted *not like a Libra but like an Aries!* Whenever she took a leap of faith—as Aries Rams do, with no Libra-like doubt or hesitation—she got lucky.

I pointed out to her that a good example was when she first moved to New York from Ohio. It had been great and stimulating for her, and she was ecstatic; she recalled feeling truly alive for the first time in her life.

In those early adrenaline days, whenever she was challenged, she got competitive. That was how she landed her first editorial position and left modeling in the dust, not caring what her "beautiful people"–friends thought of the move. Whenever she got excited and inspired by the fire of Aries, passing up the air (thought) of Libra, and went with her (Aries) leader side instead of her "nice" (Libra) follower side, *she got lucky and got what she wanted.* That in effect was how she snagged her first boyfriend. Whenever Joanna acted like an outstanding Aries, she succeeded like one.

You could say this type of behavior shift can work for a lot of people. Or that it's a particularly female challenge—not to sacrifice your strength in order to be liked. Or, what can you expect, she lived in New York City for eight years—of course she had to learn to be aggressive and assertive! All the above may have their truths.

However, it's different for everyone. For example, an impatient, outspoken Aries woman with her Jupiter in Scorpio would get lucky acting like a patient, keep-it-to-yourself Scorpio, not like an Aries. A self-sufficient Capricorn with her Jupiter in family-oriented Cancer will succeed better by working and living with others rather than by being a completely independent Capricorn. A play-it-by-the-rules Virgo with her Jupiter in buck-the-system Aquarius gets more successful being irreverent than appropriate!

What if you're a Taurus Sun, and you love nothing more than being a cute couch potato and eating chocolate (which Bulls are very partial to), or a Cancer Sun for whom staying home all weekend makes you all content as nothing else? But your Jupiter is in adventurous Sagittarius?

Your Sun sign's aversion to venturing boldly forth into the world won't make you lucky nor will it give you the greatest, most exciting life. For that, you need to express your fun Sagittarius lucky self. Take care of your Sun and Moon needs, by all means, but if you allow those parts of you to swamp the rest, the overall result will not be good.

Your motivation comes from the Sun (your Sun sign), but *how you succeed in bringing it about* comes by way of Jupiter (your Jupiter sign).

You'll still want what you want, but knowing and using your Jupiter sign is a much more effective way of guaranteeing that you'll get it. For every Jupiter sign, it is different. Behavior that is lucky for one Jupiter sign is unlucky for another.

In 92 percent of all cases, the Jupiter sign is completely different from the Sun sign. This is breaking news in astrology, giving us even greater freedom of choice than we ever thought possible. To act from an unfamiliar part of yourself that will bring you luck and that may even conflict with your known personality is like discovering a new land, one loaded with many treasures! It's your choice—your luck!

Within these pages you will find more than general advice on how to attract the luck and success of Jupiter; you will also find advice on how to create the greatest luck with lovers, coworkers, mates, children, family, and friends. You will find tips on your best look, your best dates and times, and the healthiest environment for you to live in. You can also look up the Jupiter signs of everyone you know! There are special sections on how to get on with the boss of any Jupiter sign, as well as your significant other, your child, and your parent of any Jupiter sign. You will find what you need to do to be successful, happy, healthy, and lucky. You will understand how opposite Sun signs and Jupiter signs can work together. You will learn to use both to your advantage instead of being torn or stifled by potential conflict. Your Jupiter sign could be close to your Sun sign, or completely different.

For some people, their Sun is in the same sign as their Jupiter. I refer to this as "double Jupiter," and a description of their traits and fates can be found in every Jupiter sign chapter.

By the simple act of taking this first terribly important step, you are already accepting your Jupiter energy and allowing Jupiter to help you reach your potential. Act out your Jupiter sign's traits, described in each chapter, *for a week,* and see what happens. When you read your horoscope, read the sign of your *Jupiter,* not your Sun. In other words, if you are Aquarius with Scorpio Jupiter, follow the advice for Scorpio, but not for Aquarius. Each Jupiter sign is different. Behavior that is lucky for one Jupiter sign is *unlucky* for another Jupiter sign. Put the Jupiter theory to the test in your life. Try it for a week! You will lose nothing and you will gain everything.

There are times when we must go down—it's part of the equation— but let's not *stay there.* Everything that goes up must come down—and vice versa!

If you follow the guidelines set out for your Jupiter sign, you will find luck magnetically drawn to you. You will find new opportunities and learn to take advantage of them. You will understand your personal path toward success and achieving your dreams.

It isn't magic; you will still have to work and plan and strive hard. But you will know one very major and highly useful shortcut—the way to harness the energy of Jupiter in the workplace, the home, in love, and in the wider world to make your dreams come true. Ready to take the Jupiter leap and make yourself lucky?

Because Jupiter famously expands everything it touches, it traditionally rules excess, so it is *possible* to go overboard with Jupiter's power—*but!* . . . Don't use this as an excuse not to use your Jupiter. It is hard to change yourself and easy to think of excuses for not changing. It is important to remind yourself how you can sabotage your own success without even realizing it. And since it is true that Jupiter rules excess, understand that too much of a good thing is not always beneficial. But let me ask you: Have you ever really had a problem with *too much happiness?*

I, like most people, have always been inspired by stories of strength and survival and wisdom, fascinated by what makes one person able to overcome and triumph and another not. I tell people I became a psychic and astrologer as a way to communicate. But really, more deeply, it was to find out for myself what it meant to have contact with the deepest, unconscious part of myself.

I'm going to share with you some inspirations that have helped me through life:

> *If you bring forth what is within you, what you bring forth will save you. If you do not bring forth what is within you, what you do not bring forth will destroy you.*—Jesus Christ, the Gospel of St. Thomas, discovered in 1945 in Nag Hamadi. The only record of a gospel written in Jesus' words.

> *Security is mostly a superstition. It does not exist in nature, nor do the children of man as a whole experience it. Avoiding danger is no safer in the long run than outright exposure. Life is either a daring adventure, or nothing.*—Helen Keller

> *Your playing small doesn't serve the world. There's nothing enlightened about shrinking so that other people won't feel insecure around you. And as we let our own light shine, we unconsciously give other people permission to do the same.*—Nelson Mandela recited these lines in his inauguration speech, May 1994,

after twenty-seven years in prison. He was quoting from Marianne Williamson's *A Return to Love.*

Jupiter power is the best-kept secret in astrology. Then again, maybe you're not really ready for a huge tide of good fortune. Maybe it's not the right time for you to get that promotion, to be swept off your feet, or to win the $33 million Lotto. Maybe you need to learn more, see more, before you close a particular chapter of your life. Maybe you could be missing something even better! But as long as you are informed and aware, and making conscious and kind decisions— you're doing the best you can.

Dig up that winning lottery ticket you were born with and bring it on in. As Dorothy's pretty witch said, "Look no farther than your-self." Claim your Jupiter power *now!*

♈♃

The Desire to Soar

From Horace, born in 65 B.C., who urged, "Cease to ask what the morrow will bring," to Kate Winslet, born in 1975, who exclaimed, "Life is short, and it is here to be lived!" Aries Jupiters have always been the trailblazers of the zodiac. They know, and so should you, that playing it safe is the most dangerous thing you can do.

You're only here for a short time, mate— get involved! Get it all over you!

—ARIES JUPITER RUSSELL CROWE

Let's say you're a comfort-oriented Cancer. You love puttering around, following your routine, and being devoted to your family—a vulnerable Crab in its nice, safe shell. But if your Jupiter is in fiery, shake-'em-up, risk-taking Aries, indulging your natural taste for the tried and true won't bring you either luck or success. Instead, you're going to have to do the opposite and go against nature, to take risks, to be first at everything you do.

It happened to the person who said, "One cannot consent to creep when one has the desire to soar."

Who said this? An astronaut? A president? A Fortune 500 leader?

Would you believe a blind, deaf, mute woman—born in rural Alabama at a time when voting, going to college, or earning equal pay for equal work was not even an option for women? (And there's yet more for you to do, Aries Jupiter! The latter is still legally denied to women: the ERA is yet to be passed by the U.S. Congress.)

Darkness was what Helen Keller knew. This and silence. A Cancer Sun, she craved familiarity and safety above all else. She risked losing this to be able to soar with her Aries Jupiter, and our world is a better place for it. Blind, deaf, and mute as she grew up, Helen's only connections to the outside world and the people within it were her sense of touch and psychic sensitivity.

Despite the overwhelming obstacles placed in her way, this courageous young woman discovered how to "speak" through touch. Eventually this world-famous Aries Jupiter—Helen Keller—became an activist for the blind, gave speeches on five continents, wrote a book that was translated into fifty languages, and received the highest honor possible for an American—the Presidential Medal of Freedom.

Helen Keller did not give in to her comfort-oriented Cancer Sun; instead, she chose to follow the guidance of her Jupiter sign and used her Aries fire to change lives. She decided to take risks and to walk the rough road of the innovator, and with Jupiter power behind her, she became one of the greatest pioneers of history. This is how powerful following your Jupiter sign can be: It lets you achieve more than you've ever thought possible.

Take a moment, look back long and deep over your life, recount as many events as you can, and ask yourself what has made you lucky or unlucky in your life. You'll notice that whenever you act like an Aries, *you create miracles in your life.* No matter what your Sun sign is, *you get lucky when you live out the Aries legacy gifted to you by the benevolent planet Jupiter.*

To take advantage of your luck, emphasize your . . .

Aries Jupiter Strengths	♈♃	*Be freedom loving, first, audacious, daring, inspired and inspiring, fearless, self-generating, optimistic, and self-promoting.*

But because Jupiter rules excess, watch out for . . .

Aries Jupiter Excesses	♈♃	*You can become an adrenaline junkie, a law unto yourself, or very un-Zen-like!*

BUT FIRST:

Understanding the Aries Energy and Influence

Aries starts on the first day of spring, March 21, the beginning of life in the Northern Hemisphere. Thus whenever you come first, you will tend to be lucky. Your sign is the first astrological sign, known as the "Baby of the Zodiac." Because of its location in the calendar and the zodiac, Aries energy is all about firsts and beginnings. To attract the power of Jupiter into your life, Aries Jupiter, you simply need to come first, to be at the head of the line, to be a pioneer.

Aries is symbolized by the seeds of life. Seeds need only water and a little sunlight to burst open and grow into living plants, trumpeting their newfound presence in the world with a riot of green. When you tap into your Aries energy, you must not be afraid to celebrate life. Instead, like a sprouting seedling or a newborn baby, unashamedly announce your presence in the world to all around you. This enthusiasm and desire applies to you, Aries Jupiter, whether you're nine or ninety!

In practical terms, this means that you get lucky when you break new ground and forge your own path. And you need to keep your belief in yourself strong throughout the long journey. Herein lies your exciting challenge and your success!

♈♃♈♃♈♃♈♃♈♃♈♃♈♃♈♃♈♃♈♃♈♃♈♃♈♃♈♃♈♃♈♃♈♃♈♃♈♃♈

The symbol for Aries is the Ram. Rams are fierce animals that charge, butt heads, and crash horns against one another to assert their authority—subtlety is not in their vocabulary. Several species of rams live near mountaintops, walking along cliffs, fearless and sure-footed; they climb up to great heights without hesitation. Rams are also highly sexual beasts that live by their impulses.

While you might not want to imitate the ram's behavior in your own everyday life, charging at *anyone* who challenges your authority or climbing up steep precipices, you should immediately ask yourself *when* some Aries-style self-assertion *might* do you some good. Aries Jupiter energy is about expressing yourself, not worrying what others think, and forgetting your fears. Can you use this power to conquer self-doubt and fear? Remember, too, that your Aries Ram can climb to great heights. Which challenging heights do you need to climb to, and in which areas of your life? Do you need to be more bold,

ARIES SYMBOL
The Ram

aggressive, and sexual? Acting like a ram will bring you Jupiter's energy and power as well as luck and success!

♈♃♈♃♈♃♈♃♈♃♈♃♈♃♈♃♈♃♈♃♈♃♈♃♈♃♈♃♈♃♈♃♈♃♈♃♈♃

ARIES PART OF BODY
The Head

The part of the body that Aries rules is the head. So, be *ahead* of everyone. Be the head of your company, class, and field. Plunge in headfirst! While Aries Suns need to protect their heads, Aries Jupiters, under the Planet of Protection, are, within certain obvious limits, more free to be as heady as they dare. (Don't get too carried away—head injuries are still responsible for more deaths than any other injury type.) So wear protective gear, like safety helmets and sport-specific clothing, and go climbing or skydiving—don't be afraid of heights or be reluctant to soar!

♈♃♈♃♈♃♈♃♈♃♈♃♈♃♈♃♈♃♈♃♈♃♈♃♈♃♈♃♈♃♈♃♈♃♈♃♈♃

ARIES COLOR
Red

Red, the boldest primary color, signifies fire, passion, and blood. It invites heat and literally makes one warmer. Bask in this warmth, Aries Jupiter, as you walk down your own red carpet—preferably wearing red as well. If you indulge a taste for red, crimson, or scarlet, you will have more energy and charisma, and you will find people are increasingly drawn to you. According to Feng Shui, a red room in your home invokes success. Add red towels, red sheets, red accents, and observe your new and fabulous life!

Aries Jupiter Strengths ♈♃	*And now your Aries Jupiter strengths:*

Be Freedom Loving

If you want to bring the power of Jupiter into your life, try following the advice of Aries Jupiter author and philosopher Ayn Rand. In her novel *Atlas Shrugged,* Rand's hero, John Galt, declares, "I swear—by my life and my love of it—that I will never live for the sake of another man, nor ask another man to live for mine." Galt's (and Rand's) philosophy sums up one element of Aries Jupiter: the love of personal liberty and freedom.

You may not talk this way or know an Aries Jupiter who does, but to access the strength and power of Jupiter, you must live as if your personal motto is "Live free or die." In your heart, you know if you're in an unhealthy relationship or living a self-defeating lifestyle. Are you cooped up in an office, working for someone else? Give yourself the luxury of being supremely interested in what you do. Are you living for the sake of another man (or woman), with no opportunity to be free? If you answer yes to either question, you are denying yourself the bounty of Jupiter. Your Jupiter energy will come when you reassess your life and break the chains that bind you to stale relationships and jobs, and give liberty and independence top priority.

Be First

If you are Aries Jupiter, being first matters, whether you're trying to win a race, be part of a pioneering movement, or starting a trend instead of just following one. The harder the competition the better—for you. In fact, you excel when you have to brave obstacles to achieve your "first."

W. E. B. Du Bois, born a Pisces on February 23, 1868, used his Aries Jupiter to do what many said was impossible. He became the first African American to earn a doctoral degree from Harvard University—at a time when African Americans were denied even basic schooling in most states. This leader, educator, historian, philosopher, sociologist, author, and poet was also one of the founders of the NAACP.

Other Aries Jupiter firsts include people who embrace freedom, who don't care about the conventional wisdom, and who become renowned pioneers. Aries Jupiter Nancy Pelosi became the first woman to head a political party when, in 2002, she was elected House minority leader. An inspiration to American women, she has urged girls to "know thy power . . . anything you want you can get, if you let your voice be heard"—advice that suits every Aries Jupiter.

Aries Jupiter Tatum O'Neal came in first by being the first child to win an Oscar—at age ten—for her role in *Paper Moon*. Many child stars are Aries Jupiter. Aries ambition is boundless, and you can pull off whatever you aspire to. What you come in first is up to you. It can be anything, but it must be first! The first woman in space, Valentina Tereshkova, was born Aries Jupiter. As was the first three-color traffic light in the U.S.A., born February 5, 1952, and the first nude beach in the U.S.A., born January 22, 1952!

Be Audacious

Aries Jupiter doesn't just reward you for the freedom to control your destiny and for having the guts to get to the finish first. Jupiter's lucky energy flows even more strongly when you dare to be different, when you come to claim your freedom from that most enslaving of all fears, *what other people think*—when you embrace the exciting state of being audacious, even shocking.

The key is, you must be audacious but remain true to yourself and what you, deep inside, really want to achieve. Shocking for the sake of it—fake audacity—won't help you. Just ask Aries Jupiter Dr. Ruth, the psychosexual therapist who began in 1980 with a fifteen-minute tape and ended up changing media psychology forever with her radio program *Sexually Speaking*. Her cheerful, squeaky voice relentlessly advising callers to masturbate more or to take a sex education class still has the power to astonish. Dr. Ruth—who believes free and open sexual communication is important for society—has changed sexuality in America, and become successful by breaking taboos and remaining true to her beliefs.

Television programs as well as people have Jupiter signs, and it should not surprise you that the most audacious show in the history of television is a fellow Aries Jupiter. "They've done it! They did the *zeitgeist*," Steve Martin raved when, in 1975, *Saturday Night Live* was born. Aired live, at a time when most shows were prerecorded with canned laughter, the Aries Jupiter Zeitgeist exemplified audacious, taboo-busting comedy. Launched on October 11, 1975, *SNL* did not quickly become the highest rated late-night show in America by being a nicey-nice Libra (its Sun sign), but instead by smashing American sacred cows in true Aries Jupiter style.

Be Daring

It isn't enough just to thwart the conventional wisdom and shock some of society's blue-hairs. To bring the bounty of Jupiter into your life, you must also be *daring*. It is important you take risks, dare to shine, and refuse to let negative people stop you from reaching your goals.

Pioneer Tiger Woods, born on December 30, 1975, in the cautious Sun sign of Capricorn, broke through the color barrier and forever changed the rules of golf. Few Capricorns would have been naturally daring enough to reach the top in a sport that was openly hostile to them because of their race, but Tiger braved insults, criticism, and even death threats to become, at fifteen, the youngest U.S. Junior Amateur Champion in golf history. He has gone on to win every major tournament, amass a huge fortune, and be regarded by many as the greatest golfer in history.

Another great Aries Jupiter risk taker is Jimmy Santiago Baca, a poor-born Hispanic who didn't learn to read or write until his twenties, while he was serving time in jail. Bucking every trend, refuting all the discouraging advice he received from his critics, Baca became a great writer and teacher, giving back to society by founding Black Mesa Enterprises for troubled teens.

Be Inspired and Inspiring

If you are an Aries Jupiter, and something inspires you, you must physically act on it. Red Cloud, born Sioux in May 1821, said, "I am poor and naked, but I am the chief of the nation." Say your truth loud and clear about whoever—and whatever—you are.

The point of Aries Jupiter is that when you act like your true self, it will be easier, not harder, to get along with people. Remember to work with people, communicate your ideas, and act like a leader (even if you don't feel like one—you will soon enough). Aries Jupiter will reward you if you use your new liberty and your devil-may-care attitude to inspire others, by leadership, by spirit, and by example. Like Dr. Ruth, Tiger Woods, and Jimmy Santiago Baca, who have together inspired millions, you can change lives.

Mary Baker Eddy defied the convention of her day when she founded the Church of Christian Science. She magnified her Jupiter power by working to inspire others and to convert them to her beliefs. In her ambitious Aries Jupiter way, Eddy did not stop at just trying to inspire people she actually met; she also founded a national newspaper (the *Christian Science Monitor*), and her followers inspired others. Her desire to inspire attracted the power and luck of Jupiter into her life.

Be Fearless

Aries Jupiter isn't just about daring and chutzpah: Another critical element of the Aries Jupiter energy is physical courage. A famous Aries Jupiter example is Clara Barton, the founder of the American Red Cross. Born December 25, 1821, in straitlaced New England, Clara saw the sufferings of those hurt on the battlefields of the Civil War and felt compelled to minister to the wounded. First at every scene following a bloody battle, she nursed the injured, with bullets often still zinging around her, earning her the title "Angel of the Battlefield." She single-handedly campaigned to found the American Red Cross at a time in history when it was considered scandalous for a woman to even give a speech in public. Let Clara Barton's long, heroic life—she died at ninety—inspire you to face your fears so you, too, can triumph.

Be Self-generating

The qualities of Aries Jupiter—free, audacious, daring, fearless, inspirational—should encourage you to be yourself regardless of what other people think. There is one aspect to this that you must pay particular attention to: If you are to bring Jupiter's lucky energy into your life, make sure that you know yourself well, inside and out. You will get lucky when you express the spirit and beliefs you hold within yourself or have worked hard to develop. You will actually lose Jupiter's lucky energy by posing, by being artificial, by embracing things *or* beliefs that you have not generated within yourself. The key is to be self-sufficient, to generate your own values and ideas, to trust yourself.

It all comes from you, Aries Jupiter! You *will* achieve your lucky destiny—as long as your power comes from within. Do not be afraid if life has given you little. Say it loud and clear and be true to yourself, embrace what's really within—and Jupiter's energy will flow to you.

Be Optimistic

On the good ship Lollipop . . .

Shirley Temple, born April 23, 1928, brought smiles and cheer to millions during the years of the Great Depression. Who can forget her dimples as she sang "On the good ship Lollipop"? She followed her Aries Jupiter qualities by embracing a charming innocence that the whole world craved (even today, a Shirley Temple is a drink "innocent" of alcohol). Following a childhood as America's Sweetheart and one of the most popular film stars in the world, Shirley Temple grew up, got married, became a respected civic leader, and eventually U.S. ambassador to Ghana.

To tap into your Aries Jupiter strength, you must be the one others will count on, the one with a cheery attitude who inspires and leads. You must cultivate this quality within yourself. Aries Jupiter energy is about winning against the odds, and—even if you have to refuse to be bound by other peoples' expectations—you must keep a positive attitude and feel sure that you are going to succeed. As Aries Jupiter optimist French president François Mitterrand said, "Nothing is won forever in human affairs, but everything is always possible."

Be Self-promoting

You've probably gathered by now that to tap into your Aries Jupiter energy you shouldn't be a shrinking violet. In fact, you need to work solidly on promoting yourself, because it will bring you luck!

Mary-Kate and Ashley Olsen, born Gemini twins on June 13, 1987, drew on their lucky Aries Jupiter to make themselves a fortune. The product? *Themselves.* Their look-alike Mary-Kate and Ashley dolls, videos, magazines, pocket video-game systems, and girls' sportswear line generated estimated sales of $1 billion in 2002. They began at nine months old, sharing the role of baby Michelle Tanner on ABC's *Full House* for $2,400 an episode. They became the first six-year-old producers in Hollywood history, and their company is aptly named for Gemini twins: Dualstar.

Notice that, even though Mary-Kate and Ashley are great at promoting themselves, they don't pretend to be something they are not. In real life, the sisters are gracious, smart, ordinary young women (the same adjectives have been used to describe Tiger Woods and Shirley Temple). Jupiter has rewarded their desire to stay true to themselves by generally refusing to do anything fake or artificial, to falsely promote themselves.

What Makes You Lucky in Love

If you're working on your Aries Jupiter energy, you may one day see that potential lovers are reacting to you very differently from the way they do now. Perhaps they sense that you're one of a kind, or maybe they start to notice your inner (and outer) glow of optimism, fearlessness, and charisma. You can help this process if you dress in your own original style, too. Remember, it's self-promotion, not fashion promotion, that makes you win.

Pick the most compelling person in the room and make a beeline in his or her direction. Your love luck will run out if you play passive or act indirectly. You need to be honest about how you feel . . . even if you'd rather not. For example, Aries Jupiters can become competitive and jealous when out of balance. This is easier to overcome with maturity, but always remember that you are you, and no one can compete with that.

You can expect dramatic highs and lows as well as unusually intense connections. If you are in touch with your Jupiter sign, you should be demonstrative; try also to express your sexuality in interesting and different settings. Long-term relationships, when relating becomes less "new," can on occasion become problematic and tricky for you.

One romantic hazard of following your intense Aries impulse is to fall in love too quickly. Surrendering to this impulse is not always pleasant, as breakups after hasty bonding can be temporarily devastating. But thanks to your Aries Jupiter, you always bounce back, and try, try again. Why? Because at heart, you are an extreme romantic and idealist. It can also, by the way, be bad luck for you to spend *every* waking hour with your new love interest. You need your solitude and time alone; remind yourself that your relationship will suffer if you don't get it. The first to popularize meditation in the West, Aries Jupiter Paramahansa Yogananda, wisely advised: "Be alone once in a while, and remain more in silence."

Your Best Career Moves

Aries Jupiter ignites ideas and puts them into action, so it's logical that boldness and individuality unlock your success potential. Competition brings luck in its wake, and the greater the challenge, the more psyched up you get. If you don't have a competitor, invent one! You also tend to work best in spurts and on deadlines. You will work even better if a project seems impossible—only an Aries Jupiter like you could pull it off!

Having your own schedule and the freedom to dress, speak, and act exactly as you want is very lucky for you. You're happiest and luckiest in a start-up company, a business of your own, or if you work freelance. You simply cannot do the same thing day after day with any success or satisfaction; you'd do much better working long into the night on something of your own creation than putting in a single hour on something that bores you. You must be passionately interested in your work, or you will go mad with boredom and lose your lucky Jupiter touch. The way to succeed in getting work you want is by presenting yourself in person—boldly, confidently, and assertively. Rarely will you succeed by following the typical send-a-résumé-and-wait-to-be-called format.

How to Reach Optimum Health

Aries Jupiters need to get outdoors and exercise vigorously. It's lucky for you to work up a sweat, and it's luckier for you to do your exercise outside, in a natural setting, rather than in the artificial atmosphere of a gym. So go running, take long hikes or climbs, or swim outside in nature if you can. You need variety and interest. The aerobic exercise you engage in will also make you drink more water—being fiery, you need more water than the average person.

You will love steam baths, saunas, and massage, and should indulge as often as possible. Follow these Jupiter guidelines and the giant planet's gift of success and luck will bless you with good health.

Your Lucky Foods, Element, and Spices

As a Jupiter Fire sign, heat is lucky for you. You'll feel better when you eat "living" foods rather than packaged or frozen; hot bubbling foods like soups and stews; also foods that are on the "male," or yang, end of the yin-to-yang food scale, like eggs, meat, and salty snacks. Eat fire-colored foods—red, orange, yellow, or gold—such as apples, oranges, pumpkin, cantaloupe, and beets! If you're a carnivore, eat your red meat often—and good luck being a vegan! Iron, your lucky metal, is an element you absolutely need more of in your diet, so be sure to eat your spinach and beans. Ginseng, cayenne pepper, and cinnamon are excellent additions to the Aries Jupiter diet, and are especially powerful for healing colds and zapping low energy. A lack of potassium phosphate, Aries' cell salt, can lead to depression, so eat plenty of tomatoes, olives, brussels sprouts, swordfish, and pumpkin to keep happy and to replenish your brain cells, too!

Your Best Environment

Because Aries Jupiter's heat theme applies externally as well as internally, you do best in a warm climate. If it's not warm outside, create the heat inside. Wherever you live, you will be happy and healthy if you use a fireplace or a wood stove, burn candles, and take plenty of hot baths. Ideally, your home should face south. The color red is crucial, heating is essential (Aries Jupiter, easily chilled, often suffers from cold hands or feet), and being able to see a mountain or hill nearby—or better yet, to live on one—satisfies the Ram's natural habitat needs.

Your Best Look

You will have the most success and luck when you dress like an Aries, in bold, striking reds, blacks, and whites that reveal the shape of your body. The ideal body type for Aries is strong and streamlined; it's lucky for you to show off your body regardless of your size. Confidently show your skin, particularly your shoulders and arms. For fun, you can go dramatic and theatrical with, for example, black velvet or gold sequins! Even if your Sun sign guide has told you pastel or neutrals are the way to go, ignore this if you're Aries Jupiter. Avoid fussiness and frills, as your luck requires you must be able to walk, move, and dance comfortably—to take center stage.

Your Best Times

The luckiest times for you are when the Moon is in Aries, which happens every month for two to three days. Find these special days in your astrological calendar. During these days, schedule crucial meetings, tell that special person you love them, ask for a promotion or raise, push the envelope, and plan to shine. In your everyday, the best time for you to exercise is very early morning or at the end of your workday. The luckiest times during the year for you are March 21–April 19 (when the Sun is in Aries); July 23–August 23 (when the Sun is in Leo); and November 22–December 21 (when the Sun is in Sagittarius).

The very luckiest time of your life—as in Lotto lucky—is when Jupiter reenters the sign it was in when you were born (Aries), which happens every twelve years and lasts anywhere from a few months to just over a year. This is called your Jupiter Return, and it launches your luckiest streak. You can count on the *extraordinary* to happen the very day your Jupiter Return begins. Amazing events can start happening up to six weeks beforehand, so plan ahead!

How to Be Lucky Financially

When they trust themselves, decide to be fearless and daring, inspire others, and lead with their optimism, Aries Jupiters make great entrepreneurs. Think of Aries Jupiters John Paul Getty, the oil baron, or Jeff Bezos, who founded amazon.com. So for you to attract the bounty of wealth Jupiter can bring, follow your own star, be true to yourself, and be daring and audacious. If you have a business idea, promote it, work to inspire others, and set up shop—you will be amazed at how Jupiter will bless you with luck and good fortune. You will have great success in selling your ideas so long as they are original and you do the footwork promoting yourself and your business.

Perhaps you'll find a house or apartment that's not in the most desired or nicest part of town, and then, years later, you'll find that you are at the center of the action and can sell your place for megabucks. Your intuition can guide you to good fortune—and listening to ideas generated deep inside brings the luck of Jupiter.

The payoffs will be big when you also bet on yourself. If you've always wanted to be an engineer or a trapeze artist—or take up any career that others consider unlikely—go for it! While such bets may be risky for other Jupiter signs, for you, bets on yourself are sure favorites to win.

Success in Family and Friendship

Compromise, it is said, is the specific ingredient that leads to success in any relationship. Unfortunately, compromise is sometimes an alien virtue to those channeling the power of Aries Jupiter.

While Jupiter will reward you for speaking up for what you believe in, sharing your feelings, and being bold and daring, family and friends may enjoy their time with you more if you try to be more tolerant and accepting of others. So, despite that impulsive Aries influence, you need to step back and *listen* in order to see the real wisdom in compromise. And if you're really brave (and you are if you use lucky Aries Jupiter), take a look at the "Relationships with Aries Jupiters" section. Then you'll see how your other halves live—with *you*.

When you go too far!

♈♃ *Aries Jupiter Excesses*

The myth of Icarus is a cautionary Aries Jupiter tale. Just like Icarus, who fashioned wings of feathers and, wax and then flew too close to the sun, you have to be on guard against burnout. You can go too far by taking on too much work, or too many different projects at once, or doing others' work for them because you think you could do the job better. Burnout also occurs when you're impatient, competitive, arrogant, and antagonistic to (or oblivious of) your coworkers. Sometimes the initial happiness and joy of following your Jupiter sign leads you to go to extremes—watch out, as this can hurt you.

As an Aries Jupiter, you also run the risk of overexerting and overexhausting yourself. Learn to take it easy! Take time out. Take deep breaths. Enroll in a yoga and meditation program when you find your life getting to be too much.

Excess #1 You Can Become an Adrenaline Junkie

It's in your nature to ignite a project rather than carry someone else's through, and you can handle pressure better than boredom or disappointment. That's fine so long as you don't use that adrenaline fix to avoid facing your feelings. If you don't take a break when you need one, it's a sure sign you are overdosing on Jupiter. While change refreshes you and allows you to bounce back, stronger than ever, constant change can harm.

Challenge and excitement are the highs that keep most Aries Jupiters going. But it's unrealistic to expect the highs to last forever. Many of you start new projects with confidence, get bored, and then abandon your grand plan. *If there's one thing that stops you from being superwoman or superman, it's that you don't finish things.* Sometimes this makes life more stressful, as it causes sudden deadlines to creep up on you. Be aware that you have a weakness for not finishing what you start, and for being an adrenaline junkie, and avoid this behavior pattern by refusing to give up. Carry on, stay the course, and chances are good more often than not that you will become successful.

Excess #2 You Can Become a Law unto Yourself

Many Aries Jupiters set out to break institutions but end up breaking themselves. If you allow your Jupiter qualities to build to excess, you can become so convinced of your beliefs, your rightness, and your self, that you can be unstoppable, and while this is great for a certain kind of career success, it can make you hell to live with, and it can sometimes lead to self-destruction.

John D. Rockefeller was a perfect example of Aries Jupiter power carried to an extreme. For many years he was universally hated and feared by the public as the most ruthless businessman in America, an avowed believer in monopolies, known for crushing the competition. Like almost all of the moguls of his day, Rockefeller was bitterly opposed to the establishment of labor unions. His most infamous "strategy," and certainly the low point of his career, was the Ludlow Massacre of 1914, in which strike-breaking tactics he authorized resulted in the deaths by fire of thirteen children and seven mothers. This was the typical Aries Jupiter excess of self-confidence and trust in *one's* own beliefs taken to the level of arrogance and inhumanity. Following an intense public outcry, Rockefeller used public relations techniques and his Aries Jupiter to begin transforming his image—with great subsequent success—into that of a beloved philanthropist.

You Can Become
Very Un-Zen-like!

Another Aries Jupiter cautionary tale: A man found a cocoon of a butterfly, with a very small opening in it. For several hours, he sat and watched as a butterfly struggled to force its body through the tiny hole. Then it seemed to stop making any progress. He peered into the hole and saw that the butterfly appeared as if it had gotten as far as it could—it could go no further. Wanting to help, the man took a pair of scissors and gently snipped off the remaining bit of the cocoon.

The butterfly emerged immediately and easily. But, to his surprise, the man saw that it had a swollen body and small, shriveled wings. The man expected the wings to expand to support the body, but that did not happen. The butterfly was never able to fly; in fact, the butterfly spent the rest of its short life crawling around with shriveled wings and a swollen body.

What the man, in his kindness and haste, did not understand was that the struggle required for the butterfly to get through the small opening of the cocoon was nature's way of forcing fluid from its body into its wings. This way, the butterfly would be ready for flight once it achieved freedom from the cocoon.

If you knew that you would become a butterfly at the right time, would your struggle be more meaningful? An Aries Jupiter excess of impatience and overconfidence can encourage you to take shortcuts, to avoid necessarily painful learning experiences. The next time you feel held back by a cocoon, stop looking for the scissors and instead look at it as the perfect opportunity to grow and to change!

Sun Sign-Jupiter Sign Combinations

♃♈♃

Y ou share this knock-out combo with hard-hitters such as Aries-Aries leaders Charlemagne and Eleanor of Aquitaine, both of whom never hesitated to do what they wanted to do or to say what they felt. The double Aries placement increases your fire and drive while removing many of the limits you would normally feel as a solo (Sun) Aries. Thus, the double power of Sun and Jupiter combined will conspire to bring you powers that are superlucky, but sometimes your luck will be so strong that it will feel out of your control. To work with your gift and to have an all-around balanced life requires discipline, discipline, discipline.

Think of your heavenly combination of Aries speed and Jupiter luck as a strange mixture of luxury and responsibility. It's your job to charge ahead, full tilt, while still observing the safety rules—at least the important ones. Your ego-sensitivity is doubled, so make sure not to let fear of rejection make you doubt yourself. Your strong views will doubtless be unpopular to some, but always trust your instincts, dear double Aries!

♃♈♃

T aurus hates discomfort, while Aries will brave any discomfort for their cause. For you Taurus Suns to be lucky, you must refuse the comfortable, luxuriate-on-the-couch lifestyle you naturally gravitate to and seek challenge and risk. Just look at Taurus-Aries Enrique Iglesias, who got unlucky recently when he acted out his Taurus side, cautiously lip-synching his songs at a concert in the Alps. An audience of ten thousand booed him. He would have done better to wing it as the Aries Jupiter he is!

True, sometimes you will need to push yourself harder or raise the bar higher than you want to, but in the end it's worth it. Taurus-Aries Al Pacino expresses this urge toward more in this observation about voting in the Academy Awards: "I look at all the nominees and think, What if they were all brain surgeons? Which one, if you needed brain surgery, would you let operate on your brain? That's the one you should give the Oscar to." Aries Jupiter has the gift of being able to seek out the very best, not the easiest or most convenient.

You may plan carefully in your life, but remember that it is acting like a spontaneous, headfirst Aries that brings you the greatest number of rewards, including the long-term security you seek.

ⴕⴔⴕⴔⴕⴔⴕⴔⴕⴔⴕⴔⴕⴔⴕⴔⴕⴔⴕⴔⴕⴔⴕⴔⴕⴔⴕⴔⴕⴔⴕⴔⴕⴔⴕⴔ

Tell your Aries Jupiter to redirect your overactive Gemini mind and it will. You can finally stop overanalyzing and thinking too much. Your luck comes through your senses and the force of your willpower. The Aries Jupiter individualism will make your eccentricity important and add a touch of healthy aggression to boost your career. Think of Gemini-Aries Mike Myers, whose twins, Dr. Evil and Austin Powers, took the world by storm! Aries and Gemini get on because, at heart, they're both groovy, baby.

You will do something very different in your lifetime. By daffily delighting, pushing the envelope, you may just deepen your Gemini zaniness into a useful, important life mission. You want to be different and to be perceived as such by those around you, and your Aries Jupiter is there to help you communicate your own unique message to the world.

Commit to the force of fire rather than the flow of air, and you cannot fail. You will also have more luck in love if you are direct, emotional, and purposeful. Try it, dear Gemini-Aries, and you'll see!

ⴕⴔⴕⴔⴕⴔⴕⴔⴕⴔⴕⴔⴕⴔⴕⴔⴕⴔⴕⴔⴕⴔⴕⴔⴕⴔⴕⴔⴕⴔⴕⴔⴕⴔⴕⴔ

Like Cancer-Aries Alex Rodriguez, you will jump upon your lucky star unexpectedly, suddenly, surprising everyone. Use your Aries fire to transform your sideways crab walk into a soaring romp— one you can be proud of. You have the ability to change careers dramatically, and your luck will come through your physical body (many Cancer-Aries achieve success through some form of bodybuilding, like Tobey Maguire training for the part of Spiderman). Cancer-Aries Jesse Ventura went from a successful wrestling career to a brief but stunning career in politics.

This combination is a very creative mix, lending some fiery Aries zip and zap to the Cancerian endurance. It also provides some Cancerian stability to the Aries restless inconsistency. Face it: Still waters on the outside and fire and heat on the inside is a hard combo to beat. You gain the most success in life when you publicize this fact. Caution and reticence will not serve as a safety net when there is nothing to catch in that net! Take a risk, both in love and career, and see how your courage and daring pay off. You will not regret it, dear Cancer-Aries.

ꇓ♈

IF YOUR SUN SIGN IS:

Leo

(July 23–August 23)

Here the Aries flame is set alight and becomes your shimmering throne. Your luck comes through your assertion of your own authority against the odds; not through fighting, but through *winning*—and ever so graciously and eloquently, as befits royalty. Take, for example, the great Leo-Aries writer George Bernard Shaw, who quipped, "I often quote myself; it adds spice to my conversation."

Shaw spent years bucking the status quo as an antiwar protester and socialist. So beware: Your Jupiter, the Great Protector, might also compel you to be a spokesperson—or to take care of everyone else in your life, leaving yourself high and dry. Being double fire, it's practically guaranteed that your life will be dramatic. At times, it could resemble a grand opera—or a soap opera! No matter which, keep in mind that it's better to *act* in one than to *live* in one. Put aside the desire to please everyone and commit to pleasing yourself more! Popularity contests are irrelevant if your luck comes from the power of Aries. Your success, dear Leo-Aries, is hard won and belongs to you.

ꇓ♈

IF YOUR SUN SIGN IS:

Virgo

(August 24–September 22)

Here your impulsive, larger-than-life Aries Jupiter is complemented by the wisdom and wit of sharp-as-a-tack Virgo. Your extreme discernment can reach the point of genius here, but you must drop the Virgo-dictated modesty and humility to completely fulfill your destiny as a successful Virgo-Aries. Virgo wants to serve, and Aries wants to lead, so don't be surprised when you feel yourself pulled in different directions. For luck, surrender to your Jupiter.

If all else fails you can always be a comedian. There are more comedians born under the sign of Virgo than under any of the other signs besides Gemini. Virgo Michael Keaton is a perfect illustration of a comedian glorying in his Aries Jupiter. He is as good playing an earnest Batman or an evil villain as he is at being outrageous in *Beetlejuice.* Let go of your self-criticism and begin to truly love and adore yourself, dear Virgo-Aries! Seeing the bigger picture brings you more success than counting every detail. Let yourself have great, big dreams again, as you did when you were a child.

When the Libra Scales are not in balance, sidling you into a pa-ralysis of doubt and quicksand, call on your Aries Jupiter! The Ram will charge right in, knowing immediately what to do and *when* to do it. "You know what you want," Ram says. "Go for it. Take it." Listen. The Ram is shouting in your ear: "You look stunning when you are confident and bold, striding forward! Didn't anyone ever tell you that before?"

Your inclination to pause at the wrong time will be burned away by the fire of Aries Jupiter, and you will be propelled forward again. This may happen in the form of a competition or even by an embar-rassment of some kind. You might be happy being a demure people-pleasing Libra, but once you see you are being left behind by the competition or passed over in some other important way, your Aries Ram will rear its head and send you charging back again.

Do not hesitate to make your views known and to ask direct— even confrontational—questions. It worked for successful Libra-Aries Melvyn Bragg on Britain's *South Bank Show* and it will work for you! Neither celebrity nor politician nor *any* authority should make you feel "lesser than," dear Libra-Aries. Remember: You were born a leader of the highest fire. Allow your inner fire to commit you to your cause.

ϙʮ

This placement gives you incredible strength and talent, mixing creative fire with artistic water, but you may end up fighting yourself. Scorpio and Aries are both intense and demanding. Your job is simply not to get discouraged by the absurdly high expectations you place on yourself.

When Scorpio-Aries Walter Cronkite famously said, "I can't imag-ine a person becoming a success who doesn't give this game of life everything he's got," he was describing half the equation. The other half he displayed on national television when he abandoned the objec-tive reporter's stance and cried as he read the death announcement for President John F. Kennedy; his Aries emotion cracked through his Scorpio reserve and transformed a moment in news to a collective moment in history.

Then there's Scorpio-Aries John Cleese, who is funniest when he tries to hold it in and fails. For your greatest success to come to you, dear Scorpio-Aries, let it out. Breathe into your Aries fire and sometimes you can even let your passion consume you. A scary thought for the Scorpio, I know, but lucky Aries Jupiter is too busy charging forward to allow any serious burns.

♃♈♃♈♃♈♃♈♃♈♃♈♃♈♃♈♃♈♃♈♃♈♃♈♃♈♃♈♃♈♃♈♃♈♃

Sagittarius
(November 22–December 21)

You have a fabulous mix! Your double fire burns hot, but it doesn't scorch you. This is because Sagittarius is essentially an adventurous kid. Your Aries takes you into unexpected places (which you thoroughly appreciate), and your Sagittarius Sun finds the humor there. As the zodiac's natural philosopher, you appreciate the opportunities for acting on pure impulse that Jupiter so happily provides. Surprising twists in your life appear from nowhere—and they're usually for the better.

The only potential downfall for you is disappointment with others. Your sunny disposition can be cast into shadow if your expectations are not met. For all your gusto, your Aries Jupiter side can be easily and deeply hurt. You are more sensitive to others' opinions than you may want to admit. To stay on an even keel cultivate stable friendships, preferably with Earth signs (Taurus, Virgo, Capricorn).

For your greatest success, dear Sadge-Aries, let yourself commit to one cause rather than several, and risk unpopularity as you carry your project all the way to the finish line. Sadges are loved and adored by everyone, so you can allow yourself some lone moments as all-powerful leader once in a while!

♃♈♃♈♃♈♃♈♃♈♃♈♃♈♃♈♃♈♃♈♃♈♃♈♃♈♃♈♃♈♃♈♃♈♃

Capricorn
(December 22–January 19)

This is the most intense combination. The determination of the Goat with the energy of the Ram leaves no room for pause or weakness. Success is essentially guaranteed. You are promised great rewards if you remain focused and concentrate on going your own way. You are not one who is always smiling or thinking cheerful thoughts; Capricorn-Aries Fyodor Dostoyevsky is a great example of this extremely serious combination. Following his mock execution at the hands of Czar Nicholas I's firing squad (so real that it made one of his fellow "condemned" insane), he wrote his brother, "To be a human being among human beings, and remain one forever, no matter what misfortunes befall, not to become depressed, and not to falter—this is what life is, herein lies its task."

On a much lighter note, in the name of love, you may wish to conquer your Aries impatience and take your own sweet time selecting the right mate. On the rare occasion when you fall for someone, you fall hard. Your chance at happiness in a relationship comes when you use your Aries Jupiter battering ram to crack your own defensive Capricorn wall. And once the wall falls down, allow your Aries Jupiter to loosen you up in private.

꜏꜏꜏꜏꜏꜏꜏꜏꜏꜏꜏꜏꜏꜏꜏꜏꜏꜏꜏꜏꜏꜏꜏꜏꜏꜏꜏꜏꜏꜏꜏

The fierce individualism of Aquarius is on the move as the Aries Jupiter Ram charges forward with its Aquarius flag. This combination brings with it an unusual need for exhibitionism, which is completely puzzling to the Aquarius side of yourself. Blame it on the chemistry of the Aquarius-Aries pairing. And your good looks don't hurt!

You will be torn at times as your Aries Jupiter enthusiasm struggles against your Aquarian coolness, but to be lucky, trust your creativity and indulge your passions and your fire. This may well require a great deal of time alone, far from the madding crowd. Writing being a solitary business, many successful writers are made from this mix. This does not exclude, however, the extra Aquarius-Aries talent for finding a good and permanent mate.

For greatest success, dear Aquarius-Aries, occasionally let go of your brilliant analytical powers and plunge in, headfirst. Trust your feelings and instincts, and act on them as immediately as you can. Try it around a certain person and see what happens. Remember: Your luck comes not from the air (thought) part of you, but from the fire (feelings) part of you.

꜏꜏꜏꜏꜏꜏꜏꜏꜏꜏꜏꜏꜏꜏꜏꜏꜏꜏꜏꜏꜏꜏꜏꜏꜏꜏꜏꜏꜏꜏꜏

Pisces, thank your lucky stars you have Aries Jupiter! You have the imagination and dreaminess of Pisces, and the Aries fire is exactly what you need to pull yourself into action. Use it to celebrate your gifts, as your Pisces self is inclined to soft-pedal them.

Out of this mix comes great visual creativity. Think of Pisces-Aries Dr. Seuss, whose book *How the Grinch Stole Christmas* marries Pisces kindness to a great Aries villain. He advocated the power of the tiny individual (Cindy Lou) and described the overwhelming strength that a community working together can voice. Socialism for Christmas, anyone? This is radical Aries Jupiter in concert with compassionate Pisces Sun.

In your personal life, you feel everything deeply. The emotionality of Aries combined with the intense sensitivity of Pisces can make you fall in love for all eternity—and beyond—if Pisces has anything to say about it.

For your greatest success, dear Pisces-Aries, wean yourself away from self-sacrifice and that occasional need you have to be taken care of. By cultivating the stronger and more deserving feelings you have inside, your self-sufficiency increases as well. You will then be the true leader of your life—and in love with every aspect of it!

Relationships with Aries Jupiters

Finally, for those of you who aren't Aries Jupiter yourselves but who have a spouse or significant other, boss, child, or parent who is, here are some great tips to help you learn to deal! This is how the luck of someone else's Jupiter touches *your* life.

Your Significant Other

You may not know this, but deep down, this person wants to be pursued. In spite of Aries Jupiter's seeming independence, your mate adores being chased. Dramatic gestures, intense romance, and rambunctious playfulness are often necessary adornments. The Aries vulnerability to being liked or disliked is very strong here. Aries' *deepest* fear, and you'd never know it, is that *you don't really like them*. They need reassurance constantly. The naughty ones will test you. Loving an Aries Jupiter can get exhausting at times, but play along—if you can take it—and they'll be ecstatic. Life with an Aries Jupiter lover is never boring. It's more like being on a roller-coaster ride! The more playful or kinky ones will adore acting out games and dramas of pursuit, capture, and rebellion.

One way to ensure permanence—if that's what you're after—is to factor *friendship* into the love equation. Be *stable* for the Ram, and the Ram *will love you forever*.

Your Boss

Are you sure you don't want to get another position? Perhaps change companies? Aries Jupiter bosses can be slave drivers. They are creative and fun and inspiring leaders, but they truly believe to their dying breath that no one can do the job better than they, and that applies to you, dear trusted employee. They often won't even let you do your job, because they'll be so busy doing it for you. At the very least, they will offer suggestions as to how you could do it better. It's nothing personal, remember.

You may do the best work you'll have ever done under an Aries Jupiter boss, but you won't be doing much else! Forget having much of a personal life. They get obsessed, and will work 24-7 on a project—and expect you to do so as well. Then, when they're done, that's it. They have little concept of continuity, balance, or the day to day—and no concept of bureaucracy or company politics. Aries Jupiters should never be paper pushers. They believe their true mission is always to lead the company on an inspiring crusade.

Let's say it's not your boss who's Aries Jupiter, but your partner, or even an employee. The sad fact is that Aries Jupiters can tend to treat anyone they're working with (or for) as their personal assistant. Again, they'll insist, it's not about *you*. "It's all about the *work*," they'll say. If you tell them that this behavior is painful and difficult, they'll be heart-broken and promise to mend their ways. Try it. It works every time. For a little while, at least.

You'll be surprised to hear them say, if you are doing a good job, as their employee, partner, or supervisor, that they *love* you and are eternally grateful for every single thing you do. They will shower you with gifts, as they think the world of you. It's emotional working for an Aries Jupiter. A typical nine-to-five job it's not.

Your Child

If you have a child who's an Aries Jupiter, don't clip their wings. They will never forgive you for it. Ever. This child's luck is born of fire, so if you pour water (read: your own emotional baggage) on their hopes, dampen their enthusiasm, or make them feel less about themselves, something inside them will die. It may take years for them to get it back, and this may create extreme wear and tear on your relationship.

Dear parent, this is only to save you from some anguish you may encounter raising your Aries Jupiter child. They feel everything intensely, even if their Sun sign is mild or noncommunicative. They are independent in mind, if not in action, and their luck rewards their independence. They are adventurous and imaginative and a delight to raise if you value creativity as much as they do. Play with them, encourage them, inspire them, and they'll be by your side for life. They are at heart intensely loyal and protective of those they love. They have a combative streak, so directly opposing or arguing with them are usually not the best ways to discipline them. Try teaching by example rather than by lecture.

Aries Jupiter children glow with pride if they come first in anything. If you give your Aries Jupiter child a generous upbringing, this self-pride will last a lifetime.

Your Parent

There is nothing normal about an Aries Jupiter parent. No white picket fences, no nine-to-five office jobs, no predictable times for dinner or bed. Instead, you probably had a wildly original upbringing, full of adventure and exposure to the outside world—perhaps earlier than you would have liked.

Aries Jupiters are not overly protective parents. They might say they don't understand why you won't stick up for yourself around the bigger kids, or confront the bullies, or beat up the tormentors. Even now, if you have colleagues or bosses to deal with rather than playground rivals, or even children of your own, your Aries Jupiter parent will expect you to be aggressive and forceful about getting what you want. Their advice, if you ask (and you may, if you're a more delicate Sun sign), is never to play politics—unless it's fun.

Aries Jupiter parents are intensely loyal and fiercely proud of you, but at times an odd sympathy for an enemy of yours will surface. This is frustrating, but somehow, you must chalk this up to eccentricity, not meanness. When your parents keep talking about a past boyfriend or girlfriend with whom you broke up years ago, one you'd rather forget, take deep breaths. And don't expect this phase to pass. You may have children of your own when Grandma or Grandpa Aries Jupiter, with your devoted husband and kids in the room, suddenly pipes up with, "Oh, I so liked your ex!" They don't stop there, either.

"Why couldn't you work things out, you two?" At this point, your children are snickering, and hopefully you are at the point where you can, too.

An Aries Jupiter parent is never easy, but then, who is? They love deeply—even if not wisely. Hopefully you'll see how totally great they are—unique and, above all, fun. On a serious note, think about telling them, as soon as you can, that you love them back.

That, in the end, is what astrology is really all about—appreciating how different we all are, and loving the differences! Nature gives an antidote for every poison. Isn't that the miracle of life? Problems appear, but always, somewhere, there's a solution, and Aries Jupiter is guaranteed to find it. Put on a smile and you can't go wrong. Sell yourself and you will win the world.

Remember, Aries Jupiter: You are the exclamation mark of the zodiac!

Madalyn's Movie
Choice for Aries Jupiter

*What's Love Got
to Do with It?*
The true story of singer Tina Turner, who, born poor and black, is raised by her grandmother in Nutbush, Tennessee, rises to stardom, braves her husband's abuse and death threats, and reaches the world with her singing, becoming an inspiration to many worldwide. Aries Jupiter passion and courage at its most vulnerable and sexy.

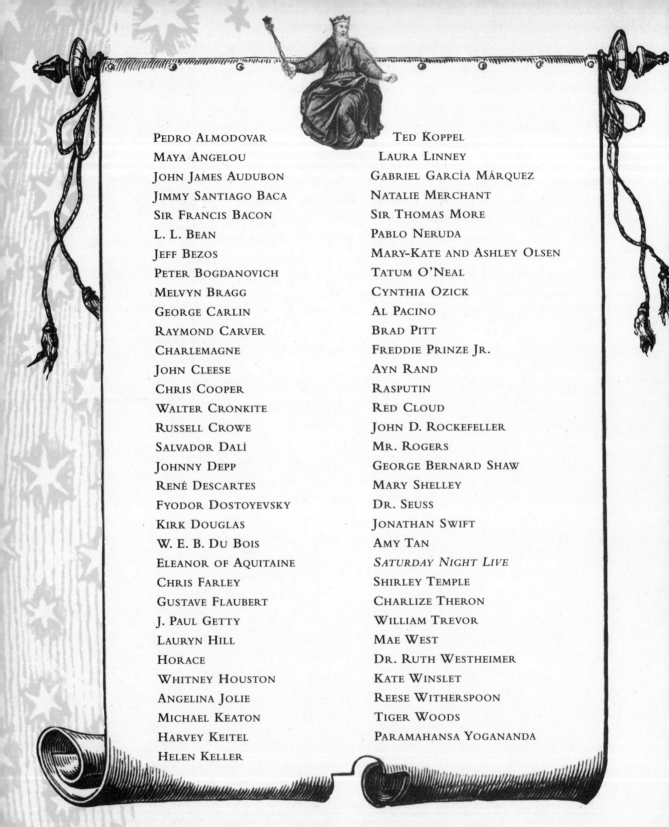

PEDRO ALMODOVAR	TED KOPPEL
MAYA ANGELOU	LAURA LINNEY
JOHN JAMES AUDUBON	GABRIEL GARCÍA MÁRQUEZ
JIMMY SANTIAGO BACA	NATALIE MERCHANT
SIR FRANCIS BACON	SIR THOMAS MORE
L. L. BEAN	PABLO NERUDA
JEFF BEZOS	MARY-KATE AND ASHLEY OLSEN
PETER BOGDANOVICH	TATUM O'NEAL
MELVYN BRAGG	CYNTHIA OZICK
GEORGE CARLIN	AL PACINO
RAYMOND CARVER	BRAD PITT
CHARLEMAGNE	FREDDIE PRINZE JR.
JOHN CLEESE	AYN RAND
CHRIS COOPER	RASPUTIN
WALTER CRONKITE	RED CLOUD
RUSSELL CROWE	JOHN D. ROCKEFELLER
SALVADOR DALÍ	MR. ROGERS
JOHNNY DEPP	GEORGE BERNARD SHAW
RENÉ DESCARTES	MARY SHELLEY
FYODOR DOSTOYEVSKY	DR. SEUSS
KIRK DOUGLAS	JONATHAN SWIFT
W. E. B. DU BOIS	AMY TAN
ELEANOR OF AQUITAINE	*SATURDAY NIGHT LIVE*
CHRIS FARLEY	SHIRLEY TEMPLE
GUSTAVE FLAUBERT	CHARLIZE THERON
J. PAUL GETTY	WILLIAM TREVOR
LAURYN HILL	MAE WEST
HORACE	DR. RUTH WESTHEIMER
WHITNEY HOUSTON	KATE WINSLET
ANGELINA JOLIE	REESE WITHERSPOON
MICHAEL KEATON	TIGER WOODS
HARVEY KEITEL	PARAMAHANSA YOGANANDA
HELEN KELLER	

"It was all from me—it was all self-generated."

—Reese Witherspoon, Aries Sun-Aries Jupiter. March 22, 1976

"You'll be the brightest light the world has ever seen."

—Natalie Merchant, Scorpio Sun-Aries Jupiter. October 26, 1963

"You've got to go in and take chances."

—John Havlicek, Aries Sun-Aries Jupiter. April 8, 1940

"Be bold, be brave, be free."

—Angelina Jolie, Gemini Sun-Aries Jupiter. June 4, 1975

"Whosoever is delighted in solitude is either a wild beast or a god."

—Sir Francis Bacon, Aquarius Sun-Aries Jupiter. January 22, 1561

"For me there shouldn't be any halfway."

—Johnny Depp, Gemini Sun-Aries Jupiter. June 9, 1963

"He who hesitates is last."

—Mae West, Leo Sun-Aries Jupiter. August 18, 1892

"Life doesn't frighten me."

—Maya Angelou, Aries Sun-Aries Jupiter. April 4, 1928

"Have no fear of perfection—you'll never reach it."

—Salvador Dalí, Taurus Sun-Aries Jupiter. May 11, 1904

"You make my day a special day—just by being yourself."

—Mr. Rogers, Pisces Sun-Aries Jupiter. March 20, 1928

♉♃

The Long Way Impels Us

From Boccaccio, the fourteenth-century author of the racy *Decameron,* to comedians like Richard Pryor and Rosanne Barr, Taurus Jupiters have never been afraid of salty, earthy, and, sometimes, explicit humor. When life gets hard, make sure to use yours!

The sooner you submit and stop denying the inevitable, the happier you'll be.

—TAURUS JUPITER JANEANE GAROFALO

Dear Taurus Jupiter, above all trust your salt-of-the-earth smarts. The great Aristotle, born a Taurus Jupiter in 384 B.C., argued that it is only through the senses that we can divine true reasoning. Since your sign rules over the senses, if you accept the bounty of Jupiter you can actually *improve* them, learning to see everything more clearly, to hear what's really going on, and to make sense of the world more acutely than anyone else. And since Taurus is the ruler of common sense and eternal truths as well as the physical senses, Jupiter power can give you deep wisdom about the world, a special ability to solve problems, and the power to work everyday miracles.

The first vision of the Virgin Mary at Lourdes, where thousands of people have been healed, occurred during a passage of Taurus Jupiter on February 11, 1858.

Your Jupiter sign won't bring you instant rewards or success. Instead, Jupiter energy comes to you through some traditional virtues: patience, endurance, planning for the long term, and good old-fashioned hard work. While it's true that shortcuts and get-rich-quick schemes won't work for you, don't see this as bad news. For you, the traditional virtues will lead to extraordinary, unprecedented luck and success.

A reporter once asked, "What do you think of Western civilization?"

"I think it would be a very good idea," replied a great Taurus Jupiter.

Born October 2, 1869, Libra by Sun, Mahatma Gandhi succeeded not by being an instantly pleasing Libra, but by being a wise and weighty Taurus—his *Jupiter* sign. He used that special Taurus Jupiter humor, which provides some of the pithiest comebacks ever, as well as Taurus Jupiter's powers of perception and insight, to find a way to make India independent of England. But, above all, he drew on the patience and perseverance that brings such luck and success to those born under this awesome Jupiter sign.

It took Gandhi twenty-three years to free India from the British—without any violence whatsoever. His tactics—patient waiting, nonviolence, fasting, refusing to fight or confront the enemy—seemed to produce very little for the longest time. But while many people thought that Gandhi's independence movement would never amount to anything, his brilliant strategy wore out the British over the long run. Dear Taurus Jupiter, set your goals high and be prepared to wait *years* to achieve them. When in doubt, think of these lines—possibly the most beautiful request ever made for faith in the face of adversity—attributed to Saint Francis of Assisi, born a Taurus Jupiter in 1181:

> *Lord, make me an instrument of your peace. Where there is hatred, let me sow love. Where there is injury, pardon. Where there is doubt, faith. Where there is despair, hope. Where there is darkness, light.*

Take a moment, look back long and deep over your life, recount as many events as you can, and ask yourself what has made you lucky or unlucky in your life. You'll notice that whenever you act like a Taurus, *you create miracles in your life.* No matter what your Sun sign is, *you get lucky when you live out the Taurus legacy gifted to you by the benevolent planet Jupiter.*

To take advantage of your luck, emphasize your . . .

Be patient, wise, determined, earthy, problem solving, hardworking, detail oriented, willing to help others, and a preserver.	♉♃ *Taurus Jupiter Strengths*

But because Jupiter rules excess, watch out for . . .

| Taurus Jupiter Excesses | ♉ ♃ | *You can become repetitive, obsessive, or self-depriving.* |

Understanding the Taurus Energy and Influence

While the first sign of the zodiac, Aries, symbolizes beginnings and the early spring, Taurus, the second sign, represents the time when crops are planted, when the hard, practical work of sowing seeds is done—work that shows no results now but produces a bounty later on in the fall. Farmers know that the best month to plant their crops is when the Sun is in Taurus, but it's also true that the best *days* to plant are when the *Moon* is in Taurus, called the Farmer's Moon. Tauruses really do have a special connection to the fertility and power of Mother Earth.

To bring the bounty of Jupiter into your life, just think of farmers sowing seeds. The results don't come quick: It takes long months of waiting, watering, and careful tending for the seeds to sprout shoots and the shoots to grow into a healthy crop. Be prepared to wait, to be patient for your rewards! The good news is that your rewards will be greater than you can even imagine. And that's true even if your Sun sign is in mercurial Gemini or dreamy Pisces: Follow the guidance of Taurus and you will enjoy success, luck, and power.

ठ4ठ4ठ4ठ4ठ4ठ4ठ4ठ4ठ4ठ4ठ4ठ4ठ4ठ4ठ4ठ4ठ4ठ4ठ4ठ4ठ4ठ4ठ4ठ4ठ4ठ4ठ4ठ4ठ4ठ4ठ4ठ4

The symbol for Taurus is the Bull, a strong, silent animal, grazing peacefully in pasture (unless he's confronted!), living life on his own terms. Your luck comes, in similar fashion, by behaving like a bull: by refusing to be swayed, by sticking to your convictions, by standing strong and silent and proud. Have you ever tried to make a bull change his mind? Case closed.

TAURUS SYMBOL
The Bull

Bulls aren't always violent or aggressive (just think of peace-loving Ferdinand the Bull), but they do usually charge when they are seriously threatened. You should be the same way: Try sticking to your ground peacefully. Successful Taurus Jupiters embody a brilliant sense of timing—knowing when to stand still, when to charge as the bull does. You, wise Taurus Jupiter, will know when the time is right. Trust yourself. Think about what Aristotle said, *We are what we repeatedly do. Excellence, then, is not an art, but a habit.*

4Ծ4Ծ4Ծ4Ծ4Ծ4Ծ4Ծ4Ծ4Ծ4Ծ4Ծ4Ծ4Ծ4Ծ4Ծ4Ծ4Ծ4Ծ4Ծ4Ծ4Ծ4Ծ4Ծ4Ծ4Ծ4Ծ4Ծ4Ծ4Ծ

TAURUS PARTS OF BODY
The Throat and the Neck

The parts of the body Taurus rules are the throat and the neck, including the vocal cords and the palate. For you, Jupiter gives special powers when you use your voice and your sense of taste. There are more singers with Taurus as their Sun sign than any other sign—so sing and enjoy! Exercise your palate as well, for your taste buds are keen. While Taurus Suns need to protect their throats, you Taurus Jupiters are free to stick your necks out safely, to sing loudly and to experience new tastes—under the power of the planet that protects.

4Ծ4Ծ4Ծ4Ծ4Ծ4Ծ4Ծ4Ծ4Ծ4Ծ4Ծ4Ծ4Ծ4Ծ4Ծ4Ծ4Ծ4Ծ4Ծ4Ծ4Ծ4Ծ4Ծ4Ծ4Ծ4Ծ4Ծ4Ծ4Ծ

TAURUS COLOR
Baby Blue

Baby blue, as its name suggests, is about babying, loving, comforting, calming, and soothing. This beautiful color resembles the color of the sky on a lovely day, and the color of robins' eggs, one of the harbingers of spring. While you're no baby, it's lucky for you to wear baby blue. Put on a baby blue shirt or sweater and see how much better you feel: stronger, more alert, *and* more calm. Your lucky color lets tension flow away from you while simultaneously giving you more energy. You will feel more positive if you wear baby blue, and you're less likely to get the blues and feel depressed.

Taurus Jupiter ♄♃ Strengths

And now your Taurus Jupiter strengths:

Be Patient

Taurus Jupiter, you must be prepared to wait for your success to come. Make sure you cultivate patience, endure hardship, and look to the long term—and lessons from the past (the past is one of the essences of Taurus Jupiter strength and wisdom)—for then luck will be your reward.

Former Superman Christopher Reeve, for example, has patiently waited eight years for his hands and feet to move. In that time, he has lobbied for bills supporting research and rights for the paralyzed—and they've passed. Taurus Jupiter Chester Carlson waited twenty-six years for his Xerox invention to be bought. Can you imagine our world now without copiers? Taurus Jupiter John Nash (the subject of the movie *A Beautiful Mind*) waited until he was in his sixties to receive the Nobel Prize for work he had published in his twenties. The Nash

Equilibrium is now used to study decision making by psychologists on behavior, biologists on evolution, and economists on international trade.

And think of the Academy Awards, also Taurus Jupiter. Every actor, actress, director, and producer wants to win an Oscar. And how much money do Oscar winners receive when they pick up their prize? Absolutely nothing—at award time. In true Taurus fashion, monetary rewards come months or years later, as awards convince more people to see movies and stars can raise their salaries.

The great Italian poet, Taurus Jupiter Dante Alighieri, born January 1, 1265, popularized the term "limbo" in his masterpiece *The Inferno*. The state of limbo is pure waiting, with no end in sight—the ultimate test of patience. "Let us go on, for the long way impels us," Dante wrote in his section on the infamous first circle of Hell. Now, you do not need to suffer in Hell as Dante indicates—not at all! In fact, as a Taurus Jupiter, you will be rewarded for your patience. You know better than other, more impatient signs that the end comes when it does, and when it is right. Your faith could move mountains—and it does.

Earthy Taurus' patient wait is never in vain. It *always* produces great results.

Be Wise

Once, while having lunch with a white friend in a restaurant, Dr. Martin Luther King Jr. was asked by the headwaiter to move to a different section of the establishment. The waiter assured Dr. King that he would receive the same service, the same good food, and the same excellent table in the other room. He could eat in absolute comfort in the other room. He would be deprived of nothing.

"I understand," Dr. King replied. "But what you are depriving me of is my right to speak to my friend, and my friend's right to speak to me. You are taking away the right of two human beings to communicate."

This Nobel Peace Prize winner declared that laws that were unjust should be ignored, as they were not truly laws. Wisdom comes from real experience, and Taurus is the most "real" sign there is. To fulfill your highest purpose, you must acknowledge all your experience, your history, all of who you are. Draw on the Earth, your own element, for energy and remembrance. Use your powers of perception to see what really matters, and go for it, no matter how long it takes.

When King made his famous "I Have a Dream" speech, he used

the enduring Taurus Jupiter future, "I may not get there with you. But I want you to know tonight that we, as a people, will get to the promised land." Many believe that King knew he would not be alive thereafter. Taurus Jupiter, in rhythm with the Earth's cycles, gives a lasting sense of time and a vision into the future. Find your luck by generating for the long term. Be the seed that grows harvests and creates posthumous legacies. Remember: That famous Montgomery, Alabama, bus boycott lasted eighteen months—eighteen months of walking to work through snow and heat, thousands refusing to board a single bus, outlasting the bus company and all who said it couldn't be done.

Like Dr. King, you must cultivate Taurus Jupiter's power to hone and sharpen your physical and mental senses. If you train yourself, you will discover you have a special gift to see to the heart of the matter, to understand what's really at stake, to perceive what's really going on. You have the ability to see the truth, even if everyone else is in denial, and the ability to speak the truth when no one else can. The child in *The Emperor's New Clothes* who cried out the obvious, "The emperor has nothing on!" is pure Taurus Jupiter. Whistleblower Taurus Jupiter Cynthia Cooper, one of *Time*'s People of the Year for 2002, told the truth about the fraud at MCI/Worldcom even when everyone else pretended nothing was going on. Taurus Jupiter, allow your senses to shine—even through others' darkness—and luck and success will be yours.

Be Determined

If you want it, your Jupiter sign can also give you unbelievable determination and resilience. If others view your sticking to what you know as stubbornness, then so be it; you can stick to your beliefs like Velcro (Velcro's Jupiter sign is Taurus) and hold fast to the courage of your convictions. Just think of Taurus Jupiter Lillian Hellman, who faced Joe McCarthy and the Committee on Un-American Activities during the communist witch-hunt of the 1950s and said, "I cannot and will not cut my conscience to fit this year's fashions."

Your determination and stubbornness come out in another way, too. Taurus Jupiter governs profit and success in business, as long as they come over the long term. You will find a unique luck will come to you if you concentrate on profits and profit making. If others tell you what you're doing is wrong, if no one agrees with your business plans, stick to your guns and be stubborn. You may have to accept short-term sacrifices in order to get long-term gains, but it will be worth it. Just look at *TV Guide*—a Taurus Jupiter. This magazine lost money in its first few years, but the owners stuck to their strategy and refused to give up. *TV Guide* became one of the most successful magazines in history—for many years the magazine with the number one circulation in the United States.

Be Earthy

Taurus Jupiters express earthy sexuality. Just think of sexually magnetic actresses Diane Lane, Faye Dunaway, Ann-Margret, and Julie Christie.

But what about earthy as in . . . just plain old Earth?

This is where Taurus Jupiter gets really wise—in recognizing the worth of the Earth. German nun Hildegard von Bingen, born in 1098—true to an overcoming Taurus Jupiter, she also became a physician, writer, and composer at a time when women were considered mere chattel—gave us the word "viriditas," or greening power, and wrote that "greening love hastens to the aid of all." Taurus Jupiter Frederick Law Olmstead gave us some of America's greatest parks, including New York's Central Park. And Seattle, chief of the Suquamish, born a Taurus Jupiter in 1786, gave us this. Sadly, he did not live long enough (and will we?) to see the results of his speech, but they live after him:

> *How can you buy or sell the sky, the warmth of the land? The idea is strange to us. If we do not own the freshness of the air and the sparkle of the water, how can you buy them? Every part of this earth is sacred . . . Every shining pine needle, every sandy shore, every mist in the dark woods, every clearing and humming insect is holy . . . If men spit upon the ground, they spit upon themselves.*

Be Problem Solving

One day, as he wandered around the field, a farmer's horse fell down into a well. The animal whimpered for hours as the farmer tried to figure out what to do. The well was deep, with smooth walls, impossible to climb, and no one could figure out how to get the horse back up onto the land.

The farmer suddenly had an idea. He asked the neighbors to shovel dirt down onto the horse. They asked him if he intended to bury the poor horse in the well, and the farmer reassured them that he did not. Puzzled, they continued to shovel, the horse began to quiet down, and the neighbors looked into the well and saw something that astonished them. As each shovel of dirt hit the horse's back, he would

✳ Plan and work long term for your future legacy.

✳ Be patient with details and small improvements.

✳ Surround yourself with white, baby blue, green, and rose—clothes, sheets, towels, home décor, and walls.

✳ Indulge your body and aid your circulation with regular massage.

✳ Eat root vegetables, horseradish, spinach, asparagus, cauliflower, celery, pumpkin, cranberries, raw nuts, and Swiss chard. Flavor with tarragon and cinnamon.

✳ Increase your intake of iodine with kelp and seaweed.

✳ Always protect and cover your throat when the weather is cold.

✳ Hang out with more Virgos, Capricorns, and Tauruses.

shake it off and take a step up, packing the earth down beneath him. To everyone's amazement, once enough earth had been shoveled into the well, the horse stepped up over the edge and trotted off.

You, Taurus Jupiter, are like the farmer (and the horse!): You get lucky when you uncover options and find critical solutions to problems. You can see what others cannot, and what is obvious to you is mind blowing to the rest of us! Just try to stretch yourself a little, and Jupiter will give you an extraordinary and unique ability to solve the hardest problems.

Taurus Jupiter Louis Pasteur devoted his life to solving practical problems in agriculture and industry, both areas ruled by Taurus. His most famous legacy, pasteurization, bears his name. He was responsible for discovering that spontaneous generation does not occur (Taurus rarely gets lucky with spontaneity!), and he also saved countless lives by discovering ways of preventing diseases such as anthrax, chicken pox, cholera, and rabies. He solved many problems using Taurus Jupiter, and so can you.

Be Hardworking

Hard work is the mainstay of Taurus Jupiter and brings great luck and success to all who pursue it. If you dedicate yourself to hard work, you will find results, and you don't have to work in dull or boring professions—just be professional in whatever you do. But be prepared to make initial sacrifices and to wait before you see results.

Think of jazz legend Ella Fitzgerald, who overcame incredible adversity. "I found that all through the years you never appreciate anything if you get it in a hurry," she said, drawing upon the luck of her patient Taurus Jupiter (her Sun was in Aries, making her *impatient*—but that's not what made her lucky). Orphaned at the age of fifteen, she was placed in the Colored Orphan Asylum in Riverdale, then transferred to New York State Training School for Girls, a reformatory institution in which state investigations later revealed widespread physical abuse. She ran away from the reformatory school and was literally living in the streets of Harlem. A year later, she sang at Harlem's Apollo Theatre Amateur Night. Her legendary career spanned six decades and earned her numerous awards, which she shared with her many nieces and nephews. No matter how hard she worked, she kept striving for more.

If you follow Ella Fitzgerald's example, Taurus Jupiter, by being the ultimate hardworking professional, Jupiter's bounty of luck and success will come to you.

Be Detail Oriented

James Joyce is a brilliant example of a Taurus Jupiter who used the gift of detail, and of producing yet more detail on his detail! Entire dissertations have been written on the various meanings and symbolisms of each word in his text. His masterpiece, *Ulysses,* painstakingly takes more than eight hundred pages to describe one day.

Both Joyce's "one day" and his literary allusions have become famous all over the world. Called Bloomsday, after the central protagonist, Mr. Bloom, and in keeping with Taurus-themed blooming of the Earth, this day is celebrated in many cities as a cultural event. The reading of *Ulysses* aloud for twenty-four hours straight by a succession of well-known actors has become tradition in London and New York.

Joyce wrote prolifically and had to wait many years to be recognized. So, dear Taurus Jupiter, concentrate, concentrate, concentrate! Do not expect instant gratification, nor to be discovered overnight. Instead, wait for your right time, and you will be more successful than you ever dreamed.

Be Willing to Help Others

Taurus Jupiter President Franklin D. Roosevelt created new, extensive social services to help Americans out of the devastating Depression of the 1930s. He reminded the country: "The only thing we have to fear is fear itself." By changing the government to help those in need, he made the United States stronger.

Decades later, President John F. Kennedy, in an effort to stop more Americans' being killed in Vietnam, ordered the withdrawal of troops—and was assassinated but only a week later. Three days after his assassination, his withdrawal of the troops was stopped cold, and Vietnam continued for ten more years. Kennedy had said at his inauguration, "Ask not what your country can do for you, but what you can do for your country."

The Taurus Jupiter physician who gave up his practice to become a revolutionary, Che Guevara, famously said: "Each and every one of us will pay on demand his part of sacrifice." Talk about Taurus Jupiter legacy! Even though vilified by many after his assassination by the CIA, Che's is the most popular and best-selling face for T-shirts than any other famous face, dead or alive.

From Kennedy to King to Guevara to Gandhi to Dietrich Bonhoeffer—executed for being a member of the German resistance against the Nazis during WWII—the number of helpful Taurus Jupiters who sacrifice themselves outnumber all other Jupiter signs.

Decide on what you are willing to sacrifice in order to get your long-term goals met. Do not sacrifice your life as these extreme examples did. Give what you are capable of giving.

Be a Preserver

Taurus Jupiter, you have another unique ability: to preserve what deserves to last! From Noam Chomsky, one of the ten most quoted sources in the world, including the Bible, to Dr. David Ho, a premier leader in AIDS research, Taurus Jupiters are amazing for what lives after them. Whether it's leaving a lasting legacy of laws, discoveries in medicine, or notable quotes, reproducing your genes, protecting your family, replicating yourself through work, collecting pop art, or preserving jam, Taurus Jupiter makes you lucky in life when you preserve.

On December 13, 1952, the Constitution and the Declaration of Independence were placed in helium-filled cases, enclosed in wooden crates, laid on mattresses in an armored Marine Corps personnel carrier, and escorted down Pennsylvania and Constitution Avenues to the National Archives. Two days later, President Harry Truman declared at a formal ceremony in the Archives Exhibition Hall: "We are enshrining these documents for future ages. This magnificent hall has been constructed to exhibit them, and the vault beneath, that we have built to protect them, is as safe from destruction as anything that the wit of modern man can devise." Naturally, this act of preservation took place during the reign of Taurus Jupiter.

Taurus, in general, loves the past. Taurus Jupiter Southern writer Carson McCullers wrote, "The voices reheard from childhood have a truer pitch."

The past is crucial for your strength and wisdom. As Taurus Jupiter Confucius, born August 27, 551 B.C., wrote, "Study the past if you would divine the future."

Collecting is also lucky for you. Taurus Jupiter P. T. Barnum collected animals and human oddities to preserve his circus tradition. The profit incentive—Taurus Jupiter rules profits and profit making—that he adopted was clear: As he said, "Every crowd has a silver lining." Equally lucky are copying and reproduction. Taurus Jupiter Andy Warhol became famous for his replications of soup cans and celebrities, and made a fortune from both his reproductions and his own art collection. Taurus Jupiter Chester Carlson invented the Xerox copier and found success making copies. Preserve and collect, learn from the past, Taurus Jupiter, and luck will follow.

What Makes You Lucky in Love

Taurus' luscious sensuality is legendary. As your Jupiter sign rules the senses, indulge them and enjoy them, even the sixth sense. Trust your instinct, your first impression, and follow it from that first introduction to . . . wherever you want to take it.

Have you ever gone on a date with someone who just could not stop thinking, who couldn't relax, whether it was because they were thinking about work or because they were nervous about how the date was going? It probably was not a date to remember . . . usually, people who are nervous and uncomfortable themselves make others around them feel the same way. This is doubly true if you are Taurus Jupiter. The key to your success in love is for you to stop thinking too much, to live in the moment, to fully appreciate and enjoy your senses. Instead, let the relationship exist and develop; stop trying to control the outcome; allow yourself to create new and unknown energy; and dive wholeheartedly into sex. Allow yourself to fully participate in the moment and enjoy each and every sensual treat. When you feel happy with someone, let go of the doubts and let it flourish!

Also, feel you can share what's troubling you. As Taurus Jupiter Dinah Shore said, "Trouble is a part of your life, and if you don't share it, you don't give the person who loves you a chance to love you enough."

This *doesn't* mean giving in to every impulse and flying on autopilot—quite the contrary. While sensuality and sex are lucky for you, so are details and planning and relationships that progress slowly. With your potential partners make plans for occasions that take time to develop: Learn new hobbies together from scratch or plan a trip that is months away. Longer relationships, and relationships that mature slowly, like fine wine, are lucky for you, so planning something that's going to take time is blessed, by Jove.

Dating should be fairly easy for you to handle. You're a genius at details, so put your gift to work! You do not need to try to outshine everyone else by making a grand splash. Instead, find happiness through the small gestures. Taurus Jupiters make fantastic wooers: They may not splash out on an amazing first date, but when they find a good partner, they make romance real (and earn luck) through all the notes, letters, and flowers they send over the months. Similarly, make sure you concentrate on the details during sex and give your partner an extra-good time. By putting your heart and soul into these gestures, you will attract the most Jupiter power and make your special relationship lucky.

Your Best Career Moves

If you draw upon your lucky Jupiter, you will become the busy bee in the zodiac. You are tireless and relentless. Your work ethic goes beyond anyone else's.

Did you know that Taurus Jupiter Virginia Woolf used to write standing up? She stood at her writing desk (in the garden shed, no less) just as an artist does at a canvas. She believed that the effort and attention this required drew out her best thinking. Not that you have to stand or otherwise deprive yourself, but you get the idea! Make your work hard and you will do it well. Young Taurus Jupiters sometimes skip over work that seems easy. An odd irony is that in school they may even get higher grades in the more difficult courses, lower grades in the easier ones. Later in life, as the love for detail sets in, this fate can be avoided.

You should think long term for your career; you will in the end build a great reputation if you carefully plan for the future. Your greatest pitfalls are worry and giving up. You should not buck trends or make too many changes. Instead, look for continuity. If you are a manager and your team isn't doing well, look for the small incremental changes that can gradually make your team better. Remember, just like Aesop's fable "The Tortoise and the Hare": Slow and steady wins the race.

Always keep your future goals in mind, and be patient, planting seeds that will take time to come through. When they do, you will have the most gorgeous orchard, better than you ever could have imagined.

How to Reach Optimum Health

Taurus has a slower metabolism than other signs, so make sure you exercise daily, drink plenty of water, and take iodine to keep your thyroid functioning well. Your throat is particularly vulnerable, so wear scarves when it is cold. Also, do stretching exercises for the neck. Cut down on salt and starches so as not to retain water.

Take long walks and breathe in the smells, sights, and sounds of nature. Walk in your bare feet on cool dewy grass. Treat yourself to regular massages, both to relax you and to aid your circulation. Above all, find a sport or exercise regimen that you like. Enroll in a dance class to show off your great sense of rhythm and to indulge your love for music. Remember: Joy and love make you healthy; worry and repressing emotion make you unhealthy. Share what's troubling you!

Your Lucky Foods, Element, and Spices

Root vegetables, such as carrots, beets, turnips, parsnips, and rhubarb, are excellent for you, because they grow *under the earth*. They ground you. Your best foods contain sulphate of sodium, Taurus' cell salt, which controls water gain. These are horseradish, spinach, asparagus, cauliflower, beets, onions, celery, pumpkin, cranberries, raw nuts, and Swiss chard.

Also, eat as many brown-colored foods, such as brown rice, potatoes, miso, and cooked grains, as you can find. (This does not include chocolate, however much you love it!)

For iodine, eat kelp, seaweed derivatives, and seafood.

Tarragon and cinnamon are your best spices.

Your Best Environment

Your best environment is in or near nature, with as many animals about you as possible. It's also lucky for you to bring the things of the Earth into your home. Add live plants to your environment. Fill your home with tile, pottery, wicker, ceramics, rocks, and crystals. Invest in a fish tank with beautifully colored fish.

Because Taurus rules the throat, you need fresh air and wind around you. Taurus Jupiter Keanu Reeves got his name "Keanu" from his Hawaiian grandparents, meaning "First Breeze over the Mountain." You need rain and water in order to be fertile. You love the seasons, even very cold winters. Remember: You are an Earth sign!

Your Best Look

Taurus Jupiters have appealingly round faces, no matter what their body type. Being plump doesn't hurt you—in fact, you may find your greatest success (and compliments) when *you put on weight*. Ooze the feeling of flesh, of Earth. Black hipless designer dresses are not befitting to a successful Taurus Jupiter. Be Venusian—soft and elegant, in flowing shapes and colors. Wear cornflower blue and see the attention you receive! In your daily life, chunky, thick sweaters look fabulous on you, worn with streamlined pants. Jeans are not your best accessory unless they are tailored. Keep your throat bare for serious sex appeal. It works for Taurus Jupiters Courteney Cox and Courtney Love!

Your Best Times

The luckiest times for you are when the Moon is in Taurus, which happens every month for two to three days. Find these special days in your astrological calendar. During these days, schedule crucial meetings, tell that special person you love them, ask for a promotion or raise, push the envelope, and plan to shine. In your everyday, the best time for you to exercise is late morning, with a nap midafternoon. The luckiest times during the year for you are April 20–May 20 (when the Sun is in Taurus); August 24–September 22 (when the Sun is in Virgo); and December 22–January 19 (when the Sun is in Capricorn).

The very luckiest time of your life—as in Lotto lucky—is when Jupiter reenters the sign it was in when you were born (Taurus), which happens every twelve years and lasts anywhere from a few months to just over a year. This is called your Jupiter Return, and it launches your luckiest streak. You can count on the *extraordinary* to happen the very day your Jupiter Return begins. Amazing events can start happening up to six weeks beforehand, so plan ahead!

How to Be Lucky Financially

Since your luck comes to you through patience, hard work, dedication, and determination, in order to gain wealth you must avoid anything that involves cutting corners or that seems too quick, too good to be true. The miracle of compound interest was made for you, Taurus Jupiter—expect your wealth to be substantial but to take a long time arriving!

Your success will come if you stick to the eternal principles. Take your own sweet time and ignore the trends everyone reads about in *Money* magazine; they should not concern you at all. Instead, put your money away in a mutual fund and make a bargain with yourself to stick with the same, conservative strategy for the next few decades before reconsidering it.

And, even more important, don't focus too much on investing in stocks, real estate, or bonds. Your greatest luck may come if you invest in yourself. Start a business—even a very small one—work long, hard hours to make it successful, stick to the course, and be very determined, and you may end up with greater riches than you could have ever found on Wall Street.

You may also find that your Jupiter energy comes when you think about higher goals than money. What about changing the world, protecting the Earth? You might want to try a socially responsible investing fund, geared to your personal beliefs, or you might want to give much of your money to charities that support your personal goals.

Success in Family and Friendship

Because your Jupiter sign is an Earth sign, you will experience greater success if you work hard on your relationships and if you keep them stable over a long time. Long-term relationships are lucky for you. Your best bet, once a course is embarked upon, is rarely to deviate from it.

You are a natural protector and Jupiter energy blesses you if you have children. If you decide not to have children, make friendships and family connections in which you can protect people who are vulnerable. This role is blessed by Jupiter and will bring you intense luck.

You also should try to be constant in your relationships and in your feelings. Stay steady and loyal to family and friends, however your Sun sign may be feeling.

The one time you should deviate from your long-term thinking is when a major relationship or friendship is unhealthy. Learn to periodically question your relationships and friendships, and if any are truly negative or unhealthy, let them go. This is not a license to be jealous or suspicious. Just be perceptive; be sure you continue to see life as it really is and act accordingly.

When you go too far!

♉ Taurus Jupiter Excesses

Aesop's fable of "The Ants and the Grasshopper" beautifully illustrates the wisdom of Taurus Jupiter planning and hard work. While the grasshopper hopped and partied all summer, the ants worked and worked, slaving away to store food for the winter. The grasshopper pitied the ants for having no fun, no life. Yet, when winter came, guess who survived and who did not?

Remember: There are many different ways you can lose in life. You can lose yourself, your sense of fun, joy—and pure silliness. You can work so much that you lose spontaneity and heart. You can plan so much that you lose the beauty of the immediate moment. "Stop and smell the roses" was certainly written for Taurus Jupiter! Savvy one, take a tip from *both* grasshopper and ant—take the good and leave the misery. You deserve to be happy, in your immediate life, just as much as the next person.

Excess #1 You Can Become Repetitive

Taurus Jupiter Pablo Picasso said to repeat oneself is to go against "the constant flight forward of the spirit." Find the perfect formula, but please keep it from possessing you! Reproducing and replicating *products* bring luck (just think of Chester Carlson or Andy Warhol), but repeating *yourself* does not. To repeat yourself over and over is very unlucky for you. It distracts you from the Earth, from sensuality, from being yourself. It can also be unlucky for you to repeat the same actions over and over; if something doesn't work, do something different or stop what you're doing!

In their worst moments, every Taurus Jupiter has been told they sound like a broken record. This happens when you begin to worry about something. The worse you feel, the more you repeat yourself, as if repetition could make the bad go away. It doesn't! When this happens, take a break. Go outside, open your senses, smell, feel, touch, and taste. Draw on that luxuriating that Taurus does so deliciously. Become *yummy* again.

Excess #2 You Can Become Obsessive

Taurus Jupiter energy is magnificently concentrated! Use it full strength on work and life goals, but please dilute in your everyday life. If you don't take that break, you can slide down the hill into obsession.

Multimillionaire Howard Hughes became lucky by saving and preserving, but was unlucky when his Taurus Jupiter preserving instincts became extreme (he was infamous for his obsessive controlling actions—he even employed the CIA to help him strike break his airlines). In his later years, his obsessive need to control extended ever further: He became deathly afraid of germs.

The actor Jane Greer recalls having dinner with Hughes once. He excused himself to go to the restroom and came back, twenty minutes later, his shirt sopping wet.

"Howard, your shirt is wet!" Jane exclaimed.

"I had to wash it several times. There's bacteria everywhere," Howard replied.

DON'TS

✳ Make impulsive decisions.

✳ Give in to changing trends.

✳ Take possession without responsibility.

✳ Live in very hot climates (cold is better for you).

✳ Forget that your actions have consequences for others.

✳ Cut yourself off from nature and the Earth.

✳ Lose your sense of humor.

There would be no life without bacteria, but Hughes was so obsessively afraid of it that he became a total recluse and ended up having no life. Don't let this happen to you, Taurus Jupiter.

A mantra for when it does—one thing you *can* safely repeat over and over!: *Ships are safe in harbor, but that's not what ships were built for.*

Excess #3 You Can Become Self-depriving

Taurus is famous for its sense of luxury, its taste for and love of beautiful things, its delicious sensuality and appreciation of abundance. Yet due to the excesses of Jupiter, these very qualities can backfire. Notice how the members of your tribe already mentioned fasted, walked, and suffered for their causes. Those who deprive themselves, either of food or other forms of self-care, are numerous. Taurus is an Earth sign, and its earthy element needs to be respected. In our modern cerebral age, it is all too easy to let productiveness become solely an activity of the mind. Resist this, strong one. Keep focused on your body, on your relation to nature and to Earth.

Look at the specific areas in your life in which you go overboard. Where do you feel deprived? Ask yourself if a moderate, more grounded Taurus-Earth approach could help. Breathe deep, into the moment. And let those you love, and the things you love, take care of themselves. They'll manage better without your worry.

Sun Sign-Jupiter Sign Combinations

Consider yourself the protector of a sacred flame—which you need to keep steady and channeled in the direction you want your life to go. The mix of your Jupiter sign and your Sun sign is an extreme combination of Fire and Earth. You need to accept responsibility (Taurus) for an unusually rare talent (fiery Aries). It's the tension of handling such creativity that can drive many Aries-Taurus bonkers.

You may struggle with this conflict, but because you are a Taurus Jupiter, you will last, and because you are an Aries Sun, you will triumph. Because Aries is the "baby" of the zodiac, you will doubtless be successful early in life. A great example of this is Aries-Taurus Haley Joel Osment, who triumphed at the age of eleven in his role of Cole in *The Sixth Sense*. Your fire will become more and more manageable as you build a safer and kinder home for it. Learn to be your very own talent agent for the unique gifts you bring to the world, and then help them grow.

You work so hard sometimes that you forget to stop. Remember: You need love and pleasure as much as the next person. Make time for relationships! Some double Taurus are guilty of seeing themselves as products to be given to the world, humbly sacrificing themselves to no end. You adore children and would do anything to make their lives better. How about your inner child who still suffers from earlier deprivation? And remember that if you let yourself get angry sometimes, it really wouldn't hurt you.

Most of you double Taurus have the fortune to be born with natural grace and good looks, but also the tendency to downplay them. Natives possessing this combo have a famous self-deprecating humor, such as writer-director double Taurus Nora Ephron. Allow others to enjoy beauty as much as you do. In this case, it just happens to be yours.

You, like Gemini-Taurus Bob Dylan (who asked how many roads must a man walk down before you can call him a man), see possibilities forever stretching in the distance. This is a blessing of foresight, or a curse of worrying how to fulfill every aspect of your potential! Choose one or two roads that make you happy, narrow things down, and work like mad. You are often politically minded. You are determined to build a better society (Taurus), and you have the razor-sharp mind and evaluation skills of a critic (Gemini).

Gemini-Taurus Katharine Graham, owner of the *Washington Post*, encouraged her reporters in breaking the Watergate scandal and won the Pulitzer Prize for her eight-hundred-page memoir, *Personal History*. In it she wrote with typical Taurus cognizance, "What is here is what I know best." Know yourself and work on what you do best.

You can succeed at anything you set your mind to, and most of you will be wealthy in old age. Both Taurus and Cancer are about preserving and saving—more than any other sign. Cancer-Taurus Bob Crane, of *Hogan's Heroes*, became famous for his self-made collection of body flicks, while Cancer-Taurus José Canseco is best known to baseball card collectors. You have a stellar career as an art collector or historian if you so choose!

The problem is your not being able to let go of relationships and circumstances that are not serving you well. Perhaps, at one time, that relationship made you very happy, but you tend to replay the lovely moments so often that you don't change what you need to change. As you are protective and caring, people don't want to let go of you. For your greatest luck, look realistically at how you want to be living in, say, twenty years from now. This gives you your answer as to how to proceed.

Your Taurus Jupiter will preserve your Leo greatness forever! Just look at the king of reproduction, Leo-Taurus Andy Warhol. This royal combination makes for childhoods in which you are already thinking in adult years and possibilities—a combination of the idealism of Leo and the eternity of Taurus. Rather than swinging from one extreme to the other, torn between fiery impulse and earthy caution, trust yourself to choose which is right at the time. Both will help you, in different periods of your life.

Leo-Taurus exhibits great ambition, but sometimes this perfectionism backfires in the realm of romance. Let love be less than perfect. At the same time, prevent yourself from marrying the first person you fall in love with. You want passion to last forever. Just remember: That raging Leo fire is best kindled and kept alive in a safe and tightly built fireplace.

ᛜᛜ

IF YOUR SUN SIGN IS:

Virgo

(August 24–September 22)

Devotion is intensified with these two Earth elements. Your work ethic is strong and you give all of yourself. You give your blood. (In the case of Virgo-Taurus filmmaker Brian De Palma, *lots and lots* of blood, as in the brilliant *Carrie*.) You will do more than your entire generation, yet you will be humble. Play a little!

You make the best writers and editors—or nitpickers—of your time. It's your choice. Once you learn to let yourself off the hook more and to enjoy life, you blossom. This is the best time, also, for you to choose your life partner. The preservation principle is superstrong in you. Until you can loosen up and communicate more openly, however, few may know of your ideals in this direction. Practice speaking aloud (and loudly) what you wish for every day. Your power to manifest is awesome with this double Earth placement, so remember to wish for what you can really live with!

ᛜᛜ

IF YOUR SUN SIGN IS:

Libra

(September 23–October 22)

You, lovely soul, are doubly ruled by Venus, goddess of love and beauty. Thus, optimistic Libra-Taurus will do very well . . . and ask for even more. The steadying influence of earthy Taurus is good for scale-balancing Libra. You will have an easier time making decisions because the strong Jupiter energy will help you come down from the cerebral clouds of the Libra sky. Taurus, the planner, will help you think ahead in concrete terms. The indulgence of each Venus sign will support each other, adding discipline. The first-ever Family Planning Clinic was born Libra-Taurus on October 15, 1916. It is always good to plan ahead, Taurus says.

Although your Libra Sun will impel you to fall in love many times and to be in love with love (many times), remember that unless you have a partner for the long term, your Jupiter luck may run out. You may struggle through a year or two of discomfort in the beginning stage of commitment, but stick with it, and you'll be glad you did.

Whether it's through creative genius or through outright ballsiness, you may be tempted to mess up the status quo. You generally create some kind of controversy in your life and become famous for it. The "rock star" poet Anne Sexton was a famous example of this Scorpio-Taurus combination, writing and singing and loving her way through "bedlam." People like you and are drawn to you, even when you want to burn your bridges, Scorpio-style, and to make a run for it. Your lucky Taurus insists on you staying, sticking it out, toward success.

This applies to your love life as well as to your career. Success is likely to come early in your professional life and later in your love life. Both Scorpio and Taurus are very fertile, so you may be blessed with many children—if that's what you want.

Here the unconventional play fire of genius meets the solid citizen. There may be no other stronger opposite. Your Jupiter will let you know which to trust at which time: when to let go of the Taurus love of possessions and inwardness and when to go for the expansive impulse of Sagittarius, and vice versa. Sagittarius-Taurus Julie Taymor, choreographer and director of the Broadway hit *The Lion King*, describes her real-life journey along the rim of an erupting volcano thus: "I shed myself of everything I was carrying and let it all fall . . . by concentrating on this line and moving slowly, I was able to get there."

If you are Sadge-Taurus, your luck comes unexpectedly and suddenly, and you must be prepared, even as a Taurus, to jump, and, then, make it last. Your Taurus makes you fear change; your Sagittarius makes you love change. The fusion of these elements creates the miracle!

This combination confers the most prestigious talents on the mortal lucky enough to receive them. You make a surprisingly charismatic and fantastic leader. Those of you who prefer to stay at home create much photographed, beautiful homes, or equally tempting world places, such as those created by Capricorn-Taurus A. A. Milne in the world of *Winnie the Pooh*. The external image is nearly always perfect: Just don't forget that the inner you needs recognition and love (and honey!) as well.

Being double Earth, work more on throwing caution to the winds when it comes to love. If you deprive yourself too much of the moment, you may impair your judgment in the moment, wait until it's too late, and then marry the wrong person. The extremes in marriage are found here: happiest ever, and unhappiest ever. Your choice.

᪾᪾᪾᪾᪾᪾᪾᪾᪾᪾᪾᪾᪾᪾᪾᪾᪾᪾᪾᪾᪾᪾᪾᪾᪾᪾᪾᪾᪾᪾᪾᪾

IF YOUR SUN SIGN IS:
Aquarius
(January 20–February 19)

Uranus, Aquarius' planet, rules the unexpected, and shocking ideas. Channeled through Taurus luck, your Aquarian brilliance will produce an invention, something new in the world. Taurus will popularize it and make sure it stays around! Just look at Aquarius-Taurus Jim Jarmusch's weird and wonderful films.

In love, too, Taurus saves you from being a complete oddball. You can retain your individuality and free spirit while having a lovely, cozy home (and partner, if you choose). Rather than being torn in two by these completely opposite types of behavior, simply train yourself to go at different speeds at different times. If you're lucky—and you often are—it will be like living one split second for decades after. Aquarius-Taurus Joey Fatone, whatever else he does, is crystallized in time forever to millions who will always associate him with their teenage worship of ★NSYNC.

᪾᪾᪾᪾᪾᪾᪾᪾᪾᪾᪾᪾᪾᪾᪾᪾᪾᪾᪾᪾᪾᪾᪾᪾᪾᪾᪾᪾᪾᪾᪾᪾

IF YOUR SUN SIGN IS:
Pisces
(February 20–March 20)

You have here the mystery of the Fish with the practicality of the Bull. The best way to use this bizarre alchemy is to produce magic. Incredibly talented actors like Pisces-Taurus Gary Sinise manage to pull this off while still sounding so normal in person. Keep your childlike Pisces hopes high while working on something practical and mundane that the rest of the world may need. Remind yourself constantly how much you are helping people—and you are.

In love, you will experience extremes. Your greatest joy will be in sharing and physical connectedness; your greatest danger is self-sacrifice and being unable to leave an unhealthy situation. Friends are especially lucky for you when they offer advice. Listen to the Earth signs (Capricorn, Taurus, Virgo) and forget the rest. You get increasingly happy, even ecstatic, as you grow older.

Relationships with Taurus Jupiters

Finally, for those of you who aren't Taurus Jupiter yourselves, but who have a Taurus Jupiter spouse or significant other, boss, child, or parent who is, here are some great tips to help you learn to deal! This is how the luck of someone else's Jupiter touches *your* life.

Your Significant Other

Taurus and Jupiter are opposites to each other. Jupiter embodies giving and expansiveness; Taurus, saving and preserving. Sometimes one takes over, sometimes the other. It's difficult being in perpetual battle, and this combination is confusing! If you are in a relationship with a Taurus Jupiter, you may have had a very chaotic period early in your love affair when your significant other faced these opposites fighting each other inside.

When you first meet them, Taurus Jupiters will charm you with exquisite manners. Don't believe that innocent look! They will have been around the block, even if they are puppy faced. They are chronologically young, experientially old. You will be wooed, caressed, and flattered. Then, when you're in hook, line, and sinker, something happens. They start to go out a lot. Perhaps they're not available as often as you like. They have a family problem and it interferes. Remember: They just don't expect a good thing immediately. Give them time. Reassure them. Work on strengthening your own life while you wait.

Taurus Jupiters get better with age. If you can hold off and meet your Taurus Jupiter when properly aged, you will have found yourself a rare vintage, like fine wine. Your relationship will last and you will be well loved; you will have a comfortable home and pleasant days. Your Taurus Jupiter may love you, but be prepared to hear the refrain, "But I've been killing myself working" or "I have to go now"—just when you want to talk about something.

Taurus Jupiters are hardworking, and they expect you to work as hard as they do. They so love to work and to make money that they will also want to do this with you. Great professional partnerships can be had with this lucky Jupiter sign, as long as you're willing to work hard too!

Your Boss

Your Taurus Jupiter boss will want you to write down what you're doing, or what you've done, every fifteen minutes. They love checklists, notes, memos, records, plans, diagrams. If you're not into paperwork, this is not the right boss for you!

These micromanagers are superconcerned about the details, and, to some, it can be crushing. I knew one Aquarius-Libra who would sit down every four months or so, write out everything she'd done in one fell swoop, then present it to her boss. She gave him what he wanted, but not at the time he wanted it. He kept an ongoing surveillance tape, which he saved at the end of every day. He liked to check up, in a timely fashion.

Taurus Jupiters are like plants—they need to be watered on a regular schedule. They work well with schedules and routines, and they prosper with detail. They worry endlessly, but they will calm down if you reassure them. They like production and tangible results (especially profits and money!), preferably every day—no matter how minute. If you need to be pushed in that direction, they will push you. If you're a flower child who prefers the loose, the easy, or the spontaneous, you might find this treatment from your boss rigid and difficult. Then, on the other hand, it might be exactly what you need. Taurus Jupiters produce some of the hardest workers around—themselves and others.

Your Child Oh wow, are you in for a treat! These children are delightful. They are the cutest, sunniest, most entertaining dolls you've ever seen. Smart, too. But when you turn to look the other way, watch out! They may bawl. They want your complete and undivided attention for now and forever more, till kingdom come. And no ifs, ands, or buts about it.

Taurus Jupiter children are really special. They perceive the world very differently from other children. They are extra-sensitive, without appearing to be so at all. Their element is Earth, and, like earthworms, they begin their lives sifting through the soil of buried family problems. You will be proud of your Taurus Jupiter children. They will do brilliantly in school (or not—but they're still near genius); they will charm thousands in their lifetime, and eventually, they will get it together and have a happy old successful age. They will not be discontented at the end of their lives—they will be satisfied—and that's more than most can say. Give them plenty of rein, sit back and try to relax as you see them stumble and founder through all that earth, and then, when you're ready to retire, you will realize it's okay. You did a fine job. This earthworm is turning into a responsible seeing human being after all. Bravo.

Your Parent

It is hard to have a Taurus Jupiter parent. I know, and I speak from experience. My father is one. They want the best for you, but they just aren't there. Their minds—and sometimes their bodies—are somewhere else. (I did not meet my father until I was seven years old.) Somehow, even though your Taurus Jupiter parent may have been making merry when you were born, worry sets in for the child. Whether you are sheltered or not, this worry can eat away at you for years into adulthood. Even if your father is a nine-to-fiver and you saw him every day of your childhood.

Even though Taurus Jupiters have all the perfect ingredients to make the very best parents—kindness, concern, genuine thought and feeling, humor, and care—they take a long time to get there. This is the challenge of Taurus Jupiters: to wait, for what seems an eternity, for their life to come into shape. Sadly, their children, spouses, and all close to them must wait with them. Some drop out along the way, exhausted by the crisis. But those who hang in are rewarded.

So, children of Taurus Jupiter parents—please give them time. They love you, but they just don't have the faintest clue how to tell you as they still don't feel quite grownup themselves. A ripe Taurus is really sixty-two years old. If your Taurus Jupiter parents are relatively old, you will find them more loving and available.

Nature gives an antidote for every poison. Isn't that the miracle of life? Problems happen, but always, somewhere, there's a solution, and Taurus Jupiter is life's ultimate problem solver! Trust your Earth sense and you can't go wrong. Have faith in your steady, unswerving progress and look forward to awesome results.

Remember, Taurus Jupiter: You are the miracle worker of the zodiac.

Madalyn's Movie Choice for Taurus Jupiter

7 UP Series
"Give me a child at seven and I will give you the man" begins each film. We revisit eleven people in the UK every seven years: They are seven years old in 1962; forty-two years old in 1997. Who by *49 Up*—in 2004—will be rich, washed up, happy, sad, successful? More real than reality TV, preserving these lives, these Taurus Jupiter results wise you up!

Dante Alighieri
Aristotle
Ann-Margret
Joan Baez
Roseanne Barr
Hildegard von Bingen
Boccaccio
Buffalo Bill
Noam Chomsky
Julie Christie
Confucius
Courteney Cox
Davy Crockett
Brian De Palma
Robert Downey Jr.
Faye Dunaway
Bob Dylan
Nora Ephron
Fantasia
Ella Fitzgerald
Calista Flockhart
Saint Francis of Assisi
Mahatma Gandhi
Janeane Garofalo
André Gide
Crispin Glover
Katharine Graham
Che Guevara
Lillian Hellman
Audrey Hepburn
David Ho
Howard Hughes
Jim Jarmusch

James Joyce
Grace Kelly
John F. Kennedy
Martin Luther King Jr.
Courtney Love
Carson McCullers
Henri Matisse
Frederick Law Olmstead
Aristotle Onassis
The Oscars
Louis Pasteur
Pelé
Richard Pryor
Vladimir Putin
Christopher Reeve
Keanu Reeves
Nicolas Roeg
Franklin D. Roosevelt
Chief Seattle
Anne Sexton
Dinah Shore
Gary Sinise
Ringo Starr
Clyde Tombaugh
TV Guide
Velcro
Andy Warhol
Elie Wiesel
Lucinda Williams
Marianne Williamson
Virginia Woolf
Mao Zedong

"Peace is our gift to each other."

—Elie Wiesel, Libra Sun-Taurus Jupiter. September 30, 1928

"As we are liberated from our own fear, our presence automatically liberates others."

—Marianne Williamson, Cancer Sun-Taurus Jupiter. July 8, 1952

*"If you board the wrong train,
it is no use running along the corridor in the other direction."*

—Dietrich Bonhoeffer, Aquarius Sun-Taurus Jupiter. February 2, 1906

*"Happiness is like perfume. You can't pour it on somebody else
without getting a few drops on yourself."*

—James Van Der Beek, Pisces Sun-Taurus Jupiter. March 8, 1977

*"Remember you have another hand—the first is to help yourself, the second is to
help others."*

—Audrey Hepburn, Taurus Sun-Taurus Jupiter. May 4, 1929

*"One does not discover new lands
without consenting to lose sight of the shore for a very long time."*

—André Gide, Scorpio Sun-Taurus Jupiter. November 22, 1869

"I have a feeling, seriously, that we pick our parents before we come into life."

—Robert Downey Jr., Aries Sun-Taurus Jupiter. April 4, 1965

"I hope the fans will take up meditation instead of drugs."

—Ringo Starr, Cancer Sun-Taurus Jupiter. July 7, 1940

"I never needed plastic in my body to validate me as a woman."

—Courtney Love, Cancer Sun-Taurus Jupiter. July 9, 1964

"Be sure you are right, then go ahead."

—Davy Crockett, Leo Sun-Taurus Jupiter. August 17, 1786

♊♃

Spreading the Word

G emini Jupiter, you were born to communicate messages the world is dying to hear! Blessed with amazing powers of expression, likability, curiosity, wit, *and* a social conscience, you have your finger on the public pulse and it is always tapping. By communicating what is honestly inside of you, you cannot help but achieve incredible results. You, the eternal messenger, will be strongly connected to words, communication, travel, education, and the future all your life. It is not surprising that Edmond Halley, born Gemini Jupiter in 1656, was the first astronomer to see and record a transit of Mercury, which is *Gemini's ruling planet* and known from ancient Greek and Roman times as the *messenger.* Whatever you do, even in the privacy of your own soul, remember that every Gemini Jupiter has a universal audience waiting in the wings!

I'm the guy who put the sin in syndication.

— GEMINI JUPITER HOWARD STERN

Five centuries later, another Gemini Jupiter, Anne Frank, began her diary: *"I hope I shall be able to confide in you completely, as I have never been able to do in anyone before . . ."* A few entries later, she was writing from a hidden attic, where, in a desperate bid to secure their safety, her family had cut themselves off from all nonessential contact with the outside world.

Can you imagine any situation less conducive to communication? Her words were discovered after her death in a concentration camp; *The Diary of Anne Frank* has since been translated into fifty languages. It is the miracle of Gemini Jupiter communication that a teenager's private diary could spread a message about the Holocaust to so many.

Speak your truth, and it will live long after you. Your words are that powerful.

Take a moment, look back long and deep over your life, recount as many events as you can, and ask yourself what has made you lucky or unlucky in your life. You'll notice that whenever you act like a Gemini, *you create miracles in your life.* No matter what your Sun sign is, *you get lucky when you live out the Gemini legacy gifted to you by the benevolent planet Jupiter.*

To take advantage of your luck, emphasize your . . .

Be verbal, communicative, truthful, versatile, a lover of learning, analytical, humanistic, witty, and inventive.	♊♃ *Gemini Jupiter Strengths*

But because Jupiter rules excess, watch out for . . .

You can become emotionally distant, superior, or controlling.	♊♃ *Gemini Jupiter Excesses*

Understanding the
Gemini Energy and Influence

Following discovery and cultivation of the land by Aries and Taurus, Gemini gathers information and comes up with ideas to help the community grow. Working as the messenger of others' ideas as well as a source of information, Gemini makes it possible for a social order to exist. Gemini's element is air, ruling intellect, and it is in this realm that language begins. Interpretation and communication are your tools, Gemini Jupiter; your passions are to think, debate ideas, and exchange information. But you don't have to be serious all the time: Naturally brilliant and funny, you are blessed with a humorous, joke-telling side, and there are more comedians born under the Sun sign of the Twins than under any other sign besides Virgo, its sister sign, also ruled by the planet Mercury. Curious and changeable, the Twins love to think, love to talk, love to be in constant motion, and can do all three at once! Multitasking comes naturally to Gemini.

The bountiful gifts of Jupiter will be attracted to you if you make yourself communicative, verbal, witty, and smart. Seek out wisdom in the form of words. Be versatile, enjoy more than one thing at once, and explore every avenue of creativity, learning, and invention.

GEMINI SYMBOL
The Twins

Interestingly, the U.S. Census Bureau records more twins born during Gemini (May 21–June 20) than at any other time. Gemini is the only sign that has more than one person as its symbol—the Twins are of course two people, and this is revealing, as whenever there's more than one person, there's bound to be communication, talking, the exchange of ideas. The Twins also symbolize the dual, divided nature of Gemini. Because Gemini governs duality and opposites, your Jupiter sign rewards those who explore their own contradictions, who hold opposing views at once, and who indulge the complexities of their personality. This is a special gift, as it also allows you to embrace dual energies in others, and to be the most open (to new people, new experiences, and new ideas) of all the Jupiter signs. Make sure you embrace the openness of Gemini, the willingness to try new things and to change your mind.

You will find your greatest luck by living in two realms at once— the cerebral and the emotional, for example, or the abstract and the concrete. Spread your hobbies and activities around so they complement one another, and use different skills. It's better for you to be someone people can't predict or pigeonhole; if possible, always act against type. Give your opposite sides equal time to exist and communicate, or you will feel out of sorts. Your luck, health, and wholeness lie in embracing duality!

The Hands, the Arms, the Shoulders, and the Lungs

Gemini rules the nerves and the parts of the upper body that come in pairs: the hands, arms, shoulders, and lungs. Open your lungs and breathe in the air from new places and new experiences. Use your hands to touch and try new activities, new sensations.

You can double the Gemini energy by using your hands, lungs, and arms to communicate: So, let yourself talk, and talk expressively—communicate with many gestures! While Gemini Suns need to protect their hands, arms, shoulders, and lungs, Gemini Jupiters, under the planet that protects, are free to be as eloquent as they want to be.

GEMINI COLOR
Yellow

Yellow is bright and luminous, inspiring energy, cheerfulness, and novelty. Wear yellow and see how much more energetic, cheerful, and interested you become in life. *Vanity Fair* declared in its "New Establishment 2002" that the signature look for Gemini Jupiter movie mogul Michael Eisner is the "yellow power tie." If yellow doesn't match your dress sense, wear yellow as undergarments, and use yellow sheets and towels—anything that comes in contact with your skin. Put yellow into your home through flowers, lamps, curtains, furnishings, and accents on walls. Try it just for a couple of weeks and see a new and fabulous you illuminate itself!

And now your Gemini Jupiter strengths:

♊♃ *Gemini Jupiter Strengths*

Be Verbal You may not be aware of it (unless you have kids!), but over the past decade a new language has spread across the globe. This language is known to millions of children: It is used at the Hogwarts School of Witchcraft and Wizardry. Its author and lexicographer has sold 200 million copies of her books,

Mental Mercury

Mercury orbits the Sun faster than any other planet in the solar system, racing around the Sun at 29.8 miles per second—about as fast as Gemini thinks. Mercury's tiny, but it's got a huge job to perform, delivering energy to the Earth by way of instant messages to the brain. It rules computers, technology, and all fast-processing conduits. Just as Mercury travels at a faster clip around the Sun than any other planet, you are fast, always moving, even if only through thought.

YOUR HISTORY

The Romans called you Mercury, the winged messenger of the gods. The ancient Greeks called you Hermes, the healer.

and at the same time managed to introduce terms like "muggle" and "squib" into the English language. The creator of this eccentric vocabulary? J. K. Rowling, born Gemini Jupiter on July 31, 1965, author of the Harry Potter books.

So, don't be afraid to let your words come out, or to play with words and invent them; they will bring you luck and success. J. K. Rowling was told that boarding schools "don't work" in contemporary fiction and that her books were too long for today's children, with their diminished attention spans. Yet these books are *long*—six hundred, seven hundred, eight hundred pages—and leave kids only wanting more! The power of lucky Jupiter extends to the Harry Potter movies—the actors who play Harry and Harry's sidekick Ron Weasley were also born Gemini Jupiter.

Rowling wasn't the first Gemini Jupiter to achieve success through language. The great Russian poet Aleksandr Pushkin revolutionized literature in his country when he wrote the first poems that used regular, everyday words—as opposed to the formal vocabulary of Old Slavonic—and, inevitably, readers fell in love with the beauty and truth of his vernacular poetry.

Ann Landers was another Gemini Jupiter who succeeded through verbal communication—she was the most widely syndicated columnist in the world, read by 90 million readers *daily*. Gemini's twin power assumed literal form with the success of her twin, Abigail Van Buren, who, of course, was the famous communicator known as "Dear Abby." Each sister harnessed the power of plainly written, homespun advice to bring comfort to millions. Don't let others' criticisms stop you! Improving your verbal skills and playing with words brings you the luck of Jupiter.

Although Gemini is about spreading the word—or many words—you do not have to print or publish to reach your full power. Muhammad Ali's Gemini Jupiter compelled him to verbalize, "I am the greatest, I am the greatest," and wisely, he trusted it. "I figured if I said it enough, everyone would really think I was the greatest." This self-titled "Onliest Poet Laureate of Boxers" continued to manifest his lucky Jupiter traits, inventing words and acting as messenger, even after retiring and being diagnosed with Parkinson's disease. In November 2002, Ali became a U.N. messenger when he traveled to Afghanistan as a messenger of peace. You can be the messenger of, or through, any form you choose. Try it, Gemini Jupiter! Say it enough and see what happens. Actualization occurs through your words.

Be Communicative

You can attract Jupiter's bounty even if you are not a wordsmith; you just have to try communicating in other ways. Just think of one of Bessie Smith's beautiful blues songs, her anguished voice revealing more about pain and sadness than the lyrics on their own could possibly convey. A Gemini Jupiter, Bessie Smith didn't write the lyrics for her songs or play with words in any way, but she did not let that stop her from communicating her pain to her listeners.

Alternately, use your skills to help others communicate. Think of Paul Allen, the Gemini Jupiter cocreator of Microsoft. He isn't famous as an author or speaker; instead, Allen has become successful giving others the tools to express themselves. Millions use Microsoft programs every day to write, talk, explore the Internet, and to develop ideas.

Who said, "Books gave me the idea there was a life beyond my poor Mississippi home," and then carried that idea into the world at large? Oprah has broken silence for millions; the star has helped people through the secret, never-talked-about nightmares of sexual abuse, unhappy marriages, and personal depression. Her book club encouraged Americans to read again, and in keeping with the Gemini Jupiter love of education, Oprah Winfrey has begun to build schools and to create scholarships for girls in South Africa. But she's best known for communicating, for talking about painful subjects that no one else on television broaches.

Such is the power of Gemini Jupiter that communication breaks through even when its twin quality, movement, is halted or prevented. Gemini Jupiter Stephen Hawking, paralyzed and mute, spread the word on space theory. *A Brief History of Time* broke all previous records for longest best-selling title, and the paperback edition reached number one in three days. Although he has to use a special machine to communicate even the most basic things, Stephen Hawking has managed—with the help of Jupiter—to spread his ideas throughout the world.

Perhaps most bizarre is the case of Gemini Jupiter Patty Hearst, who may never have become known if she had not been held hostage by the Symbionese Liberation Army. It was her communication and her shift in personality as she spoke and acted as a terrorist—here again the twin sides of Gemini—that made her famous and, ultimately, gave her a story to tell and a film career.

Be Truthful

You must speak your truth, Gemini Jupiter, no matter how unpopular it may be! "That voice that cries out doesn't have to be a weakling's, does it?" Gemini Jupiter John Osborne, the original "angry young man," wrote in the early sixties. A decade later, Gemini Jupiter Betty Ford shocked a nation by breaking with First Lady protocol and speaking out about her alcoholism. Her truth turned into a campaign, and then to her founding the Betty Ford Clinic, which still continues to save lives. And at the Academy Awards ceremony in 2003, documentarian Gemini Jupiter Michael Moore spoke out against the war in Iraq. He specifically attacked the president for fighting a "fictitious" war for "fictitious" reasons. Tell your truth, stick to nonfiction, and Jupiter will reward you.

If you do not, no matter how many people you reach, your luck will turn against you. Because Gemini's ruling planet, Mercury, was known in ancient times as a trickster, you must be particularly careful not to tell lies. Jupiter creates abundance and succeeds *only* with the positive. Whereas a Gemini by Sun may take shortcuts without misfortune, a Gemini by Jupiter may not. Martha Stewart has expanded into print with two best-selling magazines and an empire due to her Gemini Jupiter, but her star power has been fading ever since government officials began to investigate her for alleged insider trading. So, in spite of the ever-increasing emphasis on the spin of PR and politics, do not be tempted to go against the luck of your Jupiter.

Be Versatile

Since Twins can do two things at once, you find luck in pursuing more than one career; growth and mental exploration are your keys to self-improvement. Because of this, life will often cause drastic changes for Gemini Jupiter—just to get you to fulfill your potential! You are such a great source of information for people and have such a wide range of skills that even if you do stick to one job, you will often find yourself demonstrating your talents outside of work. Aren't you constantly asked to do jobs that aren't yours?

As Gemini Jupiter Jacqueline Kennedy Onassis famously advised, "Don't learn to type unless you want to be a secretary." Her versatility manifested itself in many different life paths and careers, from First Lady to editor at a publishing house. The Duke of Windsor, another Gemini Jupiter, went from King of England to traveling socialite. John Ashcroft changed from a senator who, in 1997, wrote the famed article "Keep Big Brother's Hands Off the Internet" to an attorney general who did exactly the opposite. Aldous Huxley, unable to pursue his career as a scientist, learned Braille, published a witty novel—*Crome Yellow* (yellow for Gemini Jupiter!)—but became most famous for his darkly futuristic *Brave New World*. Actor Jeffrey Wright switches roles—even *races*—with brilliant Gemini Jupiter versatility: In recent years his roles have included a Dominican drug lord, Dr. Martin Luther King, and the artist Jean-Michel Basquiat.

What do *you* want to do, Gemini Jupiter? Make sure you do both, or all, of them!

If your message offers a vision that is at odds with the accepted norms and standards of the present time, chances are even better that it will be successful. From Jane Austen's critiques of English courting rituals in the 1700s to Clifford Odets's social protest plays following the American Depression, Jupiter gifts and blesses all forms of original thought. Even *Playboy* was born Gemini Jupiter!

Be a Lover of Learning

No other Jupiter sign benefits so much from learning as Gemini Jupiter. The messenger hungers for new information and knowledge to share and explore. New facts expand your world, and you get lucky reading and studying different schools of thought—the more independent and futuristic, the better. Whether it be gossip, scientific study, learning how to write that detective novel, or speaking a different language, the learning process attracts the blessings of Jupiter and will improve your life.

Upon graduating from medical school in 1896, Gemini Jupiter Maria Montessori became Italy's first female physician, and in her practice she began to observe children and how they learn. Her realization that *children teach themselves* inspired the Montessori method of education, based on her passion to further the self-creating process of the child.

Remember my warning on shortcuts? Another "progressive" educational tool that spreads the word even further—and other writers' words at that—is CliffsNotes! First born in schools on May 23, 1953, it is Gemini by both Jupiter *and* Sun. Because it is Gemini by Sun, it is okay for it to be a shortcut, and its Gemini Jupiter has made it a renowned commercial success.

Even silent, your messages are powerful. The planets Jupiter, Neptune, and Pluto were all discovered—born—Gemini Jupiter. *Someone* gave me the message to write this book about Jupiter. Who says the planets don't have something to tell us?

Be Analytical

Never be without your sharp and quick mind. As Sherlock Holmes always reminded his dear assistant, "Cold logic and analysis, Watson!" Sir Arthur Conan Doyle established a virtual archetype of Gemini Jupiter through this fictional detective: the objective mind that analyzes and investigates, undeterred by subjectivity or sentiment.

Doyle was born a Gemini Jupiter on May 22, 1859, and his communication was so potent that even today, seventy years after his creator's death, the famous Sherlock Holmes receives more than two hundred letters a month at his fictional London address, 221B Baker Street! (It is okay that this be fictional, as Doyle is also Gemini *by Sun!*) True to his lucky Gemini versatility, Doyle pursued other careers: a medical doctor, a surgeon on an arctic whaling vessel, a supervisor of a civilian hospital during the Boer War, and an introducer of snow skis to Switzerland. For the last fifteen years of his life, he practiced psychic communication. He lectured internationally about spiritualism and other related topics, and for four years served as president of the College of Psychic Studies in London (where this astrologer holds a lectureship!). The diversity of his interests, the pattern of drastic career change, the idealization of the analytical process of deduction—all came together, for Conan Doyle, to bring him the luck and power of Jupiter.

Be Humanistic

Charles Dickens is one Gemini Jupiter who will always inspire us to look at our past, present, and future actions through the perennial favorite *A Christmas Carol*. Dickens's parents and siblings went to jail for debt when he was eleven years old. He was the only one judged fit to work, and work he did—at Warren's Blacking Factory. His factory experience haunted him all of his life—he could speak of it only to his wife and his best friend—but he translated it for others to understand in *David Copperfield* and in many other novels that touched the nineteenth-century social conscience. Dickens campaigned against child labor when it was legal in Britain and advocated the abolition of slavery in the United States. In spite of these unpopular views, he was the most read author of his day. He wrote possibly the most famous beginning of any novel:

It was the best of times, it was the worst of times; it was the age of wisdom, it was the age of foolishness, it was the epoch of belief, it was the epoch of incredulity, it was the season of Light, it was the season of Darkness, it was the spring of hope, it was the winter of despair . . .

Dickens got lucky with Gemini-double opposites! The novel is—appropriately for the Twins—*A Tale of Two Cities.*

Another author with a strong social conscience was Gemini Jupiter Harriet Beecher Stowe, author of *Uncle Tom's Cabin.* In spite of the fact that slavery was legal at the time, *Uncle Tom's Cabin* sold more copies than any other book except for the Bible; Stowe managed to be a messenger for good *and* a commercial success.

Or consider Lebanese author Kahlil Gibran, who believed humanity was best served by contact with different cultures. His most famous book, *The Prophet,* brought together two twins of thought, Western materialism and Eastern mysticism. In twentieth-century America, only the Bible has sold more copies than *The Prophet.*

Be Witty

Gemini Jupiters find success through verbal wit and smarts. You don't succeed through pratfalls as much as through telling the truth in a particularly pointed, often sarcastic way. The humor can be cold and caustic, as with Gemini Jupiter Dorothy Parker. Reviewing Katharine Hepburn's first performance, Parker wrote, "Miss Hepburn's performance ranges the gamut of emotions, from A . . . to B."

Or the humor can be wry, humanistic, and hilarious. Just think of Gemini Jupiter Jerry Seinfeld, who titled his first book *SeinLanguage* (remember Gemini's affinity for language?) and, in his jokes, turned many of our everyday experiences, like going to the movies or buying soup, into moments of unforgettable hilarity. He has amassed an estimated $500 million fortune simply through these "translations."

Gemini Jupiter, there's no one more funny than you when it comes to summing up a particular situation through sarcasm. Use it to success, Messenger of Wit!

Be Inventive

Alexander Graham Bell wrote in 1876 in a letter to his father, "The day is coming when telegraph wires will be laid on to houses just like water or gas—and friends will converse with each other without leaving home." For Bell, it was the first of many glimpses into the world of the future. Is it a surprise that this famous inventor was Gemini Jupiter, or that his greatest invention, the telephone, revolutionized the world of communication?

While working on the telephone, Bell mentioned to his assistant,

---

DOS

✳ Express yourself through writing, speech, or the media. Always tell the truth.

✳ Develop multiinterests, multicareers, travel, and live in different countries for periods.

✳ Surround yourself with yellow (sunny) colors—clothes, sheets, towels, walls, and home decor.

✳ Take time out every day to rest, stretch, and do deep-breathing exercises.

✳ Eat wild rice, cauliflower, spinach, asparagus, tomatoes, green beans, almonds, fish, raisins, apples, and all foods that are yellow. Flavor with garlic, cumin, and curry powder.

✳ Play tennis and Ping-Pong to strengthen your arms.

✳ Always protect and cover your chest when it is cold or windy.

✳ Hang out with more Aquarius, Libras, Geminis.

Mr. Watson, that their next project would be a flying machine. Bell was testing helicopter models as early as 1891—way ahead of his time, as Gemini Jupiters always are.

The telephone wasn't the only invention that revolutionized communications in the nineteenth century. Before the telephone was the telegraph, the invention that allowed us to communicate all over the world. Is it a surprise that the telegraph was also invented by a Gemini Jupiter, Thomas Alva Edison?

To attract Jupiter's energy, Gemini, you must allow yourself to be inventive, indulge creative thought, and allow yourself to explore new ideas. You will achieve particular success if you work to find new ways for people to communicate or to express themselves.

The United Nations, born a Gemini Jupiter on January 1, 1942, was made up of representatives of twenty-six different nations who pledged their governments to communicate with one another to create common understandings and laws. Versatility, Gemini Jupiter, is essential for your success. Icelandic singer Björk communicates in guttural primal sounds, and director Jane Campion communicates to us the passions of a woman who cannot speak in her award-winning film *The Piano*. However you translate the world, we hear it, Gemini Jupiter!

What Makes You Lucky in Love

You have no shortage of admirers. Your modern-day dance card is always full! People are impressed by you, enjoy your company, and they usually want you to stick around. So why do so many of you think you're *unlucky* in love? It's the Twin Dilemma: You're looking for a *soul mate*. Even if you're not ready for one, even if you don't *want* one. Here's where the abundance of Jupiter in Gemini makes it difficult for you. The need for your twin, your second half, is so strong that it makes it hard, if not impossible, for most mortals to measure up. You are selective, and many suitors don't make it past the second or third date. If it's not the real thing, you may just say, just like fellow Gemini Jupiter Greta Garbo, "I want to be left alone." Yet you go on searching. Your Jupiter won't let you stop hoping and trying.

You probably gave up a lot for your first great love, and suffered for it. Look at the list of fellow Gemini Jupiters in this chapter and you'll see more multiple marriages and divorces than you can count—and in Henry VIII's case, other types of separations, too.

Ready for the good news?

You will find your soul mate. You will. You always do.

Look at that list again. All those divorces ended up with the partners finally finding someone who was perfect for them. You'll notice those who find their soul mates early tend to work with their mates, or at least work in the same profession. Pragmatically shared goals and ambitions bode well for your partnerships. As does your giving them a chance! You'll be glad you did.

Your Best Career Moves

Many Gemini Jupiters work in media or communications and it's no accident. But to achieve success you must first and foremost cultivate the quality of honesty. You will be successfully able to lead others *only* if you are honest. Your great advantage in the workplace is your ability to generate ideas. Be sure to verbalize and spread your ideas generously. Others will then talk to you as a valued ally whose insight and creativity will lead them to the top.

As a Gemini Jupiter Twin, don't be afraid to work alone. Your challenge often comes through working in partnerships. Your partners can frequently drive you crazy, but you can learn to deal. You will enjoy most verbal exchanges of ideas, but you still may breathe a sigh of relief when you have the office to yourself. If you don't speak aloud or talk on the phone as part of your daily job, you may need to check your birth date again to see if you really are a Gemini Jupiter! The same applies to involvement with computers, which your planet, Mercury, rules.

You also have a tendency to thrive when you work in more than one job at a time. Don't be afraid to combine roles if you have the time and energy and *balance* to make this work to your advantage. If you are happy in just one, it will usually be because it gives you many different roles to play. For instance, a Gemini Jupiter I know represents opera singers. In this chosen profession, he gets to play agent, manager, personal counselor, businessman, music coach, and critic—and attend operas on a weekly basis, travel to Europe and all over the United States, meet and talk with people every day, go to galas and parties at night, and be responsible for some thirty life-and-death careers (they are professional prima donnas, after all!). All under the cloak of one job title. Did I mention that he owns his own company, used to be a star pianist himself, and still gives singing lessons occasionally? (Oh, and he wants to write detective novels.) Only a Gemini Jupiter could do all that and still find time for more. Routine, repetitive tasks will drive even the most patient of you nuts. You want to be kept busy, and you *never* want to be bored, so you better give up those fantasies of paper pushing!

How to Reach Optimum Health

Gemini runs on nerves more than other signs do, so relaxation is not just a luxury, it is a necessity. Stress affects your breathing and circulation; yoga will completely change your life for the better, I guarantee it! Practice deep breathing and avoid tobacco, coffee, soda, and all stimulants. Exercise had better be entertaining or it won't hold your interest for long, so make sure you can hear music, or talk, or both, while you are—preferably—playing a game that requires fast coordination movements, such as tennis. And please stop snacking on the run and then going all out for dinner at that hot spot you love! Eat several nutritional meals throughout the day; one or two large meals are terrible for you. Finally, your chest is particularly vulnerable and needs to be well covered when it is cold, as you can be prone to bronchitis and upper respiratory problems.

Your Lucky Foods, Element, and Spices

Gemini's cell salt is potassium chloride, which keeps the bronchial tubes unclogged (see above) as well as the lungs clear. A lack of this mineral causes problems with circulation and blood clots. So, you want to eat, in order, the following: wild rice (not white or brown), cauliflower, spinach, asparagus, tomatoes, green beans, apricots, peaches, and oranges; almonds, fish, raisins, and apples; and bananas, lemons, grapefruit, yellow squash, Swiss cheese, eggs, pears, and corn.

Curries are fantastic for you—I hope you like Indian!—as are garlic, cumin, cardamom, paprika, yellow onions, and any and all spices that make you take huge lucky gulps of air! Think *lung* foods, nerve foods, and *yellow* foods to stay your healthiest!

Your Best Environment

You may long for life in the beautiful countryside, but your energy will always work best in an urban setting, in the most cultured city you can find, exciting your intellect and stimulating your curiosity. For those who opt for rural settings, make sure you have easy access to busy social milieus. To live or visit, San Francisco, Minneapolis, Montreal, Melbourne, London, and Brussels should keep you singing, happy, and healthy.

Your Best Look

Dress *city*. Even if you live in the suburbs, tie a jaunty scarf round your throat, some accessory, something that moves as you speak. The spotlight needs to be on your mouth—always. Women, wear your bold red lipsticks as much as you like! Fun dress-up is good for you, and spots, stripes, and circle designs bring you luck, as do rings, bracelets, and anything that accessorizes your hands.

Your natural body type is thin and nervy, which is perhaps why you're always dieting. Strong colors around the shoulders and black is a must. Gemini Jupiters are born with a beautiful bone structure and it is a sin to hide it.

Your Best Times

The luckiest times for you are when the Moon is in Gemini, which happens every month for two to three days. Find these special days in your astrological calendar. During these days, schedule crucial meetings, tell that special person you love them, ask for a promotion or raise, push the envelope, and plan to shine. In your everyday, the best time for you to exercise *and* rest is midafternoon. The luckiest times during the year for you are May 21–June 20 (when the Sun is in Gemini); September 23–October 22 (when the Sun is in Libra); and January 20–February 19 (when the Sun is in Aquarius).

The very luckiest time of your life—as in Lotto lucky—is when Jupiter reenters the sign it was in when you were born (Gemini), which happens every twelve years and lasts anywhere from a few months to just over a year. This is called your Jupiter Return, and it launches your luckiest streak. You can count on the *extraordinary* to happen the very day your Jupiter Return begins. Amazing events can start happening up to six weeks beforehand, so plan ahead!

How to Be Lucky Financially

Take advantage of your adaptability when you look for financial opportunities. You may find it easier to make money through ideas, invention, and creativity rather than through investing and financial planning—try to make money betting on yourself as opposed to betting on the markets. Can you see any way for you to start a consulting business that's focused on communications?

Careers in writing, graphic design, corporate communications, broadcasting, public relations, and strategic planning will be lucrative for you.

Invest in yourself and explore multiple careers. Change and transition are lucky for you; staying in any one job for too long is unlucky for you. Whatever your field, remember: The more you communicate, the more money you make.

In the financial markets, choose diverse companies in completely different industries, but pay particular attention to companies in media and communications. Make sure you communicate a lot when you're making financial decisions: You will get great ideas from Internet chatrooms, TV networks, reading the newspaper, and comparing ideas with friends. Working alone and not talking to others is a bad idea.

Because you are—literally—mercurial, don't stick with any particular investment for too long. Quick changes are, as you know, lucky for you. Trade, trade, trade!

Success in Family and Friendship
Because you have an Air sign for your fortune, connection with family and friends by telephone, e-mail, or letters needs to be constant. Your family relationships will bloom under Jupiter's bounty if you stay in touch, send postcards, e-mail all the time, or even set up a family website. Gemini Jupiters even have that rare ability to stay friends with past lovers—so long as you keep in touch and keep communication open.

Many of you choose to live far from your original home, but Gemini Jupiter loves any excuse to travel, and this allows you to make more trips *home!* You appreciate distance, and absence does indeed make the heart grow fonder. You have friends from all over the world (more places to visit), and even in your own hometown are likely to have at least one friend who is from another country. Keep it up—having many interesting friends is lucky for you.

Gemini Jupiter Excesses ♊♃	*When you go too far!*

Frankenstein is an excellent Gemini Jupiter cautionary tale. It is Mary Shelley's story of a brilliant doctor who uses his mastery of science to bring a dead body back to life, creating a monster who kills those the doctor loves most—his brother and (on his very wedding night) his wife.

Take a lesson from Dr. Frankenstein, who tried to play god: Be wary of living life as if it were an experiment that you can control in your own "laboratory." Watch when you fall into the habit of over-analyzing and manipulating life as if it were a chess game. Sometimes, in your efforts to help others, you plot parts of their lives. Your motivation may be kindness, or, as in Dr. Frankenstein's case, an unconscious wish to create a second self (Gemini twin!), but accept that no one has complete control over life. When you find your brain taking over your heart and soul, make an effort to get back in touch with your feelings.

Excess #1 *You Can Become Emotionally Distant*

Like Sherlock Holmes, who famously avoided intimacy, Gemini Jupiters can become cut off from those they love, as well as from themselves. Gemini is an Air sign and Air rules the intellect. Furthermore, its ruling planet, Mercury, is concerned only with facts, information, and pure reason. Mercury has no connection with emotion.

Your intellect is such that it can be a delight to analyze, correct, and criticize, but reel yourself back in before you forget what you're actually feeling. It is important to remember that emotions become overwhelming if suppressed or denied. We can appreciate this when we see the fear that cerebral types can have toward their feelings. Gemini Jupiter, you need to allow your brain and feelings to coexist. Let your emotions multitask along with your thoughts. Remember: The Twin is not *complete* without its dual or opposite side!

Excess #2 *You Can Become Superior*

It is then hardly surprising that Gemini Jupiters feel superior—especially when they seem to be so much faster and brainier than others. Our current society encourages and rewards faster and faster transmission of messages. Speed seems to have become a virtue in itself. As our modern lives develop a shorter attention span, media speeds up even more, thus creating even *shorter* attention spans. It is important to remember that human feelings are still processed slowly, and true wisdom takes a long time to develop.

DON'TS

✱ Lie. Ever.

✱ Stay silent, cut off from people, or deny your self-expression.

✱ Use your charm for manipulative purposes.

✱ Stay in the same job forever.

✱ Deny yourself an education.

✱ Smoke tobacco or be in an environment that is restrictive to your lungs.

✱ Stop having fun.

So, don't make the mistake of thinking that just because you transmit and receive information faster than most other people, that ability automatically makes you wiser or more emotionally mature. Pay attention to your emotions; give them a chance to catch up.

Excess #3 *You Can Become Controlling*

From an excess of smarts comes a need to use them for one's own entertainment. Geminis rarely manipulate out of malicious intent, but more because they are *bored*. They have a need for their minds to be doing something interesting at all times!

The famous myth of Hermes (Mercury) stealing the cows from his brother Apollo illustrates a moment when the wish for control becomes denial. When his crime is discovered, Hermes charms and manipulates Apollo into not punishing him, and thus receives no consequences. Because you are so far ahead of everyone else, you will often have the upper hand, particularly in a debate. But take care, Trickster Gemini Jupiter, not to abuse a gift! Consider the long-term effect that your current feeling of control may have on your life. Might it not isolate you and ultimately deprive you of yourself as well as others? Ask yourself what truly lies underneath your sense of power. It might be momentarily reassuring or strengthening, but in the end it gives you—and leaves you with—very little.

Sun Sign–Jupiter Sign Combinations

Your Aries desire to be first provides kick-ass energy to your Gemini who *knows* you're first. These two work great together as a team! Think Aries-Gemini Sarah Michelle Gellar, who slayed vampires for seven seasons to show the world what a smart, cute, and funny actress she is.

You're sexy, too, as long as you keep it offbeat. Variety and adventure make you even hotter—*Gemini air fans those Aries flames!* You often find yourself wanting more, more, more; sometimes life just isn't enough for you. You are practically allergic to routines (except in stand-up comedy), and ambitious desires make the world turn. This combo can be dangerous though, with accidents and wandering into situations where you're not wanted, so travel with chaperone-friends to lighten the ride. You break a lot of hearts before you marry.

You long for security, homes, and money, but these are best obtained through Gemini traits of noncommitment. Be many-sided, impatient, travel, hang your hat everywhere, don't hoard, and resist routines! It's kind of like speaking Spanish to learn French, getting what you want by hook or by crook. Enter into the hook-or-crook role with complete gusto and believe that it will work. It will. Modern dance doyenne Martha Graham was born a Taurus-Gemini, and with her troupe, she changed the language of dance and movement the world over.

You are such a hard worker! Make yourself take time off for love (or the search for it) and playful moments whenever you can grab them. People love you when you're funny; remember this when you schedule all your important meetings.

You're no pushover. You find success in being exactly who you are. Except, well, you *are* double Twins, and you change who you are about every four hours. Your moods float through you like light and shadow. If you accept this—this restless rippling in your being—you will make yourself a chameleon *star* at work and irresistible in love. This combo is found in actors like Gena Rowlands,

John C. Reilly, and Alfred Molina—magnetic beings who change in a flash, revealing mysterious depths to their characters.

The search for your soul mate is all-consuming, even while others are falling hard for you. When you find someone as brilliant as you, then you fall hard, too. You need to be able to share your work and ideas, as much of the attraction will be intellectual. Your absolute ideal mate would be another Gemini Jupiter in the same profession—as Gena's John Cassavetes was.

You are sensitive and devoted to family, but if you're smart (and you are when you cease to mourn the past), you'll resist the urge to settle down early. You can see far enough into the future that you know you'll satisfy that need for a comfy crab-shell home someday, but just not now. You have a great life plan but be wary of making it so tight that it casts a stranglehold on love relationships. Either you resist love, or you take your love off into exile, as the Cancer-Gemini Duke of Windsor did. All to yourself!

It is likely that part of your past was painful and loss is a subject you know much about. Express this through creativity as much as you dare. Writing and collecting bring great success, but only if you make your talents public! Relax, and show off your private self more.

How hard can it be for a Leo to communicate with the world how great he or she is? No problem! Leos love to be great and Geminis love to talk—so you guys should have a ball. Your ideas will spread like wildfire. Just look at Leo-Gemini Ken Burns and his history of jazz, or at Leo-Gemini Sam Mendes's plastic bag blowing here, there, and everywhere in his film *American Beauty*. This is a near-perfect combination, with each side of you complementing the other. Tone down the Leo exaggerations and tall tales—remember, this is such bad luck if you are Gemini Jupiter.

Your presence is overwhelming, and people will both love you for it and be overwhelmed. In truth, it's hard to find the perfect love, as no one really seems to measure up! Expect less, sacrifice less, and your partnerships will shine with joy as much as you do.

Virgo and Gemini are both ruled by Mercury, so you get extra helpings of brilliance, potential psychic ability, and a capacity to heal everyone's problems. The latter may not be so good for you as it is for everyone else! What about your own healing?

It's hard for Virgo-Geminis to ask for help, because they are private and responsible individuals, more demanding of themselves than anyone else. Greta Garbo, who famously said, "I want to be left alone," was one, as is actor Shannon Elizabeth. (When *Rolling Stone* asked about her sex life, she became insulted and ended the interview.) In love, tell partners what you need, or you may search forever. Be careful of your intellectual superiority, which can keep you sheltered under a calculating coolness. However smart a kid is, he or she should still always be able to play with the other kids! Your superlative work must extend into *all* aspects of your life, including your happiness.

Oh, it's fun to be you! You may complain that people don't take you seriously, but don't you see how much joy you spread when you're acting silly and goofy? You are a brilliant, natural comedian, and for your greatest success *don't hide it. American Pie*'s Sean William Scott is an example of this young-at-heart Libra-Gemini appeal. At the other end of the world, Ken Saro-Wiwa, the Nigerian writer, environmentalist, and human rights activist, goes after his arguably more dangerous goals with the same largesse and joie de vivre.

You bubble over with ideas, quick as lightning, and seem to have a charmed life. But you are actually very sensitive and need to learn to brush off stinging comments, of which there could be a few. After several early disappointments in love, you get happy. You work best in ensembles and partnerships, and do well letting others take credit.

Scorpio is deep and intense; Gemini lightens. Scorpio is private; Gemini loves to gossip. This is not conflict; this is team effort! For instance, you are frighteningly sexy, but your lucky Gemini doesn't want you to frighten anyone. You have a ton of secrets that you invite the world in to see but—well, just the tantalizing invitation is enough. You like to be center stage, in all your gorgeous mystery, surrounded by a fine and graceful aura. Grace Kelly beautifully communicated this elegant coordination of Scorpio and Gemini. Some would say she had a more enjoyable time playing a princess than actually being one.

You are brilliant at ferreting secrets out of people. Use your natural detective skills in whichever field you work, and they will help you rise to the top. Add some frivolity to your emotional life, and you'll find your perfect mate. And no more performing in private!

Travel is your lucky touch, as is transforming—traveling within— yourself. Make sure you do a lot of it. The unconventional way makes you happiest, and your greatest success comes some way down the road. Sagittarius-Gemini Kim Basinger, in spite of her conventionally beautiful looks, won her Oscar at the unconventionally Hollywood age of forty-seven.

If you must stay at home, bring foreign languages into your life, cultivate contacts at a distance, and marry someone whose mother tongue is different from your own. Speaking of marriage—your growth lies in exploration and mind-expanding experiences, so you may wish to exercise your freedom rights at will before tying the knot. You go so fast, barreling through life, that many just cannot keep up with you. The last thing you want is to leave a trail of broken hearts behind you—you're too generous for that.

You could sell heaters in the middle of a New York summer, that's how powerful your persuasive powers are! In terms of communications, you could carry off interviewing, writing, and making presentations and proposals practically in your sleep. This combo of massive influence allowed Capricorn-Gemini Alan Rabinowitz to start the first-ever jaguar preserve. You will be the great success you want to be when you allow your talents to reach as many people (or animals) as possible, and when you decide to not keep *everything* under your sleeve. Let your gifts pour out, your words spill out, and leave your worries behind you.

Your critical faculties are brilliant, but criticism does not bring you luck—especially around the office—unless it's performed as social satire. If education was restricted as a child, you become self-educated and teach others. Your best mate is light, gentle, and fun.

Y ou have a spectacular combo! You are a visionary who knows how to have a good time, a creative force who shocks everyone's socks off with your irreverence. Use your circular and deductive thinking to nail down your successes. You like acting "ordinary," which is ridiculous because you're as close to genius as they come, but it works for you! Aquarius-Gemini Ida Lupino, one of the first women directors in Hollywood, used to call herself "Mother" to make those around her feel more comfortable and less threatened— and, she said, as a way to get more work out of them! And she was loved for it. In fact, you're one of the most popular combos ever. Gene Hackman and Babe Ruth are two examples of much-loved Aquarius-Geminis.

There's a harmony within your nature most people would give their left arm for. You groove on love, but emotional scenes leave you baffled. Imitate your partners to best understand how this love thing works.

╫╪╫

Y ou are as cute as they come and enchantingly childlike all your life. You are also a brave soul to be born with the Pisces-Gemini challenge. Sensitive Pisces needs to be braver in the outside insecure world, because that's where your Gemini luck wants to take you! Develop a caring, tough skin to let you succeed—and you will.

You are receptive, sexy, and often outright gorgeous. Cindy Crawford is a good example of this Pisces-Gemini mix. Due to her Gemini Jupiter, she became more than just a pretty girl; she communicated her looks through four hundred magazine covers (and still counting)—the greatest number of covers for any model.

It's easy for you to fall in love—and sometimes with more than one person at a time. Gently remind your Pisces side that dishonesty brings you terrible luck as a Gemini Jupiter, and all will be well.

Relationships with Gemini Jupiters

Finally, for those of you who aren't Gemini Jupiter yourselves but who have a spouse or significant other, boss, child, or parent who is, here are some great tips to help you learn to deal! This is how the luck of someone else's Jupiter touches *your* life.

Your Significant Other

Oh boy, were you charmed by this one when you first met, or what? Gemini Jupiters are the best flirts in the business. They can roll words and ideas like no one else. Let's see how they appear to you on the fifth or sixth date. (If they show up by then.) A Gemini Jupiter is a difficult bird to catch. They are constantly disappearing. It's almost like light in the clouds, how quickly and beautifully and evasively they can fly away from you. They will wish to give themselves away to many people—if you're lucky, to the public—for many years before they will want to be with just one person. It's as if they've been given an excess of the Twin mindset; they just can't stop themselves. A hard, healthy, adaptable constitution is recommended if you plan to pursue this Golden Being.

Once, however, they have decided you are their soul mate (twin), they don't ever let go. They will fuss over you, play with you, fix your life if you let them, and generally treat you like you were their favorite twin doll. Which you are. It's wonderful if you love them back and if you crave security.

Your Boss

You will learn a great deal from this type of boss. They will want to teach you *everything,* sometimes even beyond what you need to know for the job! They also want to be your friend and will frequently invite you out after hours. They will then give you advice on everything, from your telephone manner to your love life. Often Gemini Jupiter bosses will say, "I only want the best for you," but you somehow feel they're plotting up some secret future plan of their own for you.

You can't figure out Gemini Jupiter bosses. They just happen to be very changeable, and you are not to take this personally. Moods may change hourly. You will feel as though you're seeing a complete shift in personality, a completely different person, from day to day. It can be a bit scary.

But not to worry. You're not crazy. If they are successful, they will have used their lucky dual natures often, and to their profit. It's always worked well in the past. No doubt, they are very successful now, if not somewhat known in their field. (These people are secretly prima donnas, though much better behaved in public.) They will have their idiosyncrasies, like everyone. Reassure and calm them, and they become true delights.

Your Child

You are probably worried about your Gemini Jupiter child this very minute. How are they doing? Do you miss them? Do you not see them that often—or often enough? Why is it they are gone so long?

I truly believe this is a difficult position for a Gemini Jupiter child. They need lots of space and lots and lots of no worry. Gemini Jupiter children love stories, and stories are the best way to discipline them, soothe them, make them go to bed, do their homework, finish their dinner, and clean their room. It is the fastest threat you can utter, to take away their bedtime storytelling—they cannot *live* without their stories. (Now that you know this, dear parents, you may thank me every day for the rest of your lives.)

Your Gemini Jupiter child sacrifices for those he loves, so go easy on the demands. You're deep in their hearts, even if they can't always express it. Gemini Jupiters are dense in emotion, quick fluid in brain. It's like an affliction, but never treated, because its side symptom can be, and often is, brilliance.

Your Parent

This parent knows everything that is good for you, how everything *should* be done, how you should be doing it, how bright you are, and how much you want to get ahead, and isn't this, no isn't *this*, the best way to do such a thing?

You would be forgiven for thinking that your Gemini Jupiter parent is a mite overprotective and has a raging anxiety disorder to go with it. You see, the poor things, they never got to be children when they were children. They have no idea what it's like. They always had to be like the grownups. Not only were they wiser than their years, but they took care of the people they loved, no matter how young they were.

Lighten them up and make them laugh. Learn to be children together. The child of a Gemini Jupiter parent doesn't know much about being a child, either! So, having a Gemini Jupiter parent turns out to be a very healing experience, after all.

That's what astrology is all about—appreciating how similar and how different we all are, and loving both similarities and differences.

Nature gives an antidote for every poison. Isn't this the miracle of life? Problems happen, but always, somewhere, there's a solution, and Gemini Jupiter will always think of two, if not many, solutions! Encourage your curiosity and follow where it takes you. The destination will be as much fun as all the things you learn getting there.

Remember, Gemini Jupiter: You are the magical juggler of the zodiac.

MUHAMMAD ALI
PAUL ALLEN
JANE AUSTEN
SAMUEL BECKETT
ALEXANDER GRAHAM BELL
BENAZIR BHUTTO
JULIETTE BINOCHE
BJÖRK
KEN BURNS
JANE CAMPION
NENEH CHERRY
CliffsNotes
CHARLES DICKENS
BARRY DILLER
SIR ARTHUR CONAN DOYLE
THOMAS EDISON
BETTY FORD
ANNE FRANK
ARETHA FRANKLIN
MARY GAITSKILL
GRETA GARBO
ART GARFUNKEL
KAHLIL GIBRAN
DIZZY GILLESPIE
MARTHA GRAHAM
GENE HACKMAN
EDMOND HALLEY
STEPHEN HAWKING
KING HENRY VIII
ALDOUS HUXLEY
JUPITER DISCOVERED
GRACE KELLY
ANN LANDERS

LENNOX LEWIS
STEVE MCQUEEN
ROBERT MITCHUM
SPIKE MILLIGAN
ALFRED MOLINA
MARIA MONTESSORI
MICHAEL MOORE
NEPTUNE DISCOVERED
JOSÉ MARIA OLAZABAL
JACQUELINE KENNEDY ONASSIS
SHUGGIE OTIS
DOROTHY PARKER
PLAYBOY
PLUTO DISCOVERED
JOHN C. REILLY
CHRIS ROCK
BETSY ROSS
GENA ROWLANDS
J. K. ROWLING
DANE RUDHYAR
BABE RUTH
ARTHUR SCHLESINGER JR.
JERRY SEINFELD
ALICIA SILVERSTONE
PAUL SIMON
HOWARD STERN
HARRIET BEECHER STOWE
BARBRA STREISAND
ABIGAIL VAN BUREN
GUS VAN SANT
OPRAH WINFREY
KEN SARO-WIWA
JEFFREY WRIGHT

"I'm so far out I'm in."

—Barbra Streisand, Taurus Sun–Gemini Jupiter. April 24, 1942

"Sometimes funny is silly."

—Chris Rock, Aquarius Sun–Gemini Jupiter. February 7, 1966

"One day the 'Don't Knows' will get in and then where will we be."

—Spike Milligan, Aries Sun–Gemini Jupiter. April 16, 1918

*"I'm just naturally drawn to mixing and matching things
because I was neither one nor the other."*

—Neneh Cherry, Pisces Sun–Gemini Jupiter. March 11, 1964

*"You have to take off one skin, then another one, and then another one,
to be as clear and transparent as possible."*

—Juliette Binoche, Pisces Sun–Gemini Jupiter. March 10, 1964

"Sometimes I really feel I need the wisdom of Solomon."

—Barry Diller, Aquarius Sun–Gemini Jupiter. February 2, 1942

"We are all born mad, some remain so."

—Samuel Beckett, Aries Sun–Gemini Jupiter. April 13, 1906

"Until we stop harming all other living beings, we are still savages."

—Thomas Edison, Aquarius Sun–Gemini Jupiter. February 11, 1847

"I'm not sure that acting is something for a grown man to be doing."

—Steve McQueen, Aries Sun–Gemini Jupiter. March 24, 1930

"I think lightness has to come from a very deep place if it's true lightness."

—Alicia Silverstone, Libra Sun–Gemini Jupiter. October 4, 1976

54

...rmonious
...Multitude

...r to be exalted in the sign of Can-
...—more successful people are born

> *We all live in a yellow submarine, a yellow*
> *submarine, a yellow submarine!*

— CANCER JUPITER "YELLOW SUBMARINE"

...ted as a species for more than 500
mi... on the successful evolution and
co... ...nding community. With Jupiter's
sup... ...s and enduring. Allow yourself to
be a... ...he passionate being: Just think of
Canc... ...o painted her inner vision while
literal... ...bodycast, or of Cancer Jupiter actress
Salma ...ayek, the determined actress and producer who brought *Frida*
to the screen.

To bring the power of Jupiter into your life, practice living, working, and coexisting harmoniously with others, whether it's within the
context of your family, your company, your team, or the world at
large. If you embrace Jupiter energy, you can enjoy a tenacious luck
that shines like a beacon for the rest of the human race, a sense of self
that resists egocentricity, and the power to integrate yourself into the
larger community in order to become part of a benevolent and harmonious whole. Jupiter will reward you for this: Not only was Jupiter
in Cancer when the Berlin Wall came down, reuniting East Germany
and West Germany, but it was a Cancer Jupiter, Galileo Galilei (born

Art intoxicates the mind,
tickles the spirit and
colors your life with
beauty and creativity.

— FROM THE BOOK
Five Good Minutes

The Dragonfly Center
Creating change through art & health

February 15, 1564), who proved that the Earth revolved around the Sun, instead of the universe revolving around *us*.

Luck will desert you only if you perversely turn your back on the rest of humanity and isolate yourself, seeking safety in obscurity. Your shell is not a place to hide—consider the life of Cancer Jupiter J. D. Salinger, who gave the world such literary gems as *The Catcher in the Rye* and *Nine Stories* only to retreat to a self-imposed hermithood in New Hampshire for the last forty-plus years, deliberately withholding his precious literary genius from the rest of humanity.

Your shell is, rather, a shield that protects you as you move through life. To see the difference, look at two Beatles songs, *Eleanor Rigby* and *Yellow Submarine*, both born Cancer Jupiter on August 5, 1966. The first is about a woman who hides inside her shell, numbered among "all the lonely people." Yet in the second song, the shell is a bright yellow submarine, a secure sea-worthy vessel that can undertake unimaginable adventures.

Take a moment, look back long and deep over your life, recount as many events as you can, and ask yourself what has made you lucky or unlucky in your life. You'll notice that whenever you act like a Cancer, *you create miracles in your life*. No matter what your Sun sign is, *you get lucky when you live out the Cancer legacy gifted to you by the benevolent planet Jupiter.*

To take full advantage of your luck, emphasize your . . .

Cancer Jupiter ♋♃ **Strengths**	*Be connected, active for the greater good, money conscious, creative, regenerative, healing, record keeping, second sighted, and hospitable but with good boundaries.*

But because Jupiter rules excess, watch out for . . .

Cancer Jupiter ♋♃ **Excesses**	*You can become overly emotional, overloaded, or panicked.*

♋♃

The Harmonious Human Multitude

Astrologers consider Jupiter to be exalted in the sign of Cancer, and with good reason—more successful people are born under Cancer Jupiter than under any other Jupiter sign. In fact, the list of famous and successful Cancer Jupiters is *double* that of any other sign!

We all live in a yellow submarine, a yellow submarine, a yellow submarine!

— CANCER JUPITER "YELLOW SUBMARINE"

Your symbol, the Crab, has existed as a species for more than 500 million years, its survival dependent on the successful evolution and continued well-being of the surrounding community. With Jupiter's support, you can be equally tenacious and enduring. Allow yourself to be a very private and at the same time passionate being: Just think of Cancer Jupiter artist Frida Kahlo, who painted her inner vision while literally encased in a shell-like bodycast, or of Cancer Jupiter actress Salma Hayek, the determined actress and producer who brought *Frida* to the screen.

To bring the power of Jupiter into your life, practice living, working, and coexisting harmoniously with others, whether it's within the context of your family, your company, your team, or the world at large. If you embrace Jupiter energy, you can enjoy a tenacious luck that shines like a beacon for the rest of the human race, a sense of self that resists egocentricity, and the power to integrate yourself into the larger community in order to become part of a benevolent and harmonious whole. Jupiter will reward you for this: Not only was Jupiter in Cancer when the Berlin Wall came down, reuniting East Germany and West Germany, but it was a Cancer Jupiter, Galileo Galilei (born

February 15, 1564), who proved that the Earth revolved around the Sun, instead of the universe revolving around *us*.

Luck will desert you only if you perversely turn your back on the rest of humanity and isolate yourself, seeking safety in obscurity. Your shell is not a place to hide—consider the life of Cancer Jupiter J. D. Salinger, who gave the world such literary gems as *The Catcher in the Rye* and *Nine Stories* only to retreat to a self-imposed hermithood in New Hampshire for the last forty-plus years, deliberately withholding his precious literary genius from the rest of humanity.

Your shell is, rather, a shield that protects you as you move through life. To see the difference, look at two Beatles songs, *Eleanor Rigby* and *Yellow Submarine*, both born Cancer Jupiter on August 5, 1966. The first is about a woman who hides inside her shell, numbered among "all the lonely people." Yet in the second song, the shell is a bright yellow submarine, a secure sea-worthy vessel that can undertake unimaginable adventures.

Take a moment, look back long and deep over your life, recount as many events as you can, and ask yourself what has made you lucky or unlucky in your life. You'll notice that whenever you act like a Cancer, *you create miracles in your life.* No matter what your Sun sign is, *you get lucky when you live out the Cancer legacy gifted to you by the benevolent planet Jupiter.*

To take full advantage of your luck, emphasize your . . .

Cancer Jupiter Strengths	♋♃	*Be connected, active for the greater good, money conscious, creative, regenerative, healing, record keeping, second sighted, and hospitable but with good boundaries.*

But because Jupiter rules excess, watch out for . . .

Cancer Jupiter Excesses	♋♃	*You can become overly emotional, overloaded, or panicked.*

Understanding the
Cancer Energy and Influence

The time of Cancer is the time when the hard work of planting seeds and cultivating the soil is done but when crops are still seedlings and baby animals are still learning to take care of themselves. The crops and animals need to be nurtured, and this is your purpose, Cancer Jupiter: To domesticate, nurture, and help everyone around you grow stronger and more purposeful.

Your Cancer element is Water, the element that rules the emotions. You have the capacity for deep and passionate feeling; your knowledge of water allows you to be sensitive to every little drift and current of emotion. Cancer time begins on June 21, at the commencement of the summer solstice, the longest day of the year, when the Sun's light seems to linger forever in the sky. Solstice means "Sun stand still," midpoint in the season, after which the days begin growing shorter. It's lucky for you, Cancer Jupiter, to hold still, and, like a brilliant wizard, to think backward and forward, from the past to the future. Your historical perspective, your ability to think about past and future simultaneously, has the happy effect of making you feel a lot more grounded in the present.

By trusting its instincts and its gift for self-preservation, the tough little crab has survived longer than almost any other species. Crabs are cautious and need to feel safe; they try out different things to see what will work, and they learn from their mistakes. The crab has a hard protective shell on the outside and a vulnerable softness on the inside. When a crab grabs hold of an object, *any* object, it will *never* let go! If a claw breaks, a new one grows in its place.

CANCER SYMBOL
The Crab

To gain the bounty of Jupiter, you must act like a crab! Be a survivor, hone a sense of history, gain knowledge about your past, and spend time thinking about what's worked and what hasn't. Let yourself feel deeply, but, Cancer Jupiter, balance it with a knack for self-protection. When you find something good, never let go.

When you are hurt, use your tenacious Cancer Survival Will: Retreat inside your shell, where you can peer out cautiously as you wait for the danger to pass. You should learn to walk sideways, like the crab, to avoid head-on confrontations. Your indirect walk, however, is at the same time highly purposeful, in that it always gets you where you want to go while protecting you from harm.

✦✧

The Breasts and the Stomach

Cancer rules the breasts and the stomach. Allow yourself to nurture, to mother those around you (whether you're male or female)! Be the one who comforts and provides. Sharpen your intuition, your "gut" reaction to new things. Make sure you keep yourself centered. The Moon rules Cancer, and as such, when the Moon is in Cancer (this occurs once a month for two to three days; take care to check your astrological calendar for the different times each month), you should beware of undergoing (or performing) any breast or stomach operations.

✦✧

Sea Green

Green in any of its forms is lucky for you, Cancer Jupiter. Sea green is the truest color of the sea, the light tone of the underwater surface, as well as the color of the unconscious, rippling deep below. Oceans crash on the shore to scrub the Earth, and the clean rain falls to wash its surfaces before joining rivers and reentering the ocean. Green also symbolizes growth and birth, the health of the planet; any work you do in the garden or the wider world to protect the Earth and the sea attracts the bounty of Jupiter.

Cancer Jupiter Strengths ♋♃

And now your Cancer Jupiter strengths:

Be Connected

Your lucky energy flows to you, Cancer Jupiter, when you are most connected to others, surrounded by your family—it can be your family of *birth* or your family of *choice*. This is why many Cancer Jupiter artists and creative types have found success by working with their own close-knit families or creating a new type of family; for example, a performance ensemble. Ingmar Bergman, Anton Chekhov, Alvin Ailey, Joel Coen, Mike Leigh, Martin Scorsese, Jean Renoir, and Jean-Luc Godard—all of whom have worked repeatedly with a small group of performers to make their art—are Cancer Jupiters, as is Denzel Washington, who credits his success to his love of family, saying, "Acting is not my life; my children are my life." And where would America be without its favorite dysfunctional family, *The Simpsons*, born Cancer Jupiter in January 1990?

As a Cancer Jupiter, you don't have to have children (although it is lucky for you to have kids!) or be in close contact with your family, but you should maintain close ties within some type of community. As Cancer Jupiter Meryl Streep says, "Motherhood has a very humanizing effect." If you have no children of your own, try "adopting" people, caring for them almost as if they *were* your children. Think of the 1920s cabaret empress and Cancer Jupiter Josephine Baker: The legendary performer adopted twelve children of different races, naming them her "Rainbow Tribe." If you're not in a physical community, try a virtual one. The virtual community of the World Wide Web, "born" a Cancer Jupiter on August 5, 1990, is often referred to as a kind of "family," created to keep the people of the world in instantaneous communication. The Web functions best (as do you) when privacy is secured and protected. Boundaries—an essential attribute of your healthy Crab shell—are required in order for you to live in a mutually beneficial way within your community. Because Cancer also rules the home, even when it is a simulated home or house, you find luck! Cancer Jupiter Coco Chanel made her success through the "House" of Chanel, inviting visitors in to view her fashions as if they were visiting her home, a practice that still brings the House of Chanel enormous profits today.

Be Active for the Greater Good

If you are Cancer Jupiter, you will gain phenomenal luck by expansively *giving,* especially when your giving nurtures and cares for those around you. Many great philanthropists and thinkers—Andrew Carnegie, who endowed the nation's public libraries; Rachel Carson, whose book *Silent Spring* led to a presidential commission and gave birth to the modern environmental movement; Paul Mellon, who funded thousands of scholarships—have been Cancer Jupiters. Your philanthropy will speak for you, reaping a harvest of goodwill tenfold for each kind deed you do. Cancer Jupiters are givers and protectors; for more proof, look at Cancer Jupiter Benjamin Franklin.

Born on January 17, 1708, the tenth son of a poor soap maker, this Founding Father began working at the age of twelve. He helped pave, clean, and light Philadelphia's streets, and agitated for environmental cleanup. He helped launch the nation's first subscription library

We can't see the Moon's gravitational force, but we can still feel it from 239,000 miles away. The Moon rules the oceans and us, too, since 90 percent of our bodies are made up of water! Inside every human and animal, 90 percent of who we are—especially our emotions and unconscious— is being rhythmically moved like the tides. The Moon orbits around the Earth once per month, and month comes from *mensa,* which relates to menstruation—the rhythmic movement of blood in women's bodies.

YOUR HISTORY
1,500 B.C. Called Luna by the Romans, Selene and Artemis by the Greeks, the Moon has always been related to Mother. There are more births at the full moon and the new moon than at any other time.

(1731), and the first fire company in the city (1736). He established the Pennsylvania Hospital (1751) and helped draft the Declaration of Independence, which he signed, and also had a hand in drafting the U.S. Constitution. He invented a new type of stove but refused to take out a patent because he wanted his invention to help the public, not his pocket! In one of his last public acts, in 1789 Franklin wrote an antislavery treatise. When he died, at the age of eighty-four, twenty thousand people attended the funeral of the man who had become known as "the harmonious human multitude."

Among his Cancer Jupiter sentiments: "A penny saved is a penny earned" and "An ounce of prevention is worth a pound of cure." His statement "We must all hang together or assuredly we shall all hang separately" is a classic Cancer Jupiter remark made to his fellow representatives at the signing of the Declaration, underscoring both his sense of patriotic benevolence as well as his need for community.

What can *you* do for the greater good? Remember that working for the greater good is an investment in selflessness with a high rate of return—if you were born Cancer Jupiter and you're using your talents in this way, you will be successful beyond your wildest dreams!

Be Money Conscious

Whether you're a Scorpio Sun (used to extreme financial ups and downs) or a Leo Sun (prone to flamboyant spending), ignore your Sun sign and make sure you have a good, regular cash flow. Take both an offensive and a defensive approach: You *must save,* but you also need to make a substantial and predictable living.

Cancer rules coins and precious metals, and has among its natives a staggering number of financial wizards and collectors. David Geffen, the first self-made billionaire, is Cancer Jupiter, as is Warren Buffett (the second wealthiest man in the United States), Ross Perot, Steve Jobs, Rupert Murdoch, and the British economist John Maynard Keynes. And even though Keynes himself said, "The moral problem of our age is concerned with the love of money," nevertheless, for a Cancer Jupiter, saving and collecting coins and precious metals will make you lucky beyond the actual wealth accrued, as will acts of simple frugality. Perhaps the most dramatic example of Cancer Jupiter luck through thrift is Theodore Dreiser, author of one of the greatest books ever written on the theme of the effects of economic inequality, *An American Tragedy*. Dreiser decided to save a few pennies by changing his berth reservation from the top luxury liner to the lesser

Kroonland, a last-minute decision that saved his life. The luxury liner: the *Titanic*.

Don't worry about appearing overly concerned with financial matters. Sometimes high art is created by people who are trying only to pay the rent or make a buck. Listen to your fellow Cancer Jupiter Paul McCartney: "Somebody said to me, 'But the Beatles were anti-materialistic!' That's a huge myth. John and I literally used to sit down and say, 'Now let's write a swimming pool.'"

Be Creative

Sometimes Cancer Jupiters, even the most innately creative among them, are so skilled at saving money and creating financial security that they will often neglect their creative sides and forgo working in the arts. Don't you often feel a creative urge, a force lying dormant inside of you, waiting to be expressed if you only had the time? Make the time—your artistic side must find its expression!

Cancer Jupiter John Grisham was a Mississippi lawyer working sixty to seventy hours a week, but he felt the urge, and he managed to squeeze in an hour here and there to write his first novel, *A Time to Kill*. It took three years to complete and was given a modest first printing of five thousand copies. Five years later, it had gone through more than twenty-five printings, spent ninety-eight weeks on the *New York Times* bestseller list, and sold more than 7.75 million copies. Overall, Grisham's works, including *The Firm* and *The Pelican Brief*, have sold more than 60 million copies!

Not that you have to strive to be a best-selling author, but in between the busy hours, try to carve out some time for yourself and your creativity. Even if you don't sell a thing, it will, I guarantee, somehow sooner or later bring you luck!

Cancer Jupiter Brittany Murphy has become lucky by using her acting talent to "let it out" on screen. She said in an interview: "It basically came from wanting to get out all that's inside me, and I get to make other people happy in the process. My life has always danced to a song." She was scheduled to play Janis Joplin, another Cancer Jupiter. You can release your creativity simply by letting out your sense of humor or expressing your whimsical side. Many Cancer Jupiters achieve success by expressing a wacky, offbeat sensibility. Just think of actor Christopher Walken, who always seems to look at life from a unique perspective, or the bizarre and silly British comedy troupe Monty Python. Three of the five original Pythons are Cancer Jupiters.

DOS

❋ Work and live for the whole, rather than for the self.

❋ Cultivate sensitivity to others while maintaining strong boundaries.

❋ Explore history and the past—your best ideas come to you this way.

❋ Pursue visions of the future; it will be to your financial benefit.

❋ Surround yourself with soothing greens, blues, and browns, and speak gently.

❋ Increase your calcium intake; okra is your wonder food.

❋ Keep records scrupulously, and save all paperwork, memos, books, and documents.

❋ Hang out with more Scorpios, Pisces, and Cancers.

Be Regenerative

Seek out all the "umbilical cords" in your life—the things that connect you to the web of life—and ask yourself if they are truly serving you by bringing you nutrients or if they are actually chaining you. If it's the latter, you need to have the courage to let them go and trust that the healthy connections will regenerate (not unlike the way a crab sheds one claw to form another).

There is no better or more enduring example of continual regeneration, after the severance of ties, than the United States itself. A typical Cancer Jupiter, America holds family values in high esteem, is self-protective, benevolent and philanthropic, economically prosperous, and diversely creative in both an artistic and a spiritual sense, a true democracy with (according to CNN, 1/31/03) "2,630 religions and counting." America's official birthday is, of course, July 4, 1776, the day it embraced the Declaration of Independence and cut the cord with Mother England.

America's belief in regeneration still resonates in the words the Founding Fathers (Benjamin Franklin among them) sent to King George III in 1776, words defiant enough to sever the connection but suffused with the confidence of a nation eager to be born:

> He has erected a multitude of New Offices, and sent hither swarms of Officers to harass our people.
> He has affected to render the Military independent of and superior to the Civil power.
> He has endeavoured to prevent the population of these States; for that purpose obstructing the Laws of Naturalization of Foreigners
> For depriving us in many cases, of the benefits of Trial by Jury
> He is at this time transporting large Armies . . . to compleat the works of death, desolation and tyranny, already begun with circumstances of Cruelty & perfidy scarcely parallelled in the most barbarous ages, and totally unworthy the Head of a civilized nation.

No matter how attached you are to a relationship or a situation, use your ancient smarts to decide whether it needs to stay or to go. Cancer Jupiter astrologer Donna Cunningham puts it perfectly: *"Birth is a trauma; yet if the baby remains too long in the womb, it will die."*

Be Healing

Cancer Jupiter Nelson Mandela, Nobel Prize laureate and president of South Africa, is no stranger to suffering or to healing. Mandela was released in 1990 after spending twenty-seven years in prison. In 2003 he began to paint scenes (letting out his creative resources) of the prison at Robben Island, where he was held, pictures in which the harbor where prisoners arrived is a gentle blue. The black of the buildings is unthreatening and diffused. The bars of his cell are a cheery orange, and the vast distance between Mandela and his freedom is a simple green field. He said in an interview, "I have attempted to color the island sketches in ways that reflect the positive light in which I view it."

Taking a cue from Nelson Mandela, and with a view to self-healing, which parts of your life do you need to view differently? Use Mandela's example: *Try painting your painful experiences,* either on paper or in your mind, and color them with bright, happy colors, and then ask yourself how you feel. I guarantee some of your pain will have lifted in the process of reconciling and uniting these two different parts, the wounded person who suffers and the healer who is wise.

Chiron is the metaphor in astrology for the wounded healer. Charles Kowal discovered—brought to us—the planetoid Chiron on the Cancer Jupiter birthday of November 1, 1977, naming it for the famous surgeon and teacher of Greek mythology, an immortal centaur who was half man, half horse. The immortal Chiron had an agonizing injury that would not heal. The gods took pity on him and finally granted him the right to die so he could be released from his pain. Chiron teaches us that only by surrendering, by letting the life out, are we finally able to be free and healed.

Be Record Keeping In *Remembrance of Things Past,* Marcel Proust's autobiographical narrator is led to the discovery of a timeless center existing beneath the transitory flux of surface events, what Proust terms the *"moi-permanent."* Involuntary memory recall is the key to this discovery. The quality of memories invoked by certain sensual triggers leads the narrator to conclude that the past is not lost to time, but exists *within.* The novel, in sixteen volumes, records those memories, written from a cork-lined room (the shell) he never left after asthma rendered the author an invalid. It's little wonder, with their gift for record keeping, that Aleksandr Solzhenitsyn, Franz Kafka, Louise Erdrich, Isabel Allende, Iris Murdoch, James Michener, and so many other Cancer Jupiters are writers. Cancer Jupiter Nobel Prize–winning writer Toni Morrison said, "I think if we understand a great deal more about history, then we can progress together more as a nation."

Cancer Jupiters understand that without records, we can't learn from our mistakes.

Cancer Jupiter Coleen Rowley, named one of *Time* magazine's "Persons of the Year 2002," is the courageous woman who recorded the events surrounding her attempts to close in on the twentieth hijacker in the now famous memo to FBI Director Robert Mueller. For weeks before September 11, the twenty-one-year FBI veteran and her team had been trying to obtain access to twentieth hijacker Zacarias Moussaoui's computer. Each request was denied by higher-ups in the FBI. Rowley, harassed at work, requested whistle-blower protection. Rowley concluded her thirteen-page letter to Mueller thus: "Let me furnish you the *Webster*'s definition of 'careerism'—the policy or practice of advancing one's career often at the cost of one's integrity.' Maybe that sums up the whole problem!" Cancer Jupiter, characteristically and rightly, does not believe in advancing one's own self at the expense of the group to which he or she belongs.

Be Second Sighted

Don't underestimate your own psychic sensitivity, Cancer Jupiter, indulge it! Cancer Jupiter and former world chess champion Bobby Fischer possesses the ability to recall precisely an opponent's *every* move, including games played in the distant past, which often allows him to predict his opponent's next move. Fischer's genius is strictly literal, the gift of total recall, which allows him to remember perfectly an actual series of intricate moves and his own mental analysis of it as it occurred. Some Cancer Jupiters possess another kind of second sight.

Nostradamus (1503–1566), the French astronomer and the most famous psychic prognosticator of all time, was a Cancer Jupiter!

Second sight also means not accepting things at first sight. It means refusing to be taken in by surfaces, going beyond the way things seem. It was a Cancer Jupiter, Bob Woodward, who refused to accept the Nixon administration's dismissal of the Watergate break-in as an insignificant third-rate burglary and looked beyond the surface of things, helping to bring down an entire administration.

Be Hospitable but Set Good Boundaries

While it's lucky for you, Cancer Jupiter, to be giving, you must also have strong and healthy boundaries to guide you and tell you when to stop.

Boundaries mean staying safe, responsible, and planning ahead, all while showing hospitality and generosity toward self and others.

Should it surprise us that Margaret Sanger was a Cancer Jupiter? Born in Corning, New York, on September 14, 1883, the sixth of eleven children of an Irish Catholic mother who died at age fifty, Margaret pointed to her mother's frequent pregnancies as the underlying cause of her premature death. Sanger came to see responsible birth control as a tool by which working-class women could liberate themselves from the economic burden of pregnancy, and opened the first birth control clinic in Brownsville, Brooklyn, on October 15, 1916. Even though she was subsequently sentenced to the Queens County Penitentiary for "maintaining a public nuisance," Sanger won her case on appeal and opened the door for women everywhere to take control of their lives by exercising responsible family planning. Sanger's work won the power of Jupiter because, by giving mothers the power to decide how many children to have, she created boundaries that allowed them to be even more giving and nurturing.

What Makes You Lucky in Love

Even if you're a Gemini Sun (that is, a serial dater), success in love will come when you settle down! So get serious! Stop looking for physical perfection and concentrate on finding a loving mate interested in building a wonderful home. Remember, home and family are lucky for you. If you just climb down from your flighty penthouse perspective and focus on making that home a warm and welcoming place, you will find a greater love than you had dreamed possible.

Despite the notion that opposites attract, you will find that it's not always a good idea to look for mates among people who are different from yourself; the boy or girl next door just might end up being the answer to your prayers. Make sure you ask your family and close friends what they think about your new significant other—you will find wisdom in their views, so trust them—it's lucky!

Jupiter's energy won't be attracted if you pick a fiercely independent mate who'd be comfortable being without you for long periods of time. Absence is unlikely to cause your heart to grow healthily fonder, and it just might cause your nervous system to go *crazy*. If you're honest with yourself, you can admit that you'll have a tough time spending a single night without the one you love. In fact, that sort of intensity is lucky for you unless it gets too extreme—so try to learn from this. To avoid prolonged periods of absence, it may be especially helpful to make sure you both have the same schedules. If not, be willing to change yours! Find what works.

Your Best Career Moves

No matter your Sun sign, your Cancer Jupiter means you may be more likely to achieve success in a family or group setting. You might consider working for the family business or in a cooperative establishment. Short of that, try a small company where you'll work with a closely knit group of people with ideals and aspirations similar to your own. Trust your intuition and be natural with others. Don't worry about office politics or putting on a disguise to hide your true self. You have a naturally tough shell that will protect you; though you are such a feeling person, you may not even know it's there.

Avoid spending too much time working in isolated settings—the hermit life is not for you. Whatever you choose to do, there's a good chance that sooner or later you'll crave a mutually supportive group atmosphere in which to thrive and flourish. You will also need, more than other people, to go deeper into your work. Being naturally psychic, you love detective work, because you're so good at it! People, especially friends, have a hard time holding a secret around you for long.

How to Reach Optimum Health

Crabs can have delicate stomachs and are prone to ulcers and other stomach problems. Born under a Water sign, you may have a tendency to overindulge in alcohol, a substance that you may not easily tolerate. It's not lucky for you, so be careful! Be sure to get enough of Cancer's cell salt, calcium fluoride (see "Your Lucky Foods, Element, and Spices" for sources of it); the lack of calcium fluoride causes varicose veins, receding gums, curvature of the spine, and cataracts and other eye problems. Last, though hardly least, Crabs (like so many of your fellow mortals) recuperate best in their beds at home.

Your Lucky Foods, Element, and Spices

Food sources for calcium fluoride are egg yolks, whole-grain rye, yogurt, beets, watercress, fish, and oysters. Italian food is always a happy choice for Cancer Jupiter. Rich, full, beautiful sauces made with oregano, tomatoes, and garlic will be beneficial to you. You tend to hold on to things and have a need to cleanse your system with some regularity, so don't forget to drink lots of water.

Crabs need large amounts of calcium, because they're prone to skin disorders resulting from a lack of calcium in their diet. Calcium is readily available in milk and yogurt; okra, also high in calcium, helps reduce stomach inflammations. Sugar and starches cause constipation, and salt produces bloat, and it is also best to stay away from extremely spicy, highly seasoned foods.

Your Best Environment

There is no one more likely to seek out luxurious interiors than a Cancer Jupiter. You will even forsake a comfortable location in favor of possessing what you consider the best possible home on the *inside*. Wherever you live, you need plenty of sky. Best of all for you is to be by the ocean. If you could, you'd play outside all day and all night long like a rambunctious ten-year-old!

Your Best Look

Tight-fitting clothing will have an extraordinary effect on the way Cancer Jupiters look. Cancer Jupiter Pamela Anderson certainly gets away with it. You can make a success out of any work or dating situation, no matter how you feel, if you wear not only tight-fitting but also *soft* clothes. Tight clothing made from fabrics that shine will show you off to your greatest advantage. Sea green looks stunning on you, and with an occasional Marilyn Monroe red, you will literally dazzle. At work, a flamboyant necktie will do nicely.

Your Best Times

The luckiest times for you are when the Moon is in Cancer, which happens every month for two to three days. Find these special days in your astrological calendar. During these days, schedule crucial meetings, tell that special person you love them, ask for a promotion or raise, push the envelope, and plan to shine. In your everyday, the best time for you to exercise is late morning, with a nap midafternoon. The luckiest times during the year for you are June 21–July 22 (when the Sun is in Cancer); October 23–November 21 (when the Sun is in Scorpio); and February 20–March 20 (when the Sun is in Pisces).

The very luckiest time of your life—as in Lotto lucky—is when Jupiter reenters the sign it was in when you were born (Cancer), which happens every twelve years and lasts anywhere from a few months to just over a year. This is called your Jupiter Return, and it launches your luckiest streak. You can count on the *extraordinary* to happen the very day your Jupiter Return begins. (Halle Berry and Denzel Washington won Oscars during their Cancer Jupiter returns!) Amazing events can start happening up to six weeks beforehand, so plan ahead!

How to Be Lucky Financially

"A penny saved is a penny earned," noted sagacious Cancer Jupiter Benjamin Franklin. So listen up! Your Cancer Jupiter means you need to be conservative with money and save wherever possible. Stick to the tried and true: Chances are you'll do best in financial markets with relatively safe conservative bonds and blue-chip companies. Avoid debt and leverage, personally and in the companies whose stocks you buy. As far as possible, stick with companies that are household names. Family businesses are lucky for you, as are companies that make household products used by families in the home.

Now, don't be shocked: This is for you Cancer Jupiters only. A little bit of sexuality never hurt anyone, and I swear that by letting the world know about your sexuality, you just may *draw gold to you.*

Success in Family and Friendship

You have a tendency to stick by someone's side, through thick and thin, until they literally make you go away. While taking this tendency to extremes is bad for you, it's lucky for you to be extremely loyal. Yet be careful, and remember your boundaries!

To attract Jupiter's energy, you should not tolerate being subjected to any kind of rough handling; it would crush your shell, which you have worked years to build and maintain. In families and friendships, make this crystal clear from the get go with anyone you let into your inner circle—once damage is done, you can't, with few exceptions, ever be friends with them again, no matter how hard you, or they, try.

You, dear Crab, may scuttle along, occasionally trying to resuscitate broken friendships until you just can't do it anymore. Sooner or later you will choose to free yourself and, sadder but wiser, move on. Friends count on you for anything and everything. Furthermore, as you probably know already, you're the best brother or sister in the world! Your best policy: Claim more stable souls as your friends.

If you do have a quarrel, misunderstanding, or falling-out with one of your siblings, it's very important that you make up right away. Whatever you do, don't leave it, thinking it will heal on its own or with time. Chances are it won't. *Fix it now.* Talk, scream, yell—do whatever you have to do. Then heal and make up. Family really is that important to the Crab. You may simply have a hard time functioning otherwise.

"Chicken Little" is a Cancer Jupiter cautionary tale. The story begins with Chicken Little in the woods. An acorn falls on her head. Instead of realizing it's only gravity, and only an acorn, she gets scared. She shakes so hard "half her feathers fall out," and she runs to tell the king. Along the way she meets Henny Penny, Ducky Lucky, Goosey Loosey, Turkey Lurkey, and gets them all running along beside her to tell the king. The last animal to join the terrified ensemble is Foxy Loxy. Remember how the story ends?

Foxy Loxy leads Chicken Little, Henny Penny, Ducky Lucky, Goosey Loosey, and Turkey Lurkey across a field and through the woods and straight to his den, and *they never see the king to tell him that the sky is falling.*

You need to avoid becoming so worried in your life that your fear reaches fever pitch and half your feathers fall out. Snap out of it, Cancer Jupiter—you've got another 500 million years to live. Don't make it worse.

Excess #1   You Can Become Overly Emotional

You are such a feeling person, Cancer Jupiter, and family and friends are so important to your success and luck. Is it a surprise, then, that you sometimes go overboard? Indulging your emotions to excess and being too dramatic are unlucky for you. Watch out also for your innate tendency to become codependent, unable to end relationships that are no longer nurturing. You can give, give, give—but remember, luck comes to you only when you know when to stop giving and when you set boundaries.

Excess #2   You Can Become Overloaded

Becoming overloaded can lead to disaster for a Cancer Jupiter. You become so burdened and exhausted that you fall asleep in the moment. At times like these, there's a real danger that you could become catatonic.

DON'TS

✳ Abandon your family, burn all your papers, and move to a remote island to be a hermit.

✳ Work like an Aries Jupiter or a Libra Jupiter.

✳ Forget to pay your taxes.

✳ Burn up money.

✳ Invade others' privacy, or invite them to invade yours.

✳ Have a tummy tuck or breast surgery when the Moon is in Cancer (two to three days a month).

✳ Refuse to share your feelings and cut off friends.

Cancer Jupiter! Stop, stop, stop! Whatever you're doing, you're almost certainly overdoing it. Tell everyone to go away, or better yet, just turn off your phone, draw back into your shell, and spend time in front of the TV, an entire night, with something chilled by your side, maybe even some ice cream or pretzels. That's all you need to get back on your feet again, for now, anyway, and it's so much cheaper than going to a spa. Remember, however, that you must keep moving—this is how Crabs stay alive. Restorative periods are their only time of stillness.

It's not always easy to be honest about your need for stillness. Because you are a caring individual, and someone who is often counted on to be the life of the party as well, you have to pretend at times you're okay, no matter how you feel. In the ancient folk story "The Raven's Tale of Honesty," a character named Black Bear learns how hard it can be sometimes. In the fable, Black Bear arrives late for a meeting and says, "I'm feeling frazzled after dealing with my cubs. What if I don't feel compassionate?"

"Fake it," Raven says.

"That doesn't seem honest," says Black Bear.

"It doesn't begin with honesty," says Raven.

Excess #3   You Can Become Panicked

Cancer is a *very* subjective sign. Your tendency to universalize from personal experience can sometimes result in distorted or limited viewpoints, leading to unfortunate emotional extremes. How to deal with the situation when it comes up? When feeling overloaded, it's useful to remember to simply *take small steps.*

Hewlett-Packard CEO Carly Fiorina, born Cancer Jupiter and one of the top female executives in this country, has said about her various and successful dealings (such as a recent megamerger with Compaq), "The path to the future is about taking one step at a time." Countless leaders in business and other fields would immediately and wholeheartedly agree. Cancer Jupiter pop artist Niki de Saint Phalle, famous for her Tarot sculpture garden in Tuscany, Italy, issued the following warning against the dangers of becoming panicked: "If we are in a state of panic or obsessive thinking, we block the flow of this great spiritual force that lies within each of us. If we tune in to the angel's force we will make the right choices. Observing and interacting with nature can help us connect to this life force. The angel's treasure is this flowing blood. Everything in our universe is in movement. Let's join the Celestial dance."

To begin, very simply, come to a full stop, however driven and ambitious you are, and cease thinking about huge unsolved problems, and in particular all the problems you *might* face in a tomorrow *that has yet to come*. The constant state of alarm into which you can plunge yourself accomplishes nothing, sets up lots of unnecessary false steps to get back to zero, and paves the way for a myriad of wrong-headed mistakes.

Sun Sign-Jupiter Sign Combinations

Combine Aries Fire with Cancer Water and you have a sizzling, smoky fire. Feel as much as you want to—therein lies your success! Aries-Cancer Norah Jones pours her feeling out in her songs yet still retains a close intimacy and mystery. Your power lies in not letting everyone in on the secret; by keeping it to yourself, you grow stronger. The shell intensifies the fire, as the fire hardens the shell.

Aries requires the Now, and Cancer wishes for the Later, and this confuses you, even as it attracts others. To love or not to love? Work and career are so much easier! You speak better than almost anyone, witty and brainy and brave. Keep close to your heart and to your original sideways view of things.

Both Taurus and Cancer *want* so much. They need, like Taurus-Cancer Liberace, to make everyday life larger and more grandiose than it really is. Both signs are also notorious for not liking change, so you tend to suffer in silence for way too long in your personal life.

You'd like to spend less time reassuring others and more time indulging yourself, but your real luck lies in giving to others and watching over them as they grow. This could manifest itself in the form of investments, creative works, or simply in raising your children. People count on you, in work, friendship, and love. No actor or actress cries on screen as totally and as convincingly as Taurus-Cancer Judy Davis, but no one laughs like her, either!

Geminis have all the fun; Cancer is full of dreamy intensity. You can combine both into a startling career, leading to fame and fortune, or you can boogie the night away.

The ones who make the best use of their Cancer determinism are the most successful. Strategies and planning and finances are crucial, however much your Gemini side suggests you just blow it off and improvise. Gemini-Cancer Laurence Olivier, considered possibly the greatest and most dramatic actor of all time, quipped, "Nothing is beneath me if it pays well."

Your brilliant wit attracts many and in love sparks fly when *you* are attracted. Remember, however, that cerebral (Gemini) connections are not nearly as lucky for you as emotional (Cancer) ones are!

⊰◈⊱◈⊰◈⊱◈⊰◈⊱◈⊰◈⊱◈⊰◈⊱◈⊰◈⊱◈⊰◈⊱◈⊰◈⊱◈⊰◈⊱◈⊰◈⊱◈⊰◈⊱◈⊰◈⊱◈⊰◈⊱◈⊰◈⊱◈⊰◈⊱◈

To function, you must examine everything, like Algerian-born French deconstructionist and double Cancer Jacques Derrida, whose analytical theories revolutionized the way books are read. My favorite British poet, Ted Hughes, examined life in its extreme moments, using wild animals as a metaphor for life forces—and us.

Be careful to avoid being alone and isolated, which might assist you in your creations, but it's not how you get lucky in real life. Keep in mind what Franz Kafka said: "A novel's purpose should be to cut through the icy sea in every man's heart." Remember this when you examine your life and when you reach out to others you care about. You have much good to do in the world. Practice reassuring yourself as much as you reassure others. Emotionally, you tend to take things personally, so take a few breaths before you speak when you are with someone you really care about.

⊰◈⊱◈⊰◈⊱◈⊰◈⊱◈⊰◈⊱◈⊰◈⊱◈⊰◈⊱◈⊰◈⊱◈⊰◈⊱◈⊰◈⊱◈⊰◈⊱◈⊰◈⊱◈⊰◈⊱◈⊰◈⊱◈⊰◈⊱◈⊰◈⊱◈⊰◈⊱◈

Oh my, what a genuine knockout you have the capacity to be! Yet you will arrive at your greatest successes by underplaying them. Use humor, sassiness, and just the right twinkle in your eye. Think of Leo-Cancers Kim Cattrall and Kobe Bryant. By dressing up your ideas, making jokes, and sexualizing your sideways indirect Crab walk, your power increases.

Always, *always,* use your originality. If you plant an idea with a potential business partner, take it seriously and follow up. Thanks to your uniqueness, your ideas stand a better-than-even chance of becoming fabulous successes. Jerry Garcia of the Grateful Dead was a Leo-Cancer, and he now has a Ben and Jerry's ice cream flavor named after him—Cherry Garcia!

⊰◈⊱◈⊰◈⊱◈⊰◈⊱◈⊰◈⊱◈⊰◈⊱◈⊰◈⊱◈⊰◈⊱◈⊰◈⊱◈⊰◈⊱◈⊰◈⊱◈⊰◈⊱◈⊰◈⊱◈⊰◈⊱◈⊰◈⊱◈⊰◈⊱◈⊰◈⊱◈

You will have the most beautiful home of all your friends. The thoroughness of your Virgo directed toward the Crab shell will simply make it shine. Your home is your haven and necessary for the *off* times. There are many days when you are utterly torn between being happy and satisfied with the people around you, and being miserable.

It's only because you are caught in a tug-of-war between your lucky Jupiter side, which insists that success is found only with others, and your Virgo Sun side, which knows that you would rather be alone. It's quite a dilemma, but you solve it admirably as long as you are true to yourself and remain disciplined. A very hard worker, you have a fine chance to succeed in whatever work you give yourself. You work best in a company, and even better if *you are the boss.*

Oh, such drama and such kindness! You Libra-Cancers are at heart sweetly innocent souls and much loved for your open hearts. Libra vibrates to every tiny breath, while Cancer intensifies the reception. You are fluid, not solid, and can weep at the drop of a hat. Remember that self-control ultimately brings you your greatest success.

Don't be overly concerned about seeming too cold or stolid, because you won't fool anybody—your feelings are already written all over your face! This combination of drama and kindness imparts to you a beautiful sense of dignity, such as is possessed by astounding Libra-Cancer playwright Harold Pinter. Noble and kind, you could, if you so desired, lead world movements. Libra-Cancer Bob Geldof is a great example of a rock star who gives generously, and speaks up at every Live Aid concert he organizes.

Oddly enough, you are the most traditional of the Cancer Jupiter combinations, which is paradoxical, because the phoenix of Scorpio is so compelled to rise again and again from the ashes, that change is an almost daily occurrence!

When Scorpio joins Cancer Jupiter, that strength to rise repeatedly turns to possessiveness, and suddenly, there is a strong extra need for control. It may channel itself through the recording aspect of Cancer, such as the famous collection of interviews about sex by Scorpio-Cancer Shere Hite. Or, as in the example of Scorpio-Cancer Bob Hoskins, give some of the best portrayals of men—usually gangsters—searching for control ever seen on film. Souls born into this combination have had extremely heavy past lives and seek to heal their injuries in this life. Many Cancer Jupiters with their Sun signs in Scorpio end up as surgeons or healers.

Take time in your love life. You have plenty of time. Passion is your middle name.

IF YOUR SUN SIGN IS:

Sagittarius

(November 22–December 21)

Y ou are gifted, and although you often hide your super powers, you can't stop them from coming out by themselves. Your Sagittarius loves fun in the sun, while your Cancer luck is on the Moon side—how can you reconcile? By partying all day and all night! Thank goodness work is also a kind of party for you, or you'd never get anything done. Sagittarius-Cancer Jimi Hendrix is perhaps the most famous example of this—he worked extremely hard and conscientiously. You have good friends whom you can trust, though if you don't trust them, do not hesitate to let them go.

Extremely lucky events come to you naturally. Sagittarius is ruled by Jupiter (no other sign is) such that with Cancer exalted in Jupiter, you have a doubly positive whammy in effect. It's the Lotto winner's life this time around. Enjoy it! *You know you can do it.*

≈⊛

IF YOUR SUN SIGN IS:

Capricorn

(December 22–January 19)

Y ou are a strong and lonely soul in possession of powerful gifts, whether it be Capricorn-Cancer J. Edgar Hoover, who reigned as director of the FBI for an obsessive-compulsive forty-plus years, or Capricorn-Cancer Robert Duvall, who showed how prophets can corrupt and become corrupted in his brilliant, amazing, and overlooked film *The Prophet*. His performance was pure genius, and so is yours!

You were born with an ability to reach and affect a great many people. Both Capricorn and Cancer love planning and are deeply ambitious, so you will likely achieve great successes. Just allow for more spontaneity in your love life.

≈⊛

IF YOUR SUN SIGN IS:

Aquarius

(January 20–February 19)

I f you do not use your gifts in this lifetime, you will regret it through many other lifetimes. Wake up, Aquarius-Cancer!

You'll breeze through life until you realize that to achieve your greatest success, you must give and continue to give, and it will always involve a helpful family member or friend—one way or the other. Benicio Del Toro is an example of distant-yet-close Aquarius-Cancer: He grew up in Puerto Rico in a family of lawyers, who expected him to become one himself. He had to leave it far behind to do what he does today—which is brilliant and deeply felt acting. Aquarius-Cancer James Dean is another example of one who drew people toward him while seeming far away in his own world.

A Crab cannot, of course, survive without its shell, while Aquarius requires distance and freedom. Both requirements are very deep and strong inside of you. The pearl comes out of the damaged clam. You will find success by compromising and happily marrying one with the other.

Pisces

(February 20–March 20)

You are the healers of the zodiac. You were born with the eternal belief, as Pisces-Cancer Mike Leigh puts it, that "if you water it enough, it might grow." You have within you golden treasures, and yet you're so shy that your gift seems to shine out of a very deep, dark, place. An example of this darkly creative combo is Pisces-Cancer Kurt Cobain.

You must take care of your gifts, and that includes your body. Of all the combinations, yours is the second most prone to being adversely affected by overindulging in alcohol (Pisces Jupiter is most prone), and so you need to find less substance-centered ways to loosen up and have fun. You will find help and support on that score from your mate. A cautionary word regarding any potential mate: Do not drive away everyone who comes to your door.

Relationships with Cancer Jupiters

Finally, for those of you who aren't Cancer Jupiter yourselves but who have a spouse or significant other, boss, child, or parent who is, here are some great tips to help you learn to deal! This is how the luck of someone else's Jupiter touches *your* life.

Your Significant Other

With a Cancer Jupiter mate, you have a sweet Venus, a Shakti and Devi to everyone, a lover of all, with exquisite heart energy and exquisite orgasm energy. Cancer Jupiters are *made* for love, though if you disappoint them, they will never forgive you (the only catch but, sadly, a major one).

Cancer Jupiters will sulk mightily when they believe they are being deprived of romance. They go into dreamy and secretive states guaranteed to drive you bonkers, but they're so darned cute that you may simply give up and hopelessly, helplessly stare at them. They will, in turn, be moody, affectionate, anxious, bossy, generous, loving, sweet, and jealous. They are frustrating but worth the extra effort. If you step away from them, there will be a line of a mile long of suitors waiting behind you to take your place, so do your best to hang in through the ups and downs.

The only real problem is that they do not easily break up with their mates. They will keep up a relationship, even if it's purely for the pretense of it, or for the sake of family. If the thrill is gone and you want to break up with or divorce your Cancer Jupiter mate, good luck. It may take a couple of decades. I'm not kidding. Decide you *really* want one *before* you hook up with one. From your Cancer Jupiter mate you will gain the luck of finding home, family, and stability. They never let go. Some of the lessons may not be easy, but remember that a lifetime of devoted partnership is a rare treasure.

Your Boss

Cancer Jupiter is *always* the boss. Cancer is ruled by the Moon, which symbolizes the Mother, and working for a Cancer Jupiter boss can be a little like working for your mother! She, or he, will see the business as a family, creating something wonderful together. Do you have a problem, need a helping hand—or loan—or the comfort of a motherly bosom when the going gets rough? Your Cancer Jupiter boss is there for you—even when you'd really rather maintain more distance and keep your personal life private. After all, if it's your problem, then it affects the family and the business, so it's her/his problem, too. "Mother knows best" may often be true, but "Mother knows all" can be a little stifling. Film director Cancer Jupiter Martin Scorsese (the boss on his movie sets) has shown again and again in his films how "da boss" in a family, even a crime family, seeks and often deserves love, obedience, and loyalty. In the movie *Goodfellas*, Joe Pesci, another Cancer Jupiter, plays a character whose "family" values go askew, until he gets on the wrong side of his bosses, a performance so powerful he won an Oscar for it.

From your Cancer Jupiter boss you will gain the luck of finding your value as a family member in a potentially indifferent environment. As part of a family team, you will learn to make choices that benefit everyone, including yourself.

Your Child

Your Cancer Jupiter child was born into a fairy tale and wishes the world would stay that way. They have great, wonderful ambitions as youngsters, dreams that shock and at the same time delight you. "How are they going to be able to do all that?" you wonder, shaking your head. Your Cancer Jupiter toddler doesn't let go of anything and, for better or worse, that includes you. Later, her needs will change as she learns to appreciate books, and as a teen, when she learns to appreciate money. You should give this child born under the sign of the Crab an allowance, and start early, because, unlike other children, a Cancer Jupiter will actually save some of it every week.

A Cancer Jupiter child could teach you a thing or two, and they usually do. They're funny, wise old souls, and from infancy they know how to make you laugh. You will probably laugh together all your lives. The Cancer Jupiter child will be lucky when he or she absolutely adores and worships his or her mother, and remains cautiously kind with the father. The mother is usually the most stable relationship of this child's life. It's very important the mother be personally happy, because her child picks up everything from day one in the womb—everything!

Your Parent

Your Cancer Jupiter parent is fanatical about family. *Fan-at-i-cal.* They will want you to work hard within the family. It is extraordinary how many Cancer Jupiter parents take their children to work with them. The best thing about having a Cancer Jupiter parent is that you get to play in the grown-up world much sooner than other kids, and it's fun, not strange or intimidating.

These kinds of parents love to play dress up and dance around with you, and they generally have a ball. They really seem to be in love, and they are, with you and with each other. You could hardly ask for a more involved, caring parent, and because of that, the ride can be a little scary at times, like riding a roller coaster with no brakes. Nevertheless, it is still, in many ways, preferable to the lives of all the other kids who have bland, noninvolved or even absentee parents. Your Cancer Jupiter parent is the most attentive and education-oriented parent you could hope to have, and if it gets to be too much sometimes, remember that nature provides an antidote for every poison. Problems happen, but always, somewhere, there's a solution. Cancer Jupiters are just the people to find those solutions. *Remember, Cancer Jupiter: You are the creative feeder and nurturer of the zodiac.*

Madalyn's Movie Choice for Cancer Jupiter

Monsoon Wedding
Set in India amid frenzied preparations for a daughter's wedding, this movie is *all* about family—its roles, rhythms and habits, its nostalgia, its financial concerns— and how far it will go to protect its children. A brilliant cast engages with one another in moments of extreme vulnerability, humor, tragedy, dance, and song.

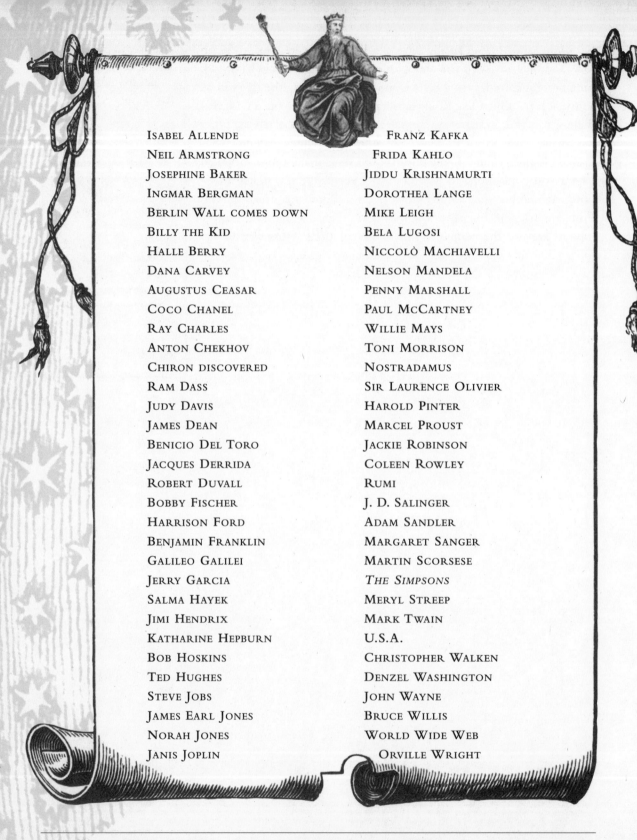

ISABEL ALLENDE

NEIL ARMSTRONG

JOSEPHINE BAKER

INGMAR BERGMAN

BERLIN WALL COMES DOWN

BILLY THE KID

HALLE BERRY

DANA CARVEY

AUGUSTUS CEASAR

COCO CHANEL

RAY CHARLES

ANTON CHEKHOV

CHIRON DISCOVERED

RAM DASS

JUDY DAVIS

JAMES DEAN

BENICIO DEL TORO

JACQUES DERRIDA

ROBERT DUVALL

BOBBY FISCHER

HARRISON FORD

BENJAMIN FRANKLIN

GALILEO GALILEI

JERRY GARCIA

SALMA HAYEK

JIMI HENDRIX

KATHARINE HEPBURN

BOB HOSKINS

TED HUGHES

STEVE JOBS

JAMES EARL JONES

NORAH JONES

JANIS JOPLIN

FRANZ KAFKA

FRIDA KAHLO

JIDDU KRISHNAMURTI

DOROTHEA LANGE

MIKE LEIGH

BELA LUGOSI

NICCOLÒ MACHIAVELLI

NELSON MANDELA

PENNY MARSHALL

PAUL MCCARTNEY

WILLIE MAYS

TONI MORRISON

NOSTRADAMUS

SIR LAURENCE OLIVIER

HAROLD PINTER

MARCEL PROUST

JACKIE ROBINSON

COLEEN ROWLEY

RUMI

J. D. SALINGER

ADAM SANDLER

MARGARET SANGER

MARTIN SCORSESE

THE SIMPSONS

MERYL STREEP

MARK TWAIN

U.S.A.

CHRISTOPHER WALKEN

DENZEL WASHINGTON

JOHN WAYNE

BRUCE WILLIS

WORLD WIDE WEB

ORVILLE WRIGHT

"Macho does not prove mucho."

—Zsa Zsa Gabor, Aquarius Sun-Cancer Jupiter. February 6, 1919

"You can build a throne out of bayonets, but you can't sit on them for long."

—Boris Yeltsin, Aquarius Sun-Cancer Jupiter. February 1, 1931

*"I'm gonna put a curse on you—all of your children
will be born completely naked!"*

—Jimi Hendrix, Sagittarius Sun-Cancer Jupiter. November 27, 1942

"I felt very, very small."

—Neil Armstrong, Leo Sun-Cancer Jupiter. August 5, 1930

"If I'm left to my own devices, I might just turn into a madwoman."

—Claire Danes, Aries Sun-Cancer Jupiter. April 12, 1979

"The most wasted of all days is one without laughter."

—e. e. cummings, Libra Sun-Cancer Jupiter. October 14, 1834

"If I had it to do again, I'd definitely pay more attention."

—Adam Sandler, Virgo Sun-Cancer Jupiter. September 9, 1966

"Do CEOs get to ask for a recount?"

—Ross Perot, Cancer Sun-Cancer Jupiter. June 27, 1930

"Man is the only animal that blushes. Or needs to."

—Mark Twain, Sagittarius Sun-Cancer Jupiter. November 30, 1835

"Love is the difficult realization that something other than oneself is real."

—Iris Murdoch, Cancer Sun-Cancer Jupiter. July 15, 1919

♌♃

If You've Got It, Flaunt It!

From Saint Augustine, born November 14, 354, who said, "Give me chastity and continence—but not yet," to actor Robert De Niro, born August 17, 1943, who said, "You'll have time to rest when you're dead," Leo Jupiters know they have to make hay while the Sun—their natural ruler—shines!

There is no end. There is no beginning. There is only the infinite passion of life!

—LEO JUPITER FEDERICO FELLINI

Dear Leo Jupiter, you were born, quite simply, to light up the planet. As Leo Jupiter William Saroyan urges you, "In the time of your life, live!" Whether you flaunt your royal power like flamboyant Leo Jupiters "Queen" Liz Taylor, Dame Kiri Te Kanawa, or Lady (Sings the Blues) Diana Ross, or let your achievements speak for themselves, like Alice Walker or Yo-Yo Ma, your luck comes from being the greatest at everything you do.

Harvard University is Leo Jupiter. "Be not afraid of greatness," urged Leo Jupiter William Shakespeare, "some are born great, some achieve greatness, and some have greatness thrust upon them." Remember it well, O Leo Jupiter!

You love being *numero uno,* because that's where you belong. Even the *Guinness Book of World Records,* the list of those who achieve top ranking in every possible human activity, was born a Leo Jupiter (August 27, 1955). And, Leo Jupiter, you're not known for passivity, inaction, or even patience. Jupiter does not reward you if you sit back, demurely, and wait for your reward to come; instead, Jupiter asks you to jump into the world and embrace all it offers.

You gain luck by exploiting tremendous energy and being prolific. Leo Jupiter Lauren Ambrose, born November 16, 1978, the youngest star of the brilliant *Six Feet Under*, said, "I love playing young women coming of age, trying to figure things out, starting to see their lives for what they are." Manifest your creativity and love for playful exploration.

You cannot be a natural lone wolf, Leo Jupiter: Jupiter rewards you if you work with people and lead a group. You need to support and listen to others, to lead through teamwork, if you hope to win the greatest luck. Every Leo Jupiter who achieves success relies upon others, listening seriously to their advice. Financial mogul J. P. Morgan, known for his ruthless manipulation of the markets, did not operate alone. In fact, he wouldn't make a single move on the stock market without listening to the advice of his psychic consultant, Evangeline Adams, first. He proudly claimed: "Millionaires don't use astrology, billionaires do."

Take a moment, look back long and deep over your life, recount as many events as you can, and ask yourself what has made you lucky or unlucky in your life. You'll notice that whenever you act like a Leo, *you create miracles in your life.* No matter what your Sun sign is, *you get lucky when you live out the Leo legacy gifted to you by the benevolent planet Jupiter.*

To take advantage of your luck, emphasize your . . .

Be confident, generous, powerful, warm, dramatic, expansive, protective, royal, and fun.	♌♃ **Leo Jupiter Strengths**

But because Jupiter rules excess, watch out for . . .

You can become self-obsessed, suffer from hubris, or use any ends to justify your means.	♌♃ **Leo Jupiter Excesses**

Understanding the
Leo Energy and Influence

Leo represents the time of year when the summer ends in a bright flash of heat and color, and we reap the harvest of vegetables and fruits that come in September. It is a time of enjoyment, of fun, so it should not surprise you that Leos are the top, the best, the biggest, the brightest! The world wouldn't be any fun without Leos—and they act like it. They are sunny, on top of their game, generous, charismatic, and work on inspiring others to follow and imitate paths they have created.

Leos have it all, love it all, and want it all—sometimes so much that reality disappoints. Leo Jupiter Peggy Lee's exquisite song "Is That All There Is?" truly sums up Leo's grandiose expectations of life. But the rest of the song says that if *that's "all there is," then we should dance, drink, and party*! Leos are not content to sit and mourn a life they don't want—they change it to what they do want.

LEO SYMBOL
The Lion

The lion, walking proud and tall, is the king of the jungle, the leader of the animal kingdom. Surrounded by a thick mane, announcing his presence with a loud roar, the lion is fearsome and impressive. Whether pouncing on a fleeing gazelle or leading their cubs to a fresh spring to drink, lions (and lionesses) are the world's natural leaders. And while the lion is clearly in charge, lions are never alone: Wherever the lion goes, so too goes his pride.

To win the luck and power of Jupiter, you must act just like a lion! Leo Jupiter, attract people to you like the lion does his cubs, and protect them by taking care of them. This is the strongest thing you, Leo Jupiter, can do to make yourself lucky: be loyal and protective of your pride! And, like a lion, try to bask in confidence and leadership, and to stay true to your real self. Don't feel like you ever have to pretend; Lions don't go about wearing masks, and if you hide your true self, you will not get lucky. And don't indulge your insecurities (how many Lions look like they have self-esteem issues?): You will be pushing away opportunity after opportunity. How long have you been doing *that?*

Leo rules the heart, the spine, and the back, the bodily elements that power the blood and hold up your entire walking carriage. The back means courage, the spine—as in backbone—means strength, and the heart pours forth deep, full emotions. Allow yourself to feel: Repressing your emotions and not taking time to destress is unlucky for you. While Leo Suns have to be extra careful to protect their back, spine, and heart, you, Leo Jupiter, should feel free to indulge and stretch these parts of your body under the power of the planet that protects.

LEO PARTS OF BODY
The Heart, the Back, and the Spine

Your lucky colors, Leo Jupiter, are orange and regal gold. If red is fire and passion, then orange is all warmth and friendliness—the friendly part of fire—as well as the true color of the "fires" of creativity. Jupiter energy will be drawn to you when you wear orange (if it doesn't suit you, remember orange purrs warmly when worn as an undergarment) or surround yourself with orange and gold colors and things. Your lucky colors send a message that is definitely sexual but with a healthy overlay of soothing friendliness. *Just be close with me. I don't bite. Not that hard.*

LEO COLOR
Orange

And now your Leo Jupiter strengths: ♌♃ *Leo Jupiter Strengths*

Be Confident

Confidence is a gift you must give *yourself.* And it's all in your own head and heart. Just listen to Leo Jupiter Eleanor Roosevelt, who said, "No one can make you feel inferior without your consent," or Leo Jupiter Whoopi Goldberg, who proudly announced, "Normal is in the eye of the beholder." You're not normal, Leo Jupiter—don't even try! You're spectacular, and you must show the world how amazing you are. Whoopi, for instance, defied all the stereotypes of how a "star" was supposed to look: "Nobody looked like me when I got famous. You have to believe in yourself in spite of what other people believe. That self-confidence is what brought me through everything in my life."

Explorer Roald Amundsen drew on the bounty of Jupiter by act-

THE RULING PLANET OF LEO

Shining Star Sun

With only *half a billionth* of its energy, the Sun sustains every life-form on Earth. All would be destroyed, however, without the ozone layer, as the Sun burns 27 million degrees Fahrenheit and is more than 1 million times the size of the Earth. Even from 93.2 million miles away we feel its god-like power. The Sun is larger than all the other bodies in our solar system *combined* and is the source of all life, warmth, and light. Just as the entire universe revolves around the Sun—so all revolves around Leo!

YOUR HISTORY

The Sun was worshiped in all ancient cultures. From the Sun god Ra of the Babylonians to the Greek Helios to the Roman Apollo, the Sun has been god of light and truth and power.

ing like a relaxed, confident Leo, calmly overcoming all errors and accidents thrown his way by the influence of his excitable Sagittarius Sun. Born July 16, 1872, Amundsen left his mark as one of the most successful polar explorers ever, becoming the first human being to reach the South Pole. His great rival, British explorer Robert Scott, failed, his expedition wracked by arguments, disorganization, and insecurity; the brilliant Amundsen seemed to sweep in, effortlessly, with blinding confidence, to pass Scott to reach the South Pole. Like Amundsen, Leo Jupiter, cultivate confidence and feel secure in your regal role.

Be Generous

Give me your tired, your poor,
your huddled masses yearning to breathe free,
the wretched refuse of your teeming shore.
Send these, the homeless, tempest-tossed to me.
I lift my lamp beside the golden door!

The Statue of Liberty, in typical Leo Jupiter style, is the largest statue *ever made* and was a *gift* of generosity from the nation of France to the people of the United States, funded mostly by French schoolchildren and workers. The great lady is three times a Leo Jupiter: her birth on July 4, 1884; her cornerstone laying on August 5, 1884; and even the writer of those immortal lines celebrating America's generosity to all who come to her shores was a Leo Jupiter—Emma Lazarus!

Consider the symbol of the "lamp beside the golden door" in its relation to *light* and especially to the Sun, Leo's natural ruler, as well as its gold representing the gift of opportunity and vision. Leo Jupiter Rabindranath Tagore wrote, "I slept and dreamt that life was joy, I awoke and saw that life was service, I acted and behold, service was joy." Like Tagore or Lady Liberty, luck will come to you if you are magnanimous, if you bestow gifts like a queen or king. Be generous and kind, not self-sacrificing or penny pinching. Give gifts that actually help people and make their lives better; practical, regal generosity is the rule.

Think of what's been made possible for women by the Nineteenth Amendment to the Constitution, born Leo Jupiter on August 18, 1920: It gave women the right to vote. Think of Leo Jupiters Dian Fossey, who gave her life to save gorillas in the wild, and of Walter Annenberg, whose philanthropy funds the arts and public television.

Be Powerful

True leaders—for example, Leo Jupiters Archbishop Desmond Tutu and Supreme Court Justice Thurgood Marshall—have power based in moral authority. For you to attract the bounty of Jupiter, earn your power and authority the old-fashioned way, by leading, by making things better for your subjects, and by being wise and fair. Luck will come if you have power and the legitimate authority that allows you to use it.

Leo Jupiter New York City Mayor Rudolph Giuliani rose to the occasion on September 11, 2001 as few public men could. Giuliani was so confident in his strength and power that he was not afraid to show emotion publicly, crying on television with David Letterman (who cried right along with him). In true Leo Jupiter fashion, his decision to make himself publicly vulnerable and explore his feelings made him *more,* not less, powerful. The general feeling was that Giuliani used his warmth and love for New York and New Yorkers to pull everyone through, displaying a manner as befitting a true royal, exquisite grace under pressure. Giuliani ended his career blessed with luck and success. Ironically, before September 11, his popularity had been under attack, as a result of his cold response to two terrible cases of police brutality, and he had been perceived as a bully, a wielder of power without moral authority. (This is the excess of Leo Jupiter: Don't let it happen to you, dear one!)

In Giuliani's book, aptly titled *Leadership,* he wrote, "I couldn't tell people, 'Be brave,' unless I was willing to walk the streets, or not to panic over anthrax unless I was willing to go to the places where it was suspected. That's what the optimism of leadership is all about. Once the leader gives up, then everybody else gives up. And then there's no hope." Remember, if the great leaders can shed tears and be loved more for it, so can you.

To keep Jupiter's lucky energy behind you, Leo Jupiter, be wary of excess pride, make sure you always listen to the advice of those close to you (even if it is a bitter pill to swallow), and always, *always,* be nice to your "subjects." (Look at what happened to Leo Jupiter Joan Crawford.) You will thrive on publicity and recognition of your success; you can be a "royal benevolent leader."

Be Warm

Leader and founder of the Steiner Schools all over the world, Leo Jupiter Rudolph Steiner, spoke of connection thus: "When human beings meet together seeking the spirit with unity of purpose, then they will also find their way to each other."

Leo Jupiter, you must ensure that your attitude to others is warm and friendly. Put this book down now and think of "warm" as a *verb*—warm to yourself, and warm to others. Warm to your beliefs: Do not fear to show your passion toward a cause or a person important in your life. Think of it: Every day, concentrate on warmth, exercise it like a muscle and you have an ability to connect instantly and deeply with others, even if you don't feel like it, and this brings you luck. Even when you're feeling distant or separate, attempt to passionately communicate what is deep and meaningful and best and most real inside you. Go with it and your warmth will become real and alive. As Leo Jupiter comic George Burns quipped in all sincerity, "Acting is all about honesty. If you can fake that, you've got it made." Warmth keeps you free and flowing; it connects you to the world; it translates into pure energy, which gives you the greatness to achieve your dreams.

Be Dramatic

Now for the fun part of how you get lucky—absolutely no minimalism welcome here. No Prada needed, thank you, throw out the ultrathin, narrow, understated, subtle. For dazzling luck to come to you—*be dramatic!* Give yourself permission to pull out all the stops. Be just downright jumbo-size exaggerated. Be a go-for-broke goofy maximalist! Spend some fun time strutting your best outlandish stuff in the hightop boots of the all-time drama queens, but let's not forget—*they're not all female*—the flamboyance of Mick Jagger, the going-to-excess of Jim Morrison, the outright swaggering, self-destructive performance of Sex Pistols Johnny Rotten. Or think of another Leo Jupiter, hippie-era freak (Mr. Natural/"Keep on Truckin") cartoonist R. Crumb, whose caricaturing of his own sexual desires and proclivities won for himself the luck of Jupiter by virtue of being superconfident enough to *totally* express himself and wildly individualist enough not to give a flying flatulating figleaf what others thought.

So even in your most mundane chores—in the shower, going shopping, buttering a slice of toast, having a casual conversation—act as if you are performing in a great play or opera: Dance your movements, sing your demands, set your thoughts to song, be dramatic! Give in to the drama inside yourself and use it in every situation to your best advantage. Like any great actor, you can put it on even when you don't feel it.

Be Expansive

Leo Jupiters get lucky by seeing themselves in the world in *mythic* terms. So *don't be afraid to mythologize yourself.* You will not get lucky by being small or overly humble. Allow your inner child to play in as large a playground as possible.

Part of being expansive is relating to others, understanding their lives, figuring out how other people live. Leo Jupiter Jacob Riis did this literally when he produced a powerful book, *How the Other Half Lives*, which exposed New York City's immigrant poor to public view and led to significant improvements in their lives. *How the Other Half Lives*, as its title suggests, provides a Leo Jupiter example of expanding one's understanding beyond yourself, beyond the accepted circumscribed world—whether that means to other genders, other colors, religions, other types of people.

People love expansion. Go expansive. Leo Jupiter Bette Davis said, "Attempt the impossible in order to expand your work." Living up to your Leo Jupiter potential means making every part of you louder and more extravagant. Let your personal expansion include your enthusiasm, vitality, humor, bravery, nobility, pride, pleasure, creativity, confidence, forgiveness, affection, radiance. Expand it to be the greatest it can be and, I guarantee, your personal joie de vivre will touch the world.

Look at the example of Leo Jupiter Blind Tom (Tom Wiggins), born blind and into slavery on May 25, 1849, who became the most famous and highly paid performer of his day, touring all over the world and giving as many as four piano recitals a day. Even while he was being exploited by his "masters," who stole most of his earnings, his joy and charisma with audiences increased his playing to a legendary stature.

Be Protective

While Pope John Paul II is another example of an expansive Leo Jupiter, stretching the borders of his office more than any pontiff in Church history, traveling to 125 different countries to call on the world's one billion Catholics, he is also an example of Leo Jupiter protectiveness. The pope threw himself into the center of diplomatic efforts to stop the 2003 war with Iraq, declaring, "Everyone has to knowingly take responsibility and make a common effort to spare humanity."

Another Leo Jupiter, Deepak Chopra proposed that he and the pope serve "as human shields to avoid bombing in Iraq." Two Leo Jupiter human shields! Add to this interesting mix yet *another* Leo Jupiter—Defense Secretary Donald H. Rumsfeld—who shows belief in protection of another kind. Named the "Pentagon Warlord" by

Time magazine, Rumsfeld was one of the leading architects of the "shock and awe" tactic of the American military campaign. An example of daring mythologized Leo Jupiter expansionist thinking, Rumsfeld warned that those serving as human shields in Iraq would be treated as war criminals. Three strong Leo Jupiter leaders, and the most dramatic roaring one wins—or does he?

Be Royal

Leo Jupiter, you may be the king but don't push it! As befits royalty, you need to act with kindness and graciousness and modesty (as the true royals know, *you don't have to prove it*). If anything, true royals act modestly, with an almost overdone humility.

One example of homegrown athletic royalty is tennis star and Leo Jupiter Arthur Ashe. Ashe, an African American, grew up in Virginia, playing on the segregated courts of Byrd Park in the 1950s. He called the coolness he projected an "adopted shield" (the shield of royalty) and never complained (true royalty never does) as he went on to become the number one tennis player in the world. Along the way he raised millions of dollars for the United Negro College Fund as well as for AIDS research. How classy an act was Arthur Ashe? When Nelson Mandela was released from prison after twenty-seven years, he was asked who in America would he most like to see. His answer: Arthur Ashe. Upon his untimely death, Ashe was eulogized as "a champion of mankind." He was surely that and more.

Indian poet and philosopher Sri Aurobindo, born in Bengal on August 15, 1872, gave his early life to the struggle for India's independence, and his later life to retreat and writing. He achieved success and luck by becoming the leader of a spiritual community of people attracted to his teachings, drawn to his writing. He protected those around him; as a result of his regal actions, Sri Aurobindo's community is still active today, many years after his death.

Be Fun

Like the lion who rolls around and *plays a lot,* be playful, romantic, childlike, and *be* your luck, Leo Jupiter. Purr in the sunshine! Speaking of which, remember Sly Stone? Of course, being a Leo Jupiter, Sly appropriately named his band Sly and the Family Stone and bestowed upon them their most famous song, "Hot Fun in the Summertime." *Just listen to it carefully,* if you never have before. Make a ritual of listening to it several times a week. It will inspire you.

If you are Leo Jupiter you can be the world's greatest party giver as well as the life of every party you go to. And it's lucky! You can earn all the luck you need by being funny and entertaining. From Madame de Staël, who after the French Revolution hosted Europe's leading intellectuals in the greatest salons of the day, to *Politically Incorrect*'s Bill Maher, who makes serious political statements while being very funny, to *Saturday Night Live*'s Will Ferrell, Leo Jupiters become successful through entertaining themselves and others. You could have a great career in all the fun arts: party planning, performing, theater.

What Makes You Lucky in Love

You will most likely achieve success in love if you approach the dating scene with quiet confidence. You don't need to be flamboyant, so don't show off. Instead, project inner peace and happiness with your achievements and, yes, don't forget to listen. You will find potential partners happy to take you for who you are, but make sure you give credit and affection to them. Be generous with your compliments. Doing this attracts Jupiter love energy like crazy! It will help you from diminishing your power and success in the bedroom as well.

Your one weakness: You are very loyal, even beyond the call of duty. If your relationship isn't working out and you suspect your partner is playing the field, move on. But when you know, you know. So pick faithful mates and you will never go wrong. And while you're at it, be sure you return the favor.

Last, but not least, make your partner *laugh!* Because you can—maybe better than anyone—and don't we love those who make us laugh the most?

Your Best Career Moves

Leos are the world's natural leaders, and if you are a Leo Jupiter, you must allow yourself and encourage yourself to run for leadership positions, communicate your charisma, and encourage others in their careers as well.

You can reach the top of a major organization if you avoid overconfidence and just behave naturally and confidently with others. Have no fear of being visible and making appearances in public and in the media. Luck will come to you from this prominence. Show your truly best work and you cannot fail to impress. Even when you are a major star, you give so much credit to everyone around you that there is never any jealousy of you. This allows you to enjoy your success all the more!

DOS

❋ Accept your role as royal leader (but not despot) and graciously support others.

❋ Express your inner light, warm sexual energy, and dramatic creativity.

❋ Be highly visible and promote your public appeal; inspire others to follow and imitate.

❋ Take serious catnaps every day; remember, lions sleep up to twenty hours a day!

❋ Eat and carry walnuts during your birthday month to ensure good health all year.

❋ Surround yourself with orange, gold, and tawny colors: clothes, sheets, towels, home décor.

❋ Each day eat *five* of the following: carrots, ginger, plums, pears, oranges, spinach, raisins, lamb, dates, coconuts, green, yellow or orange vegetables, or Thai food!

❋ Hang out with more Sadges, Aries, and Leos.

Issues of self-worth simply cannot be brought into this arena. Save that for the therapist. Your career moves are all about how smooth you are, how easy you make it look, how much you can laugh—and make others laugh—while you're doing it. Insecure, envious Leo Jupiters earn very bad luck as it's not their natural style. This can only be learned behavior from childhood or the effect of a parent; you must work this through and get over it. In the meanwhile, go out there and seize the world! It's yours, it's your time.

How to Reach Optimum Health

Take a tip on beauty sleep from lions in the wild: They sleep as much as twenty hours a day! Practice yoga and cut out coffee. You're wound up tight enough as it is. Your mane is your shining asset and does best with warm weather, plenty of sleep, brewer's yeast, B vitamins, and regular scalp massages.

Lucky lions eat all fresh things, so as much as you love to eat rich food, know that it isn't good for you. It's either the best restaurant for you, or a partner with whom you can cook gourmet meals at home. Your health, by the way, depends very often on the condition of your mate. You both need to be in good shape. Big or small doesn't matter, just mutual good shape.

Cut down on fatty foods (bad for the heart, which is ruled by Leo), avoid lifting heavy objects, and make sure you get some sun, no matter where you live. Leo Jupiters do best with serious breaks and exotic holidays!

Your Lucky Foods, Element, and Spices

Why Leo Jupiters diet more beautifully than any other Jupiter sign: The Lion's cell salt is magnesium phosphate, which activates the digestive enzymes. You will, most likely, never see a shapeless Leo Jupiter!

For maximum health, eat heart-healthy foods and those that help circulation in the arteries. Plums, pears, and oranges are the best foods to allay heart strain. Other excellent foods are those with high iron: spinach, raisins, lamb, and dates. Use coconut or coconut juice in dishes, and eat mixed greens and cooked green, yellow, and orange vegetables every day. Thai is your best palate as it contains your favorite spices! Your luckiest food of all is the walnut. Eaten or even worn as a talisman protects the heart.

Your one weakness is a fondness for sweets, which I urge you to avoid. They go straight to your heart, and the consequences far outweigh the temporary fleeting pleasure of a sugar surge. If you find yourself craving seeds and nuts, give in, by all means.

Your Best Environment

You might be forgiven for thinking the king or queen needs to live in his or her palace. Really, all that is necessary for Leo Jupiters to be lucky is that there are enough people around. It is difficult for you to live purely in nature (unless it's some great camping exploration) or in the countryside. Try to set up your environment like a royal ancient court, with courtiers all around!

The highest cultural milieus, pace, and fun, especially when traveling, are where Leo Jupiters are most likely to be encountered. You may be constantly striking up conversations with strangers on airplanes, in hotel lobbies, elevators. This can be exhausting, but you only notice it after you've been at it for seventeen hours straight. And then, dear Lucky Lion, you need to go to a fantasy island and just snooze on the beach. No other kind of holiday will do. You'll come back tired otherwise, and what's the point of that?

Your Best Look

Leo Jupiters always look great—in this area you are naturally lucky! All you really need do to enhance this natural luck is make sure you get plenty of sleep. (Leo Jupiters show it more than most when they've missed sleep.) You have a way of always looking fastidiously turned out, which makes a nice business impression, but you look very sexy when you let yourself be just a little messy—an orange or yellow T-shirt hanging over your jeans or shorts, for instance—and when your hair is not always combed or styled. Of course, you dress for glamorous parties better than anyone (besides Libra Jupiters), and that kind of look—high power diva or mogul—you can pull off in a second.

Your Best Times

The luckiest times for you are when the Moon is in Leo, which happens every month for two to three days. Find these special days in your astrological calendar. During these days, schedule crucial meetings, tell that special person you love them, ask for a promotion or raise, push the envelope, and plan to shine. In your everyday, the best time for you to exercise *and* rest is late afternoon. The luckiest times during the year for you are July 23–August 23 (when the Sun is in Leo); November 22–December 21 (when the Sun is in Sagittarius); and March 21–April 19 (when the Sun is in Aries).

The very luckiest time of your life—as in Lotto lucky—is when Jupiter reenters the sign it was in when you were born (Leo), which happens every twelve years and lasts anywhere from a few months to just over a year. This is called your Jupiter Return, and it launches your luckiest streak. You can count on the *extraordinary* to happen the very day your Jupiter Return begins. Amazing events can start happening up to six weeks beforehand, so plan ahead!

How to Be Lucky Financially

Luck will follow you when you generously invest your leadership skills and your money. Beware your tendency to economize too much. Also, make sure you put yourself in situations where you will be in a group and be the leader of that group.

Remember, you do have natural luck on your side when it comes to finances—your ruler, the Sun, pouring its warmth into your deep coffers—so as long as you relax and don't freak out, you'll be more than fine. Once you start to worry, you may will your money to disappear. Relax! Leo Jupiters are the only natural winners of Lotto. I never give advice to anyone concerning Lotto, except for Leo Jupiters. Most types of gambling will be too individualistic for you; instead, why not get a group together and figure out lucky numbers for Lotto together.

You will achieve your financial dreams when you work with others who appreciate you; avoid, if possible, lone careers. Politics suits you, as does leading your community through work. Choose the group path and make sure you don't overly worry about your reputation and keep yourself from taking needed risks. Also, make sure you aren't too selfish about money, as that will hurt you.

Success in Family and Friendship

If you are given a great number of responsibilities early in life, relax, you're lucky. Accept this graciously like the true royal you were meant to be.

Perhaps to your surprise, taking the lead, even within your own family, will have a very positive effect on friends and relatives. They *want* you to be a leader. With your siblings, it is superlucky for you to compete. This may seem surprising, but it is actually lucky both for you and for them if you're as competitive as you can be. Leo Sun sign siblings run into trouble if they are too competitive. However, as a Leo *Jupiter,* you will find luck in competition and it will make you the apple of your parents' eyes. It's also lucky for you to be loyal, generous, and protective. You will not regret it.

People depend on you perhaps more than you know. Be big, large hearted—roar when you need to. Roaring is, in fact, a very good, lucky device for Leo Jupiter. You might be surprised to hear that even roaring in *anger* often helps. Your family and friends actually *look to you* to clear things up, and also for advice, wisdom, solace, and fun. The worst luck for Leo Jupiter in a family or any other situation: *Do not isolate or distance yourself* and *never play a passive role.*

When you go too far!

♌♃ Leo Jupiter Excesses

Macbeth, a darkly beautiful play of misplaced ambition gone wrong, serves as the perfect cautionary tale for Leo Jupiters who may become victims of their own ambition. Macbeth, a Scottish chieftain, welcomes the king, Duncan, to his castle as a guest. Goaded on by his wife, Macbeth slays the king in his sleep, gets away with the crime, and himself becomes king. But soon his foul deed leads him onto a path of a whole succession of bloody deeds, and in order to remain king, he murders everyone in his way and presides over a ruined country drenched in blood. Lady Macbeth goes mad and takes her own life, while Macbeth is finally killed by his enemies. Macbeth himself states the play's theme:

> *I have no spur to prick the sides of my intent, but only*
> *Vaulting ambition, which o'erleaps itself and falls on the other.*

The moral: Leo Jupiters are famous for taking a chance on anything, but there's a limit! Avoid the need to fulfill your partner's ambitions and beware of desiring success and power to an exaggerated degree.

Excess #1 You Can Become Self-obsessed

Microsoft founder Bill Gates, born a confident Leo Jupiter if there ever was one, is the richest man in the world, with an estimated $40.7 billion. He gives away more than $1 billion a year, donating generously to education, public libraries, and vaccine research, among many other causes. Yet, when he departs from the role of regal benefactor and becomes selfish, Jupiter's luck and success can desert him. Remember when Gates tried to wipe out his competitors and ended up spending years in court (and millions on legal fees) as a result? "Success is a lousy teacher," he said. "It seduces smart people into thinking they can't lose."

Problems ensue, Leo Jupiter, when you overreach your ambitions. Avoid tyrannical behavior, sudden mood swings, outrageous demands. It's also unlucky for you to entertain ambitions that are completely selfish or self-obsessed. Wallis Simpson, the Leo Jupiter wife of England's King Edward VIII, persuaded her husband to leave the throne (a very bad move for a Leo Jupiter) so that they could marry. In exile in Paris, she announced, "You can never be too rich or too thin," and frittered away her life in a stream of parties. By entertaining such self-obsessed, unqueenly ambitions, Simpson denied herself the power and bounty of Jupiter.

Excess #2 You Can Suffer from Hubris

The prime and criminal Leo Jupiter excess is hubris—exaggerated pride or self-confidence, wanton arrogance. Believe me, this will result in nothing but disaster and leave you sore and disappointed for years afterward. Sometimes excessive arrogance is a reflection of insecurity. If confidence was taken from you as a child, you may have to, perhaps with professional help, reteach yourself true confidence, which comes with utter acceptance and relaxation.

Don't let an excess of confidence (or insecurity) lead to arrogance and egotism. Excessive pride is your enemy. Leo Jupiter President Lyndon B. Johnson was one of the proudest presidents America has ever had—in many ways a good thing—but his pride was so strong that he could never admit making a mistake. Although he knew that the Vietnam War was going badly, his pride prevented him from acknowledging defeat and withdrawing American forces. He once said, "I will not be the first American President to lose a war." Yet, the result of Johnson's hubris? Personal defeat. He refused to run for another term because he realized his war policies had doomed his chances for reelection.

Excess #3 · *You Can Use Any Ends to Justify Your Means*

"I never cared but for one thing, and that is, simply to know that I am right before my Father in Heaven," said the great Mormon leader Brigham Young. Both he and his protégé, John D. Lee, were Leo Jupiters, and it is said that Lee's belief in their "Father in Heaven" allowed him to use any ends to justify their means. This is very dangerous for a Leo Jupiter. Never allow your ambitions to cross the rules of fair play and good, neighborly behavior.

One tragic result of this was the Mountain Meadows Massacre, in which a wagon train of 103 men, women, and children were slaughtered as they traveled across Mormon territory in Utah in September 1857. John D. Lee had gone to the wagon train and assured its members that they had safe passage to proceed, but once he had gotten them to lay down their arms and lured them out into the open, they were killed. Only seventeen very young children, considered too young to have any memory of the incident, were permitted to live and were later adopted and converted by Mormon families.

Mormon hostility to outsiders and the church's belief in polygamy had made Young (who himself had twenty-seven wives and fifty-six children) and his followers highly unpopular, but while John D. Lee was executed for the crime twenty years later, he refused to implicate Brigham Young, and Young's role was never definitively determined.

Sun Sign-Jupiter Sign Combinations

You are a natural superstar, so enjoy this awesome combo! Divas we love sparkle with this double fire: Ashley Judd, Joan Crawford, Celine Dion, Diana Ross, and Bette Davis are all Aries-Leo! You do it big or bust. Don't even think about going at it humbly and modestly. Approach all with your heart and you will be wildly successful. In work, you want plenty of control and power, and when your dreams pay off big (which they do), then you get it, too. You're impossible to resist!

In love, you require someone equal to your daring and strength; the mates who last are the ones who are very powerful in their own right. Don't listen to that "balancing" advice and get yourself an earthy homebody—it will never work (hey, Ashley's with a *daredevil race car driver,* for heaven's sake!) You are smoldering where it counts and deserve the best lover—and true love—you can find. You have natural luck, so be as picky and choosy as you like. Don't despair if there's no one in your life right now. Aries-Leo, you can go a long time waiting for the right one, but to be lucky keep socializing, don't isolate. Aries fear that people don't like them enough; Leos are confident that people *do.* Try to find the correct balance. Don't be a people pleaser. Rather, you should be worshiped or at least admired by others.

You have incredible intuition and must use this in work and love for your greatest success. You tend to find out more about others than they find out about you, and while you think this *may* give you an advantage (safety for the Taurus), remember that your Leo actually brings you more luck if you are open, giving, flamboyant—even dazzling! Mine your own creative and communicative powers to rake in the most cash. Then trust yourself to make the right investment, even if it's on pure impulse.

In love, you require real class—a true royal who isn't pretentious. Fly-by-night floozies will leave you feeling sad. You need security, and when you have your chosen mate, you want to plan the whole future. However, when you have found a mate you can respect completely, let yourself relax and allow more spontaneity into the picture. This goes for your work and dealings with your professional colleagues as well. Have a laugh!

Even though you know the greatest successes grow from adversity, you can let yours be more fun, thanks to your lucky Leo. With these two fixed signs, you can overcome anything with your strong desires and simultaneously have the wisdom to see life for what it is. Look at fellow Taurus-Leo *Star Wars* creator George Lucas, whose terrible car accident just after his high school graduation ended his dream of being a professional racecar driver! "Good luck has its storms," he said.

This combo is known as the Great Virtuoso, bringing powerful work that you believe in, a love that feeds your mind, body, and soul, much professional acclaim, and—you don't even have to try for this one—high-tiered social success. The mystery, dear Gemini-Leo, is in the timing: All of your universal feedback may come at once, or one deserved reward takes minutes while another takes years! For more luck, relax your quick Gemini rhythm and step more to the pronounced pleasure of Leo.

You are a natural communications wizard, and imagination and intuition is where your magic truly comes to the fore. You will translate this into different fields, but for your greatest luck, *you must* focus on one—that which you feel most passionate about and that does the most good. Your charisma attracts many high-powered helpers; don't bury it under a ton of Gemini analysis! Royal roars are good too, as people *love* you to be their leader.

People *clamor* for you, but do resist the Gemini urge to flit hither and thither—Leo leadership and intensity bring you more luck. Gemini-Leo idols Peggy Lee, Rudolph Giuliani, and Harvard University are not fly-by-night trends, and neither are you.

The best part is that it gets better as you get older—unusual for the eternal-youth Gemini, but your good looks turn sizzling.

In love, your search for the best mate (and we're talking very best here) can take you as long as it takes. Before then, there's plenty of fun, soaking up the sun-kissed adoration of *many,* but make yourself happy by letting your intense romantic streak and desire for faithfulness show. There's a hidden sentimentality that brings you success as well—when you've found the right person with whom to share it.

Leo makes it so worthwhile!

IF YOUR SUN SIGN IS:

Cancer

(June 21–July 22)

Tremendously creative, you ride a fine balance between the vision of artist Cancer-Leo Amedeo Modigliani and comedian Cancer-Leo Will Ferrell. Push your individuality! Your greatest luck comes through the element of *surprise* and, sometimes, aggression; try it, in your career and in your relationships, and you will be laughing all the way to the bank—or the altar.

Sweet and constant, let your lucky Leo lighten you up and dance you out of your shell occasionally. You feel your life so very deeply and need only to *enjoy* it more for complete success. A born romantic, you will attain your dreams more easily by bringing an element of playful performance into your courting—and into your sex life. Becoming a parent is also extremely good for you. Do not allow yourself to be a victim to unrequited needs and desires; *speak* when you feel hurt and ask for something when you want it. Public speaking or acting classes would be an excellent investment for you, both therapeutically and in terms of expanding your work possibilities.

᧤᧟

IF YOUR SUN SIGN IS:

Leo

(this makes you double)

(July 23–August 23)

You get lucky by being larger than life—literally. People who are not Leo-Leo tend to fade next to you; your exciting energy and talent often steal the show. You are the very best at what you do, and your confidence can at times approach legalized hubris, so know when to recognize forces—like time—that are greater than yourself. Double Leo Peter O'Toole initially rejected an offer of an honorary Academy Award as he "might win the lovely bugger outright," but 2003 was his Jupiter Return (Jupiter in Leo will not happen again for another twelve years), and thankfully, he changed his mind.

Know when to step back, when to hang your head modestly, and when to keep quiet in times of extreme drama. The power of your fate is so great that you will need to be physically taught how to do this—most probably from good friends. Your circle is huge (everyone *loves* the double Lion!), and you may pick and choose at different times.

In love, you want to be adored unconditionally. An extreme romantic with the highest standards of a mate, you become lucky with a soul mate when you can let go of control. Study the cat, not the lion, for your inspiration—in the love area, less is more.

Your career is not in jeopardy: You are a brilliantly hard worker, astute and creative way above the call of duty, you take on more than almost anyone else, and you don't quit. What you could allow more of, however, is your enjoyment of the thing. Luck is earned through your Leo, which for you means more fun, less burden. You feel responsible for the whole world sometimes and want to save it all!

You *want* to be hard on yourself (and on others), but if you know what's good for you, you'll follow your lucky Jupiter and encourage yourself to be as expansive as possible. No more going about with your head bowed and collecting doubts! Now you must shine, be a star, maximize, and show off all your greatest qualities. Virgo-Leo Alison Lohman of *White Oleander* is a perfect example of one who has to *force* herself to feel beautiful even though she already is.

In love, you climb a Mount Everest of hope and desire thinking that it probably won't even be worth it when you get there. Move to sea level! Your analysis is making you starved for oxygen. Find a solid Taurus or a relaxing Leo Sun sign, slide in, and have a good time. You receive many proposals of marriage—*don't worry about it!*

People *love* you! In fact, this is the best-liked combo of all Leo Jupiters. Your natural charm and charisma turn you into everyone's favorite prince or princess. The luckiest action you can take is simply to soak up all that admiration. Libra-Leos Catherine Deneuve, Mickey Mantle, and Luke Perry are examples of mass worship who really didn't have to do anything—but be themselves. The problems may come in your personal or work life, however, when getting by or winging it just doesn't cut the mustard. In relationships, it will hurt in terms of not being able to achieve deeper intimacy, and you may end up torturing yourself as to how to act. In your work, when you aren't credited appropriately, it will be infuriating—and often unfair. How many times have you heard that the power of your looks makes everything so easy for you?

When in doubt, which Libra frequently is, stride forcefully forward and act confidently. Push yourself to the top, smiling and laughing all the way. Draw on the lucky power of your Jupiter and work your creativity hard, and you *will* be rewarded. You have an immense talent for love, and with faithfulness comes your true soul mate.

IF YOUR SUN SIGN IS:
Virgo
(August 24–September 22)

IF YOUR SUN SIGN IS:
Libra
(September 23–October 22)

IF YOUR SUN SIGN IS:

Scorpio

(October 23–November 21)

*Y*ou *always win*. It's a natural outcome—though you do work very hard for it. From Scorpio-Leo Billie Jean King, who beat Bobby Riggs in her famous "Battle of the Sexes" tennis match in 1973 (inspiring many women to change their lives) and said, "Champions keep playing until they get it right," to Scorpio-Leo Dan Rather, the only journalist to score an interview with Saddam Hussein in February 2003, you quietly triumph over every competitor in your field. More at home with extremes, massive pressure, and deadlines, you are uncomfortable with the quieter "paperwork" times. Learn to meditate and be dull—it may save your life! A tip? Try *not* to win in love. You are wildly sexual but often push people away for the sake of control. Give in—and enjoy!

For your greatest luck with a lasting relationship, call on your Leo to relax the intensity of your expectations and judgment. You may be giving those you love a hard time without realizing it! You deserve the best, remember—and that includes the best coming from you.

IF YOUR SUN SIGN IS:

Sagittarius

(November 22–December 21)

*W*hat a blessing it is to be you! All you were born to do is *play*, to stretch your talents to the limit, and to enjoy the rewards that come from the burning gift of double fire. Sagittarius-Leo singer Nelly Furtado and Senator John Kerry are examples of those who draw people toward them without having to try. The only blemish is this perfect paradise that is you is that you need more *structure and control of your impulses*. Sadge-Leo Jim Morrison played himself right into destruction. Self-mastery and self-discipline (Leo leader of yourself) also brings you happiness in love—*when* you decide you want it!

A natural needer of space and possessing high standards in your choice of a mate means that you get luckiest the longer you look and the older you are before you decide to settle down. You must have a mate with whom you can share your ideas and intellect, or it won't work, no matter how much relief they may provide at the time.

Do not wait to bring out your creativity, however! The double fire is one of the most talented combinations there is, and it is no exaggeration to say that you would be depriving the world of a very important voice if you resisted your own inspiration.

Capricorn

(December 22–January 19)

Here the spare delicate steeliness of Capricorn is emphasized and beautifully lit by Leo fire. Capricorn-Leo actor Sarah Polley refuses to wear makeup, and Capricorn-Leo filmmaker Richard Lester minimizes *his* profession thus: "Filmmaking has become a kind of hysterical pregnancy." In your inimitable own way, you make it to the top. You know what you want and more often than not, you get it.

You may fight yourself at times, and even in natural situations you are in, such as your chosen profession! For your greatest success, keep going on your own individual path and do not bow down to anyone. Call on your lucky Leo, though, to make it more enjoyable and to torture yourself less. Let people love you and let everyone in more. Your luck comes from your social aura; however, you often turn that into a wall—albeit a fascinating and funny wall.

In love, when your restraining self wants to put the brakes on your wild Leo luck (people find you irresistible), take a deep breath and expand to the tune of the universe. For long-lasting relationship success, be forgiving, extravagant, and laugh more. Your mate is dying to make you happy; let him or her know how it is.

Aquarius

(January 20–February 19)

You guys are the *cool* kids on the block. Talk about being lucky! Always keep your sense of irony—but real, as in passionate curiosity, not pseudo-have-nothing-to-say indifference. Fellow Leo-Aquarius include Milos Forman, François Truffaut, Alice Walker, Mena Suvari, Tatyana Ali, Brandy, and Watergate-breaking Carl Bernstein, called "the ultimate feisty newsroom brat" by a colleague. Cure others' resentment of you *and* your stormy love life by calling on your Leo fire to melt that coolness into stronger warmth and to show how vulnerable you really are.

For success in long-lasting relationships, you will get lucky by softening your stubbornness and toning down your idealism or making it more realistic. No one can measure up to the height of your dreams (no one alive or real). Pour this into your work, call it ambition, and you will do brilliantly (of course!). But in love, call on your lucky Leo to turn you into a purring lion, and keep the purr going whatever the drill—as low as you like it, but keep it going. This is how the other half survives.

IF YOUR SUN SIGN IS:

Pisces

(February 20–March 20)

You were born intriguingly part-fish, part-lion and create your luck by coaxing yourself out into the world. Shine your Leo light onto your creativity. *Shock us* as Pisces-Leo actor Aaron Eckhart does as he brilliantly swims from role to role: from gentle biker in *Erin Brockovich* to gross brute in *Nurse Betty* to academic in *Possession*! Even in business, medicine, or other professional fields, you will always be a bit of an actor and gain your greatest success through exploring different sides of yourself with others. You are very well liked by your colleagues; let them in more if you can.

In love, your sensitivities run to screaming highs and lows, and you must be aware of your changing needs and not blanket them with alcohol and drugs. Temper addictive behavior by refusing to go into retreat from the world. For success with your soul mate, resist dependency and step back when you find you are giving too much of yourself. You can be the savior in love and *it will only exhaust you*! Stay highly visible and your life will be a happy one.

Relationships with Leo Jupiters

Finally, for those of you who aren't Leo Jupiter yourselves but who have a spouse or significant other, boss, child, or parent who is, here are some great tips to help you learn to deal! This is how the luck of someone else's Jupiter touches *your* life.

Your Significant Other

Leo Jupiters love romance. Rarely will you find a Leo Jupiter who is single, so snatch them up while you can. The best way to snare one is to be playful and fun. They don't go for old sticks-in-the-mud.

Want to meet a desirable Leo Jupiter? Go to an upscale party. Chances are, a Leo Jupiter will stand out. They love to go to parties where they can make people laugh, surrounded by admirers. If you get a date with a Leo Jupiter, be prepared to dress up and go out to one of the most chic eateries in town. They do not prefer the earthy, modest boy or girl next door; they want royalty and someone who stands out.

Never be cheap with your Leo Jupiter date. Potato chips on the couch watching TV, even on a fifth date or twenty-fifth, is not a Leo Jupiter's idea of a great time. They want to have fun, to show off both you and themselves. And last but not least, they *adore* sex, so if you really do want to stay at home for an evening . . . well, you get the idea!

They are great wooers, so be prepared for a big romance—or if that's the direction you're going in, a big lifetime commitment. Their mate is very important to them. Commitment can take some time to gel. Why give up all their admirers? But once they take the plunge, they are loyal and protective and will love you beyond belief.

Your Boss

"Nothing could be worse than the fear that one had given up too soon, and left one unexpended effort that might have saved the world," said Leo Jupiter Jane Addams, the great fighter for women's rights. *This is true for all Leo Jupiter bosses!* You will find yours, in the manner of great leaders, attending to *everything,* coming in early and staying late. With endless energy and enthusiasm, successes are achieved, one after the other. Your Leo Jupiter boss is fervently and completely devoted to his or her beliefs, acting on and promoting them, and wishing you to do the same. That's the way they attract others to help expand their world. In this full-scale production of your Leo Jupiter boss's work, nothing is left undone. If you're not a believer in this boss's system or world view, I might recommend that you find another position.

Your Leo Jupiter boss will be exuberant, fun, dramatic, flamboyant, and probably generous to you. You will find him or her extremely sympathetic, understanding, and able to talk to you about your problems. There is sometimes no greater listener than Leo Jupiters as long as they are the ones to provide the advice, reassurance, and pep talks, and, to be sure, they will insist that they be the one to whom you go to for all of the above. They will be unhappy if you go to another staff member first.

They will push you to do the best you can and, at that same time, have fun. They are funny as *anything,* guaranteed to have you rolling in the aisles with laughter. If you're quiet and modest and shy, they will bring you out your shell. They are your best cheerleader. Whatever you do, don't ignore them, don't be critical, and don't dismiss their advice. That is their most vulnerable point. They need to be your protector, your mentor, the one to whom you naturally go.

Your Child

Encourage your Leo Jupiter child's dreams, confidence, and leadership quality, as this is how he or she will become successful in later years. The best form of discipline for your royal cub works through appealing to their vanity. To say "Don't do that because I told you so" won't work. Say instead, "It doesn't look very attractive when you do that." That will instantly make them change their behavior, as they do not want to appear unattractive.

They can be risk taking and reckless, so while encouraging their adventures and explorations, do be on the lookout for accidents! They want their freedom but they feel crushed if they are ignored. Always encourage their creativity. It's almost useless to make them focus on boring details. They excel at going for the gold and great triumphs and want to be the class leader; they may be the *ringleaders* too. It is worth it to encourage their confidence whatever the cost and to *never* overlook them for one of their siblings.

It might be good to teach them about money early on. Discipline them in terms of an allowance because Leo Jupiters can be extravagant. You'll find they need more money than most other children. Not only do they appreciate the finest quality from infancy, they want it to keep their friends around them and exert a feeling of control and mastery!

Their home is their castle and they are proud of their family.

They're royal after all—so this means you are, too! They have all that natural exhibitionism and energy to give out, so reassure, compliment, hug, and provide plenty of affection. Be sparing with criticism as this can leave an indelible wound. A Leo Jupiter child who is not encouraged properly in early years will turn fearful and passive later on in life. Your child's success and well-being depends on being treated as a royal cub, right from birth.

Your Parent

Life with a Leo Jupiter parent is never dull. These Lions and Lionesses make life fun and entertaining, will playact with you no matter how old they (and you) are, and will make you laugh with their stories and funny imitations. They love groups and parties, and want you to have lots of friends yourself. They may have been concerned if you weren't the most popular kid in your class. They can get downright anxious if you're not doing your best and dismantle their Lion mane to become a clucking hen.

If you have a parent who is a Leo Jupiter, you had, no doubt, also what they say was an *"interesting"* childhood. Children of Leo Jupiter parents often become aware of grown-up vulnerabilities at a very early age, due to experiencing their parent's ego successes and woundings in the world to a palpable degree. You may have learned, for example, not to beat them at chess when you are only eight years old. Instinctively you will feel their need to be best and wanted and may find yourself reassuring them. Now, in terms of *you*—because Leo Jupiter parents think they are best, they want you to be best, too. Remember, all Lions possess a kingdom, and that includes you! Sometimes they act so proud of you that it's embarrassing, especially when they want to show you off in front of their friends or colleagues.

You may have been aware of a great romance going on in your household. Leo Jupiter parents never lose their bond with their mate and are not shy with their affections. With this comes storms and dramas too—possible walk-outs—and a wish to include you in the family negotiations. Because Lions cannot live without their pride, your Leo Jupiter parent will do anything to keep marriage and family together— for better and for worse.

If it was for worse—well, that's what astrology is all about. Appreciating how different we all are, and loving both similarities and differences. Nature gives an antidote for every poison. Isn't this the miracle of life? Problems happen, but always, somewhere, there's a solution, and Leo Jupiter will always try to find the *very* best one. Encourage your confidence and follow where it takes you. The journey will be as much fun as all the prizes you receive when you reach the destination. *Remember, Leo Jupiter: You are the charismatic leader of the zodiac.*

Madalyn's Movie
Choice for Leo Jupiter

8½

Fellini, who is himself Leo Jupiter, made *8½* with such grand style, such lavish color and fellini-esque images that, like a true Leo Jupiter, the film has inspired many imitations, including one by Woody Allen. Watching it is like stepping into a hallucinatory circus fun-ride as one follows the plot of a director attempting a production—and purely winging it!

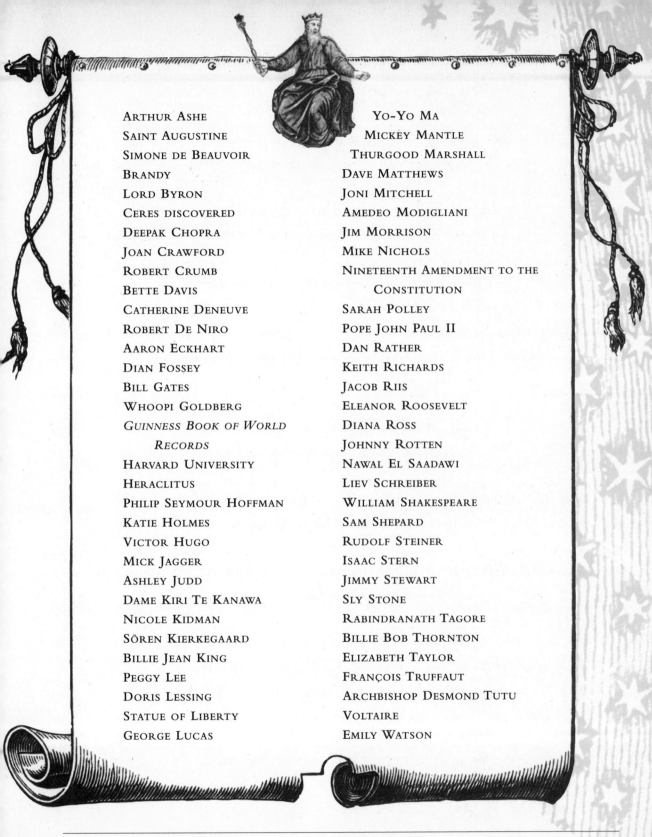

ARTHUR ASHE
SAINT AUGUSTINE
SIMONE DE BEAUVOIR
BRANDY
LORD BYRON
CERES DISCOVERED
DEEPAK CHOPRA
JOAN CRAWFORD
ROBERT CRUMB
BETTE DAVIS
CATHERINE DENEUVE
ROBERT DE NIRO
AARON ECKHART
DIAN FOSSEY
BILL GATES
WHOOPI GOLDBERG
*GUINNESS BOOK OF WORLD
RECORDS*
HARVARD UNIVERSITY
HERACLITUS
PHILIP SEYMOUR HOFFMAN
KATIE HOLMES
VICTOR HUGO
MICK JAGGER
ASHLEY JUDD
DAME KIRI TE KANAWA
NICOLE KIDMAN
SÖREN KIERKEGAARD
BILLIE JEAN KING
PEGGY LEE
DORIS LESSING
STATUE OF LIBERTY
GEORGE LUCAS

YO-YO MA
MICKEY MANTLE
THURGOOD MARSHALL
DAVE MATTHEWS
JONI MITCHELL
AMEDEO MODIGLIANI
JIM MORRISON
MIKE NICHOLS
NINETEENTH AMENDMENT TO THE
CONSTITUTION
SARAH POLLEY
POPE JOHN PAUL II
DAN RATHER
KEITH RICHARDS
JACOB RIIS
ELEANOR ROOSEVELT
DIANA ROSS
JOHNNY ROTTEN
NAWAL EL SAADAWI
LIEV SCHREIBER
WILLIAM SHAKESPEARE
SAM SHEPARD
RUDOLF STEINER
ISAAC STERN
JIMMY STEWART
SLY STONE
RABINDRANATH TAGORE
BILLIE BOB THORNTON
ELIZABETH TAYLOR
FRANÇOIS TRUFFAUT
ARCHBISHOP DESMOND TUTU
VOLTAIRE
EMILY WATSON

"We're not arrogant, we just think we're the best band [Oasis] in the world."

—Noel Gallagher, Gemini Sun-Leo Jupiter. May 29, 1967

"What's the point of doing something good if nobody's watching?"

—Nicole Kidman, Gemini Sun-Leo Jupiter. June 20, 1967

"I'm always thinking about how to get to the next level."

—Usher, Libra Sun-Leo Jupiter. October 14, 1978

"Violence is the last refuge of the incompetent."

—Isaac Asimov, Capricorn Sun-Leo Jupiter. January 2, 1920

"Life is a hard teacher; she gives the test first, the lesson after."

—Sören Kierkegaard, Taurus Sun-Leo Jupiter. May 15, 1813

"I say, Ashley . . . Girl, you just endure."

—Ashley Judd, Aries Sun-Leo Jupiter. April 19, 1968

"What separates the great actors from the average actors is hard work."

—Philip Seymour Hoffman, Leo Sun-Leo Jupiter. July 23, 1967

"My satisfaction comes from my commitment to advancing a better world."

—Faye Wattleton, Cancer Sun-Leo Jupiter. July 8, 1943

*"By hanging in there and working hard,
I could make the time and the place my own."*

—Barbara Walters, Libra Sun-Leo Jupiter. September 25, 1931

*"You could not step twice into the same river,
for other waters are ever flowing onto you."*

—Heraclitus, Pisces Sun-Leo Jupiter. March 535 B.C.

♍♃

Do the Right Thing

From Petrarch, born July 28, 1304, who wrote, "It is more honorable to be raised to a throne than to be born to one; fortune bestows the one, merit obtains the other," to Tom Hanks, who told a reporter almost seven hundred years later, "If it wasn't hard, everyone would do it. It's the hard that makes it great," Virgo Jupiters have always known, and so should

You've really got to start hitting the books because it's no joke out here.

—Virgo Jupiter Spike Lee

you, that your success depends on your very best work.

Winning the Lotto is not the way for you if your Jupiter is in the sign of *the highest standards there are.* You get lucky by earning it, dear busy bee, and the harder it is, the better you do. Now before you sigh at this idea of Jupiter boot camp (hey, didn't you get this book to make your life *easier*?) think of it this way: Your Midas touch brings you as much gold as you want. It's even good for you to worry (just not to excess). There, doesn't that make you feel better?

The Virgo Jupiter way is simply contrary to the popular myth of overnight success—exactly as it should be. Virgo is an Earth sign, remember? You, smart one, must think on your own, uninfluenced by changing trends, and trust what you know from real and concrete experience. Ever heard of that overnight success that took twelve years? That's the way it *really* happens. It's not good for Virgo Jupiters to believe in fairy tales.

As opposed to the flashy Fire signs—the lazy Libra or the I'm-too-sensitive-to-deal-with-the-world Pisces (sorry, but this chapter is about *the most* critical sign of the zodiac here)—you are anchored in the honest here and now, and you were born to set it right. If there was ever a time you were called a "goody-goody," wear the label proudly; it means you drew on your Jupiter fortune successfully!

You embrace your power in the world through conscientiousness and integrity, by being a straight shooter, dedicated and humble, by striving to remain true to yourself, and by improving, improving, improving! Remember, if you're looking for a quick fix or an Ed McMahon at your door—you'll be thwarting your helpful Jupiter and will only bring yourself *bad* luck! Or, if not bad luck, then none at all, and Virgo Jupiters—who are always working and doing, thinking and chewing over solutions—hate *that*. As fellow Virgo Jupiter Susan Sontag said, "Avoiding boredom is one of the most important purposes."

Take a moment, look back long and deep over your life, recount as many events as you can, and ask yourself what has made you lucky or unlucky in your life. You'll notice that whenever you act like a Virgo, *you create miracles in your life.* No matter what your Sun sign is, *you get lucky when you live out the Virgo legacy gifted to you by the benevolent planet Jupiter.*

To take advantage of your luck, emphasize your . . .

Virgo Jupiter Strengths ♍♃	*Be committed, of service, a purist, hands on, structured, humble, graceful under pressure, an integral part of the whole, and the real thing—yourself.*

But because Jupiter rules excess, watch out for . . .

Virgo Jupiter Excesses ♍♃	*You can become self-effacing, imprisoned by rules, or fanatical.*

Understanding the
Virgo Energy and Influence

Following the golden hot days of Leo, when all was ripe to bursting, now comes the hard work of the harvest, the time of Virgo. This time, from late September to early October, is when, in the old days, the men and women of the village joined hands and worked hard to bring home the bounty of crops from the fields.

Virgo's symbol, a woman holding strands of wheat, reminds us of advanced farming techniques and cultivation (wheat cannot flourish in the wild), and of Virgo's brilliant skills of discernment and analysis. To attract the power of Jupiter into your life, Virgo Jupiter, trust yourself to separate "the wheat from the chaff," the good from the bad, to discriminate intelligently, saving what is valuable in your life and discarding what is not. Your purpose is not judgment, but a clear assessment of your environment and everyone's place in it: *Is everything working as it should be? How can I help make it better?*

Tap into your lucky energy by tapping all around you with your innate antennae. Like the busy bee that senses pollen through its legs, you can sense the truth by using your lucky sixth sense. And because Virgo is an Earth sign, the purpose of your sixth sense is *pragmatic* rather than mystical: simply to see what is right and to improve where necessary. You get lucky not by being the self-aggrandizing guru (as one of the Fire signs may!), but by being the modest hands-on fixer that the world cannot live without. Help yourself, Virgo Jupiter, by following routines that bring order into your life. Order is your time-saver, and remember, your luck comes from accomplishing as much as possible!

♃♍

Earthy and passionate Virgo is hardly a virgin. For you, Virgo Jupiter, who obtains success by setting the record *thoroughly* straight, we must clarify! Virgo was the Virgin Astraea, who preached her pure ideals on Earth during the Golden Age. Virgin in the strictest sense means *purist,* as in virgin olive oil—pure, organic, natural, and unpolluted, no toxins, genetic modifications, hormones, or pesticides added. As we are not exactly honoring these Virgo ideals, pure Astraea vanished from the Earth, just as, today, we make so much that is pure vanish from the Earth (over twenty species *daily*). But Astraea, so the

VIRGO SYMBOL
The Virgin

story goes, being the helpful Virgo she is, still watches over us from the sky as the constellation Virgo.

Get lucky by embodying Astraea's power to stay alert and clear eyed. Dear Virgo Jupiter, look at your own life and see, really *see,* what is pure and true for you. Can you recognize what may have become a toxic relationship or work environment? Can you see an original ideal you had that you might have felt compelled to water down, adulterate, even completely alter? We are not talking a necessary loss of naïveté here—Virgo is born wise and skeptical—but about calling upon the power of your lucky Jupiter to see absolutely every impure additive, shortcut, pretense, falsity, illusion, or delusion you have accepted into your life. Remember, Virgo Jupiter doesn't make you lucky through shortcuts, coverups, or great-sounding PR—not through self-deception by any name! Mercury is an emotionless planet, and Virgo purism can be harsh, ruthless even, in its drive to detoxify and to purify. But the reward is a truthful life.

VIRGO PARTS OF BODY
The Intestines and the Nervous System

Virgo rules the intestines and nervous system—all that is affected by thought and analysis: The intestines literally pick through what you feed into your body, assimilating and digesting. So let yourself be as picky as you need to be, absorbing that which is nourishing and flushing away what is of no use. Virgo rules the fine connection between the mind and the body, so be aware that what you are thinking and feeling manifests itself physically, and work on making your thoughts and emotions positive.

VIRGO COLOR
Cool Green

You do best around colors that are both soothing and refreshing—bright and bold colors can actually disturb your thinking and equilibrium—and if you indulge in soft earthy colors (green, blue, brown) as well as classy neutral colors (white, gray, black), you immediately will see the quality of your life improve. The best of all for you is cool green (think of jade); use it to nourish your soul and to relax. Add sheets, towels, china, and wall accents in this color, and observe how much more energy and clarity you have.

Be Committed

To be lucky with this Jupiter sign, you need to be fiercely committed every step of the way until you have reached completion—whether it be with a work project, an area in your relationship, signing on your new home, planting a vegetable garden, finishing your degree, or putting together that family photo album you started years ago. As the great Virgo Jupiter poet Robert Frost said, "There's no way out but through." Just think of some of your hardworking fellow Virgo Jupiters: tennis stars Venus Williams, Steffi Graf, and Martina Hingis (all of whom trained and strained for years to achieve success); determined, tough baseball star Sammy Sosa; and Julia Roberts, one of the hardest-working actors in Hollywood. These are not people who mess around—they're committed.

For those with the gift of Virgo Jupiter, with the perfectionism and worry of the purist, life is never made easier. For all his dedication to music, and his great success, jazz giant Charlie Parker, the inventor of be-bop, had a pretty hard life. He was so committed to his art he once said, "If you don't live it, it won't come out of your horn." But in your perfectionism lies all that you need to be lucky. The phrase "to make a virtue of your imperfections" must have been written by one of your tribe. Your strengths and weaknesses exist to be made into Jupiter guideposts to light the road toward your own individual success, whatever form that takes. It is no surprise that many of you experience hardship and even illness in your early years—Virgo rules the sixth house of health, after all—and herein lies the miracle of your Jupiter gift. You were born with a genius to learn and to overcome, and you wouldn't be you otherwise!

Be of Service

Many of the good things Americans enjoy today come from Thomas Jefferson's draft of the Declaration of Independence, which proclaims—Virgo Jupiter–style!—that the government is the *servant,* not the master, of the people. This revered Virgo Jupiter, born April 13, 1743—the third U.S. president and author of the Declaration of Independence and of the Statute of Virginia for Religious Freedom—served his country for more than five decades. With his strong beliefs in the rights of man and a government derived from the people, in freedom of religion and the separation of Church and State, and in education for all, Thomas Jefferson struck a chord for human liberty two hundred years ago that resounds through the decades. He built the stunningly beautiful University of Virginia, an institution that has inspired thousands of minds in Charlottesville. President John F. Kennedy recognized Jefferson's accomplishments when he told a gathering of American Nobel Prize winners that they were the greatest assemblage of talent in the White House since Jefferson had dined there alone.

To bring the power of Jupiter into your life, *make your talents matter and count.* As a powerful Virgo Jupiter does, go deeper than the rewards of fame and money, go to where it really counts, life and death. Virgo Jupiter Miep Gies did. She risked her life *daily* to bring food to Anne Frank and her family and those who hid with them. Described affectionately by Anne as "our workhorse," Miep never sought praise or reward. "The past go always with you your whole life," Miep wrote, and she honored her sense of responsibility as a gift of life for eight people, risking her own life for theirs for more than two years. When she appeared at the Oscars in 1996, I gasped—I have never gasped at a celebrity before—and saw that most of the star audience did not even know who she was. Yet she had done more than most of them. A lesson for you, Virgo Jupiter?

Be a Purist

To be a purist is to go for the absolute essential. Abstract purism is needed for truth, and physical purism, the urge to purify our bodies and the Earth, is needed for survival. To gain the luck of Virgo Jupiter, strip away all that is unessential and prioritize, and as this famous Virgo Jupiter advised, "Wash your mind clean." John Muir was born April 21, 1838, now known as John Muir Day—April 22 is Earth Day, both consecutive days being Taurus (April 20–May 20), which is the other Earth sign of Earth preservation—and Muir's Virgo Jupiter legacy is everywhere in the United States. Two hundred sites are named after him, including Muir Woods, Muir Beach, and the John Muir Trail through the High Sierras. His work laid the foundation for the creation of the U.S. National Park Service, which opened during his lifetime Yosemite, Grand Canyon, and the Petrified Forest. He founded the Sierra Club, for protecting wilderness and the human environment. It is no surprise, then, that all of Muir's books are still in print! (As befitting a brainy Virgo Jupiter, he was also extremely intelligent, starting school at the age of three.) Born in Scotland, he became American, but his allegiance was to Earth above all. In fact, he wrote his address as "John Muir, Earth-Planet, Universe."

Other famous Virgo Jupiter Earth purists include George Schaller and Chico Mendes. Mendes gave his life to stop mining interests and cattle ranchers from destroying the Amazon, where he worked as a *seringueiro* (rubber tapper) and saw many of his fellow Brazilians lose their livelihoods. In the name of profit, the miners tore down trees as quickly as possible, leaving behind a ruined desert where once stood a rain forest more than 180 million years old. Chico was jailed, fined, and threatened by government officials who were profiting from the destruction themselves; he was finally murdered in 1988 by order of a rancher named Alves de Silva. His purism was not in vain, though: Chico's death caused an international outcry, and the Brazilian government and mining companies were forced to make much-needed reforms.

(continued) pictures, Hermes carries a caduceus. This curly symbol is still used today by the medical community as the symbol for healing.

Be Hands On

Hard work is so lucky for you Virgo Jupiter! The concrete and the real are crucial for the success of any Earth sign, but you, even more than Capricorn or Taurus, need to get right in there and get your hands dirty! Naturally skeptical about impractical dreams and unproven schemes, you specialize in results and in getting things done. Not only do you stimulate incredible opportunities for yourself when you assume more responsibilities, but your confidence level goes sky high when you experience things firsthand, and this includes experiencing with your senses and being "present," even if not physically there. For all these reasons, it is best for you to be personally involved in every detail rather than be a front or a detached director.

Virgo Jupiter Peter Weir, director of *Dead Poets Society*, *The Year of Living Dangerously*, and *Witness*, among many other films, says: "Give me things that are unusual or difficult, what are called 'broken scripts.'" Weir's brilliance is stimulated by his directly working with and fixing scripts. His work does not benefit, he says, from being given a finished, perfect product; he needs a script with flaws, one that he can dig into and get his hands dirty with.

And who could be more hands on in a literal sense than the most famous magician of all time, Virgo Jupiter Harry Houdini? Born in Budapest on March 24, 1874, Houdini moved to New York with his family at the age of thirteen and became interested in magic. Houdini soon discovered that his talent at the sleight of hand of the magician was outweighed by his even more hands-on ability to escape from a bewildering array of locks and restraints. Using his hands (and his teeth, feet, and toes!) Houdini escaped from handcuffs, leg irons, straightjackets, jails and prison cells, a mail pouch, packing crates, a giant paper bag (without tearing the paper), a giant football, an iron boiler, milk cans, coffins, and his famous Water Torture Cell. In most of these escapes, upon later examination, there was never a sign of how Houdini accomplished his release. He used his Virgo Jupiter hands-on power to work miracles.

Be Structured

Your planet Mercury drives you to control mind over matter. For Gemini, also ruled by Mercury, it may be just for the enjoyment of thinking and analysis! But for you, dear one, it is often to secure structure, to make life more ordered and predictable. You do not function well in chaos or with the muse of intermittent messy inspiration. This is not the old soybean that all Virgos are tidy and neat, but on the deepest level they need to know where everything is. Strength comes for you through knowledge and transforming the unfamiliar to the familiar. Don't let this limit you to ignoring what you can't—or don't want to—know, but bear in mind that your luck comes through routines, order, and repetition of all kinds.

Virgo Jupiter actor Cate Blanchett is famous for her enjoyment of making lists and crossing items off as she accomplishes them. This star is so determined to be a normal "proper mum" that she refuses to hire a nanny to take care of her child Dashiell.

Virgo Jupiter, while structure and record keeping are lucky for you, note from your compatriot Amelia Earhart that your life doesn't have to be boring or dull! Earhart was one of history's wildest risk takers and was deeply committed to the joys of flight, a purist in her dedication to her art. Yet, like a good Virgo Jupiter, she grounded her accomplishments in hard work, routine, and organization, funding her record-setting flights through Virgo writing and speech making. She said, "It's routine now. I make a record and then I lecture on it. That's where the money comes from." Her husband, the publisher G. P. Putnam, described his wife's routine as "sheer, thumping hard work of conscientious heroing." Emulate Amelia, Virgo Jupiter, and the world is your oyster.

Be Humble

Meg Whitman, the fantastically successful CEO of eBay, has no qualms about admitting mistakes—*whatever it takes to do things right*. Born a Virgo Jupiter on August 4, 1956, under the Sun of Leo, Whitman has chosen to go with the luck of her Jupiter rather than be compulsively channeled by the protective pride of her Sun sign. "The key thing is," she says, "if you make a mistake, have the courage to fix it quickly. I do think the price of inaction is far higher than the cost of making a mistake." Sacrificing your ego for the (best) work always makes you lucky if you are a Virgo Jupiter. When you are wrong, defending and justifying yourself, or even worse, saying no mistake was made, brings terrible luck.

Humility is very attractive in you, Virgo Jupiter. In the extreme, sacrificing your ego can lead to greater art, work, and even a deeper spirituality. Virgo Jupiter Elia Kazan (director of *On the Waterfront*, *A Streetcar Named Desire*, *East of Eden*, and other treasures) famously said, "Sometimes you have to go underground for a while, to know a personal defeat, to kill yourself." He should know, as he brought himself terrible luck when he acted against his ethical Jupiter and disclosed names to Senator Joe McCarthy's Committee on Un-American Activities, thereby changing the lives of some of his colleagues, who were later blacklisted and could not work for years. He might have been able to live it down if he had been born under *another* Jupiter sign—many others did—but to this day, this is an act that still marks him.

Be Graceful Under Pressure

Virgo Jupiter Ernest J. Gaines, author of *A Lesson Before Dying* and *The Autobiography of Miss Jane Pittman* (which became a TV-movie adaptation starring Cicely Tyson), could not put your Jupiter gift in more perfect Virgo terms: the everyday, routine, work, and becoming a better (Virgo Jupiter's fave word!) generation:

"When I speak to black students about Hemingway, they often ask me what I expect them to learn from 'that white man.' I tell them: 'All Hemingway wrote about was grace under pressure. And he was talking about you. Can you tell me a better example of grace under pressure than our people for the past three hundred years? Grace under pressure isn't just about bullfighters and men at war. It's about getting up every day to face a job or a white boss you don't like but have to face to feed your children so they'll grow up to be a better generation.' "

Be graceful under pressure, Virgo Jupiter—it becomes you.

Be an Integral Part of the Whole

Virgo clarity is the clearest sight there is, cutting through all! You, dear Virgo Jupiter, must not care about changing trends or mass opinion, but must trust your own assessment. Virgo Jupiter Ralph Waldo Emerson pinpointed it exactly: *Trust thyself . . . Nothing is at last sacred but the integrity of your own mind.* He also wrote, "We live in succession, in division, in parts, in particles . . . every part and particle is equally related." Virgos, as integral parts of the whole, like individual worker bees, work selflessly and with integrity for the whole. Selflessness is ultimately for survival of the self; Virgo Jupiter is not into purposeless sacrifice!

"I am here," said the Virgo Jupiter child, "to speak for all the generations to come." And Severn Cullis-Suzuki, at twelve years old, spoke for six minutes to dignitaries from all over the world at the 1992 United Nations Earth Summit in Rio de Janeiro, urging them to protect our Earth. At nine years old she had founded ECO (the Environmental Children's Organization), and as a Yale graduate eleven years later, she described her ongoing goal in *Vanity Fair*'s Hall of Fame as, "quite simply, to improve the planet." Remember that Virgo Jupiter improvement? *You were born with it.* Another Virgo Jupiter, Kory Johnson, of Phoenix, Arizona, began Children for a Safe Environment—*also at the age of nine.* The wells where they lived, she said, were contaminated, and a lot of the kids there had leukemia. Kory was awarded the Goldman Environmental Prize in 1998, before becoming a freshman at Arizona State University. She said she had been motivated to become an activist so early by the death of her sixteen-year-old sister and the deaths of her sister's friends.

Be the Real Thing—Yourself

You, Virgo Jupiter, become successful for being straightforwardly yourself. Not for you the glamour of image mirages! If you forget your true self or try to disguise yourself, even superficially, you will never achieve your full potential. To bring yourself luck, be a straight shooter, even blunt with your assessments and criticisms. Virgo Jupiter Ruth Simmons, the first African American woman president of Brown University, said in her no-nonsense speech that "elite universities (must) remain steadfastly and resolutely the province of excellent minds and not just fat purses." And she was interrupted by an outburst of massive applause from the audience—no doubt many of whom had "fat purses"!

British actor Virgo Jupiter Helen Mirren, "the ultimate thinking man's sex symbol," sexy and gorgeous at fifty-eight (no face-lifts, thank you very much), credits her talent to "old-fashioned and traditional" hard work and professional training. Mirren's brilliant portrayal of Detective Chief Inspector Jane Tennison in *Prime Suspect* should be studied avidly by all ambitious Virgo Jupiters!

Nine-time Wimbledon tennis champ Martina Navratilova, also winner of 167 singles titles and 165 championships in doubles, is another Virgo Jupiter who insists on being herself. Born October 18, 1956, under the people-pleasing Sun sign of Libra, Navratilova was nevertheless able to be honest about her sexuality years before other celebrities felt comfortable enough to do so.

There is no one who works as hard, gives as much, and practices their trade as seamlessly and efficiently as perfect Virgo. Measure your gifts by the degree of their service and help; otherwise, you are frittering them—and Jupiter's immense luck—away.

What Makes You Lucky in Love

You must trust your first instinct about someone and go with that as far as it will take you. Illusions and projections into the distant future are not good for you. Just go with the here and now. Yet you are luckiest with a strong physical connection that is lasting, not sweet and fleeting. You will find success only with someone with whom you can say anything and be completely yourself and comfortable. This includes your worries, for better or for worse.

Virgo Jupiter, as the "Virgin," or solitary one, often finds luck with their mates, and having someone to share their life with makes them live longer, too! (Interestingly, Gemini Jupiters—both ruled by Mercury—as the twin or partnered one, often do better alone.)

Dignity and aloofness for some reason can bring you great luck, but don't be afraid to make sacrifices to make the relationship work so long as you are with an ethical, caring partner. You should stick to your core values at every turn. When you find yourself being taken for granted or being taken advantage of, get out and don't look back. Make sure you pick a partner who shares your beliefs. You get lucky with the long haul.

Your Best Career Moves

No matter what your Sun sign is—whether a fast starter Aries prodigy or a creative old mystic Cancer—if your Jupiter is in Virgo, you have one basic route to success: honest hard work. Don't oversell yourself when applying for a new position; be clear, articulate, truthful, quietly confident, with concrete examples and impeccable references. These qualities will bring you a ton more success than charm, humor, flattery, or a strong personality, presence, or looks. (And yes, Virgo Jupiter, be cleanly dressed, tidy, and straightforward.) Minimalist understatement is your key to good fortune here. Remember, God is in the details. Rather see the branches and the twigs than the whole forest.

Commit all of yourself, summon your sense of responsibility and ethics, and you will find the world shines on you for your incredible dedication and brilliant reputation. What you must not do is cut corners or do things the easy way; this is bad luck for Virgo Jupiter and always backfires. Prepare for the long haul, help out as much as you can, and try to resist personal ego gratification (another backfiring inevitability). Virgos are the best critics and mimics in the world, but do be careful of backbiting and gossip; success will shun you then. Remember, you are working for the community, whatever level or stage you work on, whether you work for yourself or for a family business.

Give advice *whenever you can*. You are the one who people count on, and you do well to be dependable, honest, and present. You rise, quietly and surely, to the top just by virtue (and it is a virtue) of your excellent work. You must follow your strong Virgo conscience and be incredibly ethical as, that way also, success comes to you. Medicine, publishing and the literary field, science, research, restaurants, service agencies of all kinds, or accounting are all open to you (Virgo Jupiter can do anything), and furthermore, they all *need you*. Who wouldn't? You're true blue, the real thing. Show it, don't shout it.

How to Reach Optimum Health

Don't go crazy with exercise and diets; moderation, in all facets of your life, is the key to good health with you. Eat and exercise regularly and gently—not too much, not too little. Resist the latest diet and food fads. Do not do that two-week grapefruit fast, for instance. One of America's greatest tenors, Virgo Jupiter Mario Lanza, tragically died an early death due to overeating combined with extreme fasting and "health" experiments.

Long walks every day—preferably with birds chirping and moderate sunshine—are the perfect exercise for you. Stop your mind at least twice a day to refresh yourself and to renew your energy. Every few months, long breaks become *crucial* for you.

Your Lucky Foods, Element, and Spices

Basically, you want to eat foods that provide good roughage, are easy to digest, and calm the nerves. Stay away from rich sauces and food that is fried or terribly spicy. Endive should be eaten *every day*: it provides potassium sulphate, which is Virgo's cell salt, as well as ideal roughage. Potassium sulphate is essential to muscle contraction, prevents fatigue, carries oxygen to cells, keeps your skin oxygenated and clear, and your hair shiny and dandruff free! Other foods high in this mineral are romaine, chicory, wheat germ oil, almonds, bananas, lemons, oats, and lamb. Brewer's yeast, brown rice, and bran are also excellent for your skin and hair, as well as for your nerves, and provide fiber and B vitamins. In addition, make sure you eat a serving of one of these every day: yogurt, melons, eggs, cottage cheese, or papaya. Substitute herb teas for caffeine and honey for sugar. Red wine is fine, but try to avoid hard liquor, chocolate, and drugs.

Your Best Environment

Your best environment is one of peace and calm. Your nerves are best soothed by space, order, and nature, and whether in a city or the countryside, trees are a necessity. You function best in a moist climate with both architecture and nature that are carefully designed and structured—for example, manicured parks and gardens. Clear whites, greens, and grays must be what you see every day! Cities ruled by Virgo include Boston, Strasbourg, and Paris; countries are Greece and Crete.

Your Best Look

The tailored look is superb on you— simple and elegant. No hippie styles for you! Dress *professional,* even if you're not. The straighter, cleaner, and sharper you look, the more success—yes, even in love—you draw your way. When you wear navy blue, black, gray, and white *suits,* you're a knockout. Jackets are particularly lucky for you. If this is too much of a switch for you, try it for just a day a week, or a couple of days a month, and see the results. Avoid tight-fitting diva styles and flamboyant hair, and be subtle with makeup and accessories. Tidy underplay and minimalism is sexiest on you. Less is definitely more.

Your Best Times

The luckiest times for you are when the Moon is in Virgo, which happens every month for two to three days. Find these special days in your astrological calendar. During these days, schedule crucial meetings, tell that special person you love them, ask for a promotion or raise, push the envelope, and plan to shine. In your everyday, the best time for you to exercise is late afternoon to early evening. The luckiest times during the year for you are August 24–September 22 (when the Sun is in Virgo); December 22–January 19 (when the Sun is in Capricorn); and April 20–May 20 (when the Sun is in Taurus).

The very luckiest time of your life—as in Lotto lucky—is when Jupiter reenters the sign it was in when you were born (Virgo), which happens every twelve years and lasts anywhere from a few months to just over a year. This is called your Jupiter Return, and it launches your luckiest streak. You can count on the *extraordinary* to happen the very day your Jupiter Return begins. Amazing events can start up to six weeks beforehand, so plan ahead!

How to Be Lucky Financially

Go slowly, save, and don't miss a single detail. Plan ahead, make lists, itemize. Resist impulsive moves or extreme dreams; through moderation, care, and caution is the way financial luck comes to you. It is not money that you love so much, but its practical purpose and what it provides. Think of your security and old age if you are ever tempted by the promise of immediate riches! Shortcuts are disaster for Virgo Jupiter. Invest in a home, property, or the Earth if you have a large sum you want to do something with.

Ethics are paramount for your success, and you will find the greatest returns through socially responsible investing funds. When you pick individual companies, look for those that have simple, understandable business models, are ethical, and value hard work, companies that have important services or actually make things. Avoid retailers and financial companies with complex or dubious business models. Don't look for immediate success; instead, be dogged and work hard on your investing skills, and you will gradually see improvement. You are not going to find luck at the gambling tables or through random chance. Follow your head (looking for what ought to be done), and you'll find luck shines on you and makes it easier to achieve than you could ever have guessed.

Success in Family and Friendship

Your greatest success comes through your care for others and your dependable nature. Your family and friends do well to get advice from you (but don't go overboard with the criticism). Above all, be yourself and be truthful. People love you to be their friend and secretly are honored that they are considered yours. You have a dignity and class that can't be bought, either through extreme shows of emotion, need, or emergency.

Dear Virgo Jupiter, you don't need to impress or please, and certainly do not make promises you cannot keep. To be lucky, *you should be the one everyone counts on.* Family and friends feel comfortable around you and can relax, as you have a very soothing quality. You are there in a crisis, you are there for the laid-back times, you are punctual and the perfect family and friend for all the right things. *Everyone* wants you at their wedding! You don't need to be entertaining, nor the one who gets the party going, nor anything at all. Just be you. That's all anyone wants from you.

When you go too far!	♍♃ *Virgo Jupiter Excesses*

The perfect Virgo Jupiter cautionary tale is William Golding's *Lord of the Flies*. It is the story of a group of English schoolboys who are stranded on a desert island and what happens to them as they attempt to organize and structure themselves in order to survive. There forms a "civilized" faction and a "primitive" faction, each trying to control and impose their order on the other, with intense fears of the unknown and sudden eruptions of violence breaking in.

The warning to you, Virgo Jupiter, is to resist ordering yourself too much! Structure and service are your lucky keys to success, but the excess of this is blocking out the rest of your human psyche, cutting off your spontaneity and your deeper emotions, all of which, at best, can give rise to uncontrollable fears of the unknown (the part of yourself that you have made unknown), and at worst erupt in damaging ways.

At the end of the novel, ironically (and intentionally so by Golding, a Virgo Sun), the boys are rescued by the "grown-ups" in ships of war and returned to "real" civilization.

Excess #1 · You Can Become Self-effacing

Smart Virgo Jupiter actress Renée Zellweger, as successful as she has become, can still say: "I'm getting better at dealing with fame, I guess. But I still feel stupid. I mean if I were Stephen Hawking then you could come up to me and tell me I'm great." Your modesty is lucky, but dear Virgo Jupiter, it is only right to credit yourself! Don't allow yourself to succumb to an excess of self-doubt or insecurity.

Beware of the extreme shown in Ken Kesey's *One Flew Over the Cuckoo's Nest*, in which the inmates have become so suspicious that rights and freedom could harm them that they remain in the asylum voluntarily, under the manipulation and control of Nurse Ratchett. Ask yourself what or whom in your life—or what part of yourself—can you see as a supercontrolling, critical Nurse Ratchett, and what part of yourself can you see as the self-doubting inmates. If you draw a blank, you're absolutely fine! If not—well, we all doubt and brainwash ourselves to some degree, but it helps to know how.

Excess #2 · You Can Become Imprisoned by Rules

Dear Virgo Jupiter, beware of giving so selflessly (even to your own ambition) that you lose control of your control, as it were, and allow a set of rules (self-created or otherwise) to become your ego, directing your every move and thought. The process of following an order so thoroughly not only results in unhealthy self-repression but also can cause you to lose important parts of your life, relationships that are dear to you, and ultimately yourself.

Virgo Jupiter Robert Swan Mueller III is a possible candidate for ignoring his human instincts and slavishly following rules and tradition. Mueller was chosen by President Bush to head the FBI in the months before September 11. Dedicated to following the bureaucracy and rules of the FBI, Mueller attempted to discredit and bury Agent Coleen Rowley's account of mistakes made by FBI higher-ups, closing her investigation of one of the hijackers *even as the second plane was going into the World Trade Center*. As Rowley has since written, an FBI supervisor called her on the phone while it was happening—she watched on television as she listened to him on the phone—and told her not to investigate the twentieth hijacker's computer under any circumstances. By sticking to protocol and procedure above all else, Mueller went against his Virgo Jupiter, at great cost to us all.

Excess #3   *You Can Become Fanatical*

The utter extremes of control and ignoring or repressing different parts of yourself can turn one into a person who believes one thing above all is right. And bingo, you're a fanatic!

Some famous examples of this way-way extreme: Virgo Jupiter Jake "Raging Bull" LaMotta became so excessive in his use of control that he ended up losing his wife, his career, and, some say, his sanity. Virgo Jupiter John Wilkes Booth became so fanatically against the rights slowly being afforded African Americans at the time that he assassinated President Lincoln.

It's rare, but it happens. Fanaticism seems to be accompanied by extreme conditions, such as isolation or, conversely, a strong mob mentality. So as long as you stay within a moderate and comfortable lifestyle, with plenty of breaks and humor, I wouldn't worry about it, Virgo Jupiter. But if you feel that fanatic within—watch out!

Sun Sign-Jupiter Sign Combinations

Aries

(March 21–April 19)

One side of you wants all the attention while the other can do nothing but self-criticize. This makes you a fierce worker bee, compelled to be the worshiped queen while silently sacrificing yourself inside. You have the highest of ideals and not everyone sees how much you suffer. If you use it to create, as Aries-Virgo Eric Clapton does, that is one form of salvation. Your work rises to such energetic heights that it produces the illusion of magic! But watch your desire for absolute perfection or it can cause actual physical suffering, as it did for Aries-Virgo magician Harry Houdini, who attempted many "impossible" feats. Aries-Virgo Elizabeth Montgomery also acted an illusion of magic—in *Bewitched*. In love, you have a devoted mate as you have a devoted following. All you need is kinder devotion from *yourself*.

Taurus

(April 20–May 20)

Except for the doubled worry, a perfect mix! Taurus lends Virgo its sensuality and luxuriousness; Virgo gives to Taurus its razor-sharp thoughts and healing powers. Nothing is impossible for Taurus-Virgo—*nothing!* It is not blind optimism that makes you succeed but pure knowledge, Earth sense, and experience in overcoming. Taurus-Virgo psychologist Rollo May wrote, "Commitment is healthiest when it's not without doubt, but in spite of doubt." It is this "in spite of" part that propels Taurus-Virgos so far forward! No matter the degree of hardship or struggle, you end up winning every time, provoking awe and surprise. And yet you are modest, hardworking, loyal, discreet—a star without any of the flashiness. Taurus-Virgo actors Daniel Day-Lewis, Cate Blanchett, and Renée Zellweger all belong in this category: fantastically talented and fantastically modest. Taurus-Virgo James Brown, the "Godfather of Soul," described his idea of perfect happiness thus: "Being able to work a lot and play a little." Of the latter, vacations in nature are essential for you.

People always want you on their side. Life always turns out well because you make it so and because you deserve it so.

Here we have the Double Mercury Special! With the two signs ruled by the quick messenger of thought, precise words and movements are your specialty, whether it's to hit the spot like comic genius Gemini-Virgo Gene Wilder or to win by being faster than anyone else, like Gemini-Virgos Venus Williams and Steffi Graf. Your talent shines through any form of communication, and you're fun to be around too! Nonstop thinking and running on your nerves makes time out for relaxation and meditation a necessity for you—don't ever skip a real vacation.

Your Gemini side, who prefers to slide into easy street, will sometimes feel as if it's being policed by ethical Virgo—and that's not a bad thing. Especially when it comes to love. Make sure you find a partner who makes you laugh, or the opening up you'll have to do with your emotions—not to speak of commitment—may just be too hard. Your brain likes to run the show. Put it to bed to allow the deeper stuff to come out. It's there!

ↀ♍

You insist on doing things *right,* and you can feel what's right or not throughout your whole body. You're a very sensitive soul, so sensitive that life can feel absolutely sticky at times as you get caught up emotionally in fixing everyone's problems. People can't help liking you because of your purity of feeling and vulnerability, but when you suddenly go into one of your dark moods, you wish everyone would just back off. What fixes you right up is food, order, and mothering, either giving it or receiving it, and this even plays a part in your love life. You need one mate who is faithful and devoted at all times—and can put up with your occasional crabbiness and criticism. Be careful with jealousy as it can seize you unexpectedly—remember the extremes of heavyweight champion Cancer-Virgo Jake LaMotta? When life's intensity becomes too much, channel it creatively, as Cancer-Virgo Gustave Klimt did so beautifully in *The Kiss.* You are devoted to your craft and do excellently when you can retain your Virgo rationality.

IF YOUR SUN SIGN IS:

Leo

(July 23–August 23)

You are fantastically gifted and should be happy as a god—if you could only learn to live with a little less! Your ideals stretch as far as the Sun and you manifest them into genuine masterpieces, natural leader that you are, but resist trying to create a masterpiece out of the one you love. Your high standards are often taken as stinging criticism, which works in your professional field as you surround yourself with similarly high-flying pressured talents—but leave it outside the bedroom door. Even though control definitely turns you on, if you express this through sexual play rather than real-life dominance, all will be groovy! You are fastidious, analytical, punctual, and organized, and need a mate who is similar or you will be driven nuts. (And remember: No trying to change them.) You can go from being a spontaneous, fun party animal to a guilt-ridden, worrying critic in one hour flat, so make sure your mate can deal with that, too. Some of our genius Leo-Virgos: Roman Polanski, Sherry Lansing, Peter Weir, and Claude Debussy.

IF YOUR SUN SIGN IS:

Virgo

(this makes you double)

(August 24–September 22)

It is no coincidence that Virgo-Virgo Macaulay Culkin is best known for *Home Alone*. Virgo-Virgos always walk alone, riding the high tightwire of precarious perfectionism, fearing that they will never be able to share their dreams and compulsions with another. Not so! It is true that you double Virgos are obsessed, in the best and the worst ways, both compulsive and utterly devoted, but you are meant for a mate, just as a mate is meant for you. (A warning: Beware of falling for one with addictions, as one of your favorite roles is that of savior. *It doesn't work*. Think of helplessly "falling to pieces" double Virgo Patsy Cline: being a savior makes you miserable.) If you can wait out the early anxious years, you end up with one who is just perfect.

Your work is beyond excellent, you put your all into it and don't even expect much back. Much of your hard work is an attempt to prove yourself, and that's fine, as long as you let go of being so hard on yourself *sometime*. There is no combo that is as hard on its native as this one. Practice self-love, surround yourself with plenty of rituals and healthy routines, and indulge your green thumb. Remember the healing powers that double Hermes has bestowed upon you, and while the plants and children grow, let yourself grow, too.

Here your Libra desire for niceness is sharpened a hundred degrees by perfectionist Virgo. Allow your lucky Jupiter to make your decisions for you—Virgo is an Earth sign, after all, and a calm, efficient, practical thinker. Life can be so much easier if you dip into the Earth and ground yourself from time to time with its simple solutions. You are meant to be happy in your real life, not meant to suffer for your work. Who can forget Libra-Virgo Melora Walters's beautiful and anguished performance in *Magnolia*, or the equally talented Libra-Virgo Naomi Watts, who said, "There's a lot of skeletons in my closet, but I know what they're wearing. I'm not gonna act all ashamed of it"? You love to share and break your back doing so: But remember, the impulse to have your good deeds recorded is *not nearly as lucky* as your Virgo Jupiter, which doesn't look for public thanks at all! Work on developing the latter, dear Libra (and I promise you will look even *better*).

Passion, passion, burning bright! You enchanting Scorpions have possibly the greatest choice of anyone: You can put it into your work or art, like director Louis Malle (if you want to see passion, just watch his *Damage*), or leap directly toward the flame, as poet Sylvia Plath did; disciplined by her Virgo, destroyed by her Scorpio.

You, brilliant one, have superdiscriminating taste, both in sex and in career. You see instantly through illusion, so it is essential that your mate is honest. To ensure faithfulness, your Virgo rationality, not your Scorpio passion, *must* choose him or her. You are loyal to the end, but if betrayed, there is no more venomous critic than you. Scorpio-Virgo Ezra Pound used his criticism *positively* to help create possibly the greatest narrative poem ever written—*The Waste Land* (if you don't believe me, just look at Eliot's draft before Pound's edit!)—but *negatively* in his personal life when he became notoriously anti-Semitic. Heaven or hell? The fork in the road is most extreme for you and creates such tension that your health can be affected. Structure your rest and eating patterns and *commit* to yoga. If you do this and call upon your Virgo clarity, helpfulness, and discipline, you will create your own nirvana on Earth.

IF YOUR SUN SIGN IS:
Sagittarius
(November 22–December 21)

You have very high ideals matched with an unwavering work ethic that allows you to accomplish miracles. Bravo, dear one! You are also generous, have a biting sense of humor, a healthy lust for life and bodies, and when you decide to pursue an agenda, whether professional or personal, you hit it right on the nail. That is, if you call upon your lucky Virgo to get the task accomplished and not follow your Sagittarius love of the pursuit with no end in sight. Button down early and make friends with discipline and structure—which your Sagittarius self loves to do without. Act like a wild woman or man by all means, but back it up with serious work and attention to detail and you will have success after success. You can't help but have fun in your life with a huge circle of friends, but attempt to limit the number of your loyalties or you will end up exhausted. And by the way, a tendency toward hypochondria is *not* unlucky for you. Remember, "boring" Virgo could save your life, as you are naturally impulsive and accident prone. A number of you make surprisingly great lawyers. Famous Sagittarius-Virgo daredevils include Nick Stahl and activist Chico Mendes. Case closed.

IF YOUR SUN SIGN IS:
Capricorn
(December 22–January 19)

Purism intensified! You know what you want and you can get it, too. Ambitious and choosy, you don't beat about the bush in your opinions, your desires, nor your loyalties. Capricorn is cautious—as Capricorn-Virgo Mel Gibson said of his fame, "I've got to put the brakes on or I'll smack into something"—but, nevertheless, both his Earth signs kept him going, going, going! And speaking of Earth, you have double Earthiness, giving you a very strong physical and sexual presence. From Cuba Gooding Jr. winning the Oscar for his most physical role (a football champ in *Jerry Maguire*) to Rod Stewart singing to great effect, *"If you want my body and you think I'm sexy,"* these Capricorn-Virgos used their bodies to great effect, and so should you. Remember: Proudly, but with the right amount of discriminating Virgo questioning: *If* you think I'm sexy. And Cuba's role was all about *asking* Jerry Maguire to prove that he wanted him.

For success in love, keep impeccable standards for your mate but give more than you normally would. Your Jupiter luck lies in self-sacrifice, not selfishness—in all areas!

Y ou guys are wonderfully weird and the world loves you for it. Forget about being or acting normal—there's no success, neither personally nor professionally in that direction. Use your lucky Virgo Jupiter to go all the way with your individuality and weirdness—to fine-tune and *perfect* it even. Don't be vague in your eccentricity. Be as succinct and as articulate as you can about all the interesting things that make you tick. From Susan Sontag's brilliant and fearless essays to Yoko Ono's sometimes-outrageous art pieces (a ten-hour film depicting nothing but bottoms, for example), these Aquarius-Virgos are exacting and unapologetic about what they have to say or be, and so should you! In fact, they go out of their way to explain the details and resist giving the impression of unreachable sloppy-artist mindset. God and good luck is *in the details!*

In love, also, you will find the most unusual mate with whom you can work *and* have sex—look at the famous examples of Aquarius-Virgos Mia Farrow and Yoko Ono—and his or her integrity must be strong throughout, or you're gone.

P isces is all feeling, and Virgo is all thinking, and each of you is the finest antennae from different planets! If you give way to your Pisces impulse, you may feel torn, moody, and often quite sorry for yourself—albeit with fantastic insight. If, however, you call upon the luck of your Virgo Jupiter, you will channel that superlative insight and creativity in a way that is helpful, productive, and clear to the outside world—a resounding success! You may even scare yourself into working harder, and, even better, become A+!

Use your lucky Virgo discipline and structure also to help you feel better and to avoid the emotional highs and lows that come with Pisces underwater territory. Whichever way you choose, you can't help but be sensitive to everything. From Philip Roth's declaration that "there are only seven serious readers left in the United States" to Nina Simone's intense dissatisfaction with the same country motivating her to become an exile in France, these ultra-talented, high-caliber Pisces-Virgos expect the best and so should you! Your luck comes through articulating and speaking out about problems, rationally and clearly as befitting Virgo, not crying about them like Pisces. This applies to your love life as well as your work life, and you will be grateful many times over, dear one, if you can also allow your lucky Virgo to make the illusions you have about your love interest more realistic. Then you will be *happy.*

Relationships with Virgo Jupiters

Finally, for those of you who aren't Virgo Jupiter yourselves but who have a spouse or significant other, boss, child, or parent who is, here are some great tips to help you learn to deal! This is how the luck of someone else's Jupiter touches *your* life.

Your Significant Other

Let's get this straight right from the get-go: Virgo is an Earth sign, and they want it real. If you are someone who prefers promises over results and dreams over reality—if unattainability is what turns you on, in other words—a Virgo Jupiter mate is not for you. As one of their native Virgo Jupiters, Robert Frost, wrote: "Earth's the right place for love: / I don't know where it's likely to go better." However, if you've made it this far—keep going!

If you thought this beautiful creature was aloof and unemotional when you first met him or her, by now you may have changed your mind. The initial distance is simply a way of protecting themselves and also serves to weed out the undesirables. Virgo Jupiters do well to be choosy.

Try not to take their compulsion for work personally as a rejection of you. You may have met them at a work-related event (or even at work, or coming home late from work). Their luck comes from making things better, and once you're part of their lives, this includes making you better, too! Not different, just better: *Coming from them, criticism means love.* To love is to fuss over; expect healthy, delicious meals as well as some great sick days in bed. They will advise you on everything, from your work to your parents to your finances to your friendships to your clothes. In return for accepting their loving criticism, get set for undying loyalty, a commitment to routine and structure, and a strong relationship with a partner you can absolutely trust and depend on. They won't ever be elusive, untrustworthy, or unattainable; if you prefer such mirage-mates to the real thing, you might not get what you want from solid, stolid Virgo Jupiter! No one will be more supportive of you—and the Earth sign can make for some pretty strong (and sometimes kinky) passion as well.

Your Boss

Virgo Jupiter bosses *never* boss. They are more like very hardworking colleagues who expect you to do your best work and pitch in extra, just as they do. They believe in concentration; as Virgo Jupiter writer Simone Weil put it: *"Absolutely un-*

mixed attention is prayer." A class act, they prefer to show who they are through their achievements rather than by turning up the wattage of their limelight (they leave that to the Fire signs!). Naturally modest, even shy, they are sharp as a tack, cannot be bluffed, and in a worst-case scenario, you may receive your notice before you receive a warning. At their best, Virgo Jupiter bosses are supportive and encouraging, giving you helpful suggestions and advice; at their worst, they can be so wrapped up *in the work* that they are unaware of their workers' need to—well, to have a boss! They don't seem to think that people should be told what to do.

Tatyana Ali describes the experience perfectly when she talked about Virgo Jupiter Will Smith's involvement in her singing career: "Will never told me he was doing this, but I think he kept his distance to let me explore and go through this whole experience alone, and I really feel like it was mine—and not like someone was holding my hand."

In an age when personality and image are rewarded, and wining and dining are key, your Virgo Jupiter boss is an oddity. Behind the scenes he or she is doing more than you know. He or she is in all likelihood a model of clarity, industriousness, modesty, and helpfulness. Drawing upon the luck of this Virgo Jupiter, you'll figure it out on your own and having earned your own self-respect, realize your Virgo Jupiter boss is one you can count on and respect perhaps above any other.

Your Child

You will learn as much from your Virgo Jupiter child as he or she will learn from you, believe me. You have given birth to a natural teacher and a wise old soul. Draw upon Jupiter's power to not let your child be lonely (a common Virgo trait, increased by Jupiter's expansion), and in the process, you as well. The Virgo Jupiter child loves deeply but cannot always show it physically, and because of their natural shyness, is often starved for hugging.

Because they are such little realists themselves, they need you to give them dreams to believe in. Virgo Jupiter children love to improve and make things better, and this includes everything and everyone they love. Be prepared for your child to offer a lot of commentary and occasional criticism! Ironically, they are oversensitive to criticism themselves, so you may learn to be more sparing in your own. Finally, if you're a free-thinking parent who believes kids shouldn't follow rules, having a Virgo Jupiter child may be your timely karmic lesson: These children prefer structure, and to make them happy (and to give them

love), the best thing you can do is make your home life a model of order, routine, and clarity.

Your Parent

Virgo Jupiter parents show their love through
manifesting the expansion of Virgo traits onto
you. This means excessive care, devotion, fuss, cleanliness, criticism,
encouragement in learning—and humor (they're funny). You may
have been the cleanest, best-mannered bookworm in your class—or
then, if you rebelled against this, the messiest, wildest lay-about
teenager or young adult. Virgo is an Earth sign—they are giving in
practical and material things, but not always in demonstrative affection;
however, whenever you needed them, your parents were the most dependable in support if you were down.

Just know that some of the behaviors of your Virgo Jupiter
parents—behaviors that may have irritated or even enraged you—are
motivated by love. This sounds like a platitude, but it's not! Their criticisms are motivated by, of course, an innate perfectionism, but also by
love. And didn't their comments sometimes help you, even if you
didn't want to admit it? What about their love of rules and structure?
This may have seemed boring or even frustrating, but remember, they
were always fair in applying the rules. Their actions really were motivated by strong, steady love, so appreciate them!

Nature gives an antidote for every poison. Isn't this the miracle of
life? Problems happen, but always, somewhere, there's a solution, and
Virgo Jupiters will not rest in their quest for all that is better. Give
yourself credit for each step up, and measure your progress as much by
how far you have gone as by how far you are still going to go.

Remember, Virgo Jupiter: You are the helpful truth-teller of the zodiac.

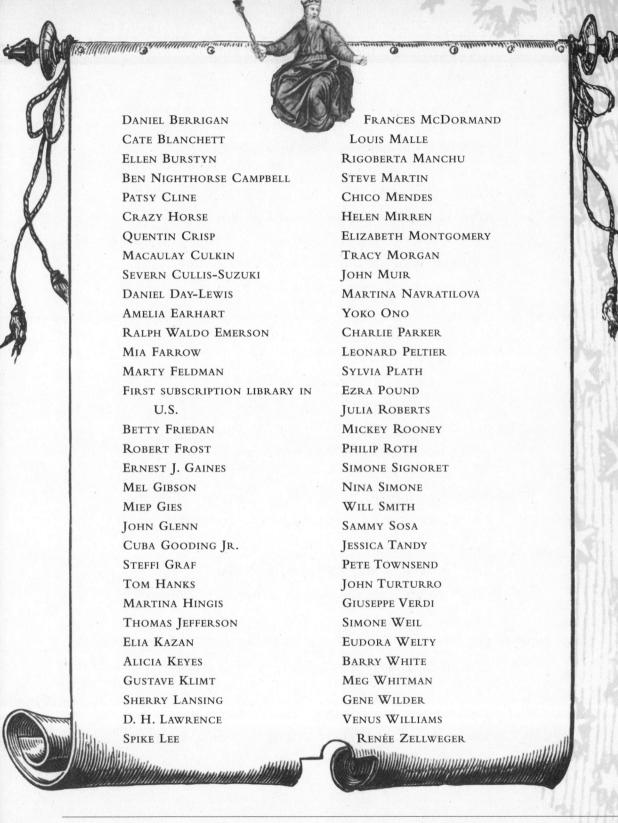

DANIEL BERRIGAN

CATE BLANCHETT

ELLEN BURSTYN

BEN NIGHTHORSE CAMPBELL

PATSY CLINE

CRAZY HORSE

QUENTIN CRISP

MACAULAY CULKIN

SEVERN CULLIS-SUZUKI

DANIEL DAY-LEWIS

AMELIA EARHART

RALPH WALDO EMERSON

MIA FARROW

MARTY FELDMAN

FIRST SUBSCRIPTION LIBRARY IN
    U.S.

BETTY FRIEDAN

ROBERT FROST

ERNEST J. GAINES

MEL GIBSON

MIEP GIES

JOHN GLENN

CUBA GOODING JR.

STEFFI GRAF

TOM HANKS

MARTINA HINGIS

THOMAS JEFFERSON

ELIA KAZAN

ALICIA KEYES

GUSTAVE KLIMT

SHERRY LANSING

D. H. LAWRENCE

SPIKE LEE

FRANCES MCDORMAND

LOUIS MALLE

RIGOBERTA MANCHU

STEVE MARTIN

CHICO MENDES

HELEN MIRREN

ELIZABETH MONTGOMERY

TRACY MORGAN

JOHN MUIR

MARTINA NAVRATILOVA

YOKO ONO

CHARLIE PARKER

LEONARD PELTIER

SYLVIA PLATH

EZRA POUND

JULIA ROBERTS

MICKEY ROONEY

PHILIP ROTH

SIMONE SIGNORET

NINA SIMONE

WILL SMITH

SAMMY SOSA

JESSICA TANDY

PETE TOWNSEND

JOHN TURTURRO

GIUSEPPE VERDI

SIMONE WEIL

EUDORA WELTY

BARRY WHITE

MEG WHITMAN

GENE WILDER

VENUS WILLIAMS

RENÉE ZELLWEGER

"All serious daring starts from within."

—Eudora Welty, Aries Sun–Virgo Jupiter. April 13, 1909

"You have to believe in yourself when no one else does."

—Venus Williams, Gemini Sun–Virgo Jupiter. June 17, 1980

"When my horse is going good, I don't stop to give him sugar."

—William Faulkner, Libra Sun–Virgo Jupiter. September 25, 1897

"Every man gotta right to decide his own destiny."

—Bob Marley, Aries Sun–Virgo Jupiter. April 6, 1945

*"A woman is like a tea bag—only in hot water
do you realize how strong she is."*

—Nancy Reagan, Cancer Sun–Virgo Jupiter. July 6, 1921

"Only I can change my life. No one can do it for me."

—Carol Burnett, Taurus Sun–Virgo Jupiter. April 26, 1933

"Whenever I get happy, I always have a terrible feeling."

—Roman Polanski, Leo Sun–Virgo Jupiter. August 18, 1933

"The wound kills that does not bleed."

—Wallace Stegner, Aquarius Sun–Virgo Jupiter. February 18, 1909

"It is easier to live through someone else than to become complete yourself."

—Betty Friedan, Aquarius Sun–Virgo Jupiter. February 4, 1921

*"It's no good running a pig farm for thirty years while saying,
'I was meant to be a dancer.' By that time, pigs are your style."*

—Quentin Crisp, Capricorn Sun–Virgo Jupiter. December 25, 1908

♎♃

Shining Bright

From Virgil, born in 70 B.C., who wrote, "Love conquers all, and we too succumb to love," to fellow poet Philip Larkin, born 1,992 years later, who said, "What will survive of us is love," Libra Jupiters know, and so should you, that love is what makes the world go round. Love will make you luckier than anything else.

> *We must trust our own thinking.*
> *Trust where we're going.*
>
> —LIBRA JUPITER WILMA MANKILLER

Since Libra rules marriage and partnership, Libra Jupiters reach their full potential by *inviting and engaging in love with others.* The more you invite love and its beauty and awareness into your life, the more happy and successful you will be.

Libra Jupiters attract more love, adoration—even sheer worship—than any other Jupiter sign. Ruled by the most beautiful goddess, Venus, who "stole the wits even of the wise," you guys are the most charismatic figures around, stealing the wits of the wise of *every single generation.* Just look at this sampling from the last five decades: from Beyoncé Knowles to Justin Timberlake; from Jennifer Lopez to Catherine Zeta-Jones; from Madonna to Michael Jackson; from Donny Osmond to The Beatles to Brigitte Bardot and Sophia Loren—we're talking some pretty worshiped people here! Then there are everyday Libra Jupiters: *Everybody Loves Raymond*'s Ray Romano, America's sweetheart Katie Couric, and everyone's best "Friend" Jennifer Aniston; adored and revered leaders such as George Washington and Winston Churchill; and those Libra Jupiters who have moved millions and changed the world through their charisma: Martin Luther, Joan of Arc, and Gloria Steinem.

Libra also rules justice and balance, and its symbol, the figure of Justice holding the scales, is always—in every court of law—blindfolded. And this is where it gets most exciting to be a Libra Jupiter—you must *see* differently! Not only must you trust your ideals but, as Justice does, trust your intuition and sense what is right and wrong, not through the obvious literal ways (her eyes are covered, re-member), but through your sense of *balance*. So, Libra Jupiter, don't just rely on your eyes: Use your extra Libra senses to feel your way through the world. Your Libra Jupiter senses give you an added bene-fit: You have many extra ways to be more sensual!

Be careful, however, to live in a harmonious, healthy, and beautiful atmosphere. As an Air sign, Libra Jupiters are vulnerable to bad air, to negative environments. Make sure your home and your workplace provide the sort of balanced, loving environment that nurtures your power and sensual energy. How often have you experienced the luck of your Jupiter in an environment that was not just so?

Take a moment, look back long and deep over your life, recount as many events as you can, and ask yourself what has made you lucky or unlucky in your life. You'll notice that whenever you act like a Li-bra, *you create miracles in your life.* No matter what your Sun sign is, *you get lucky when you live out the Libra legacy gifted to you by the benevolent planet Jupiter.*

To take advantage of your luck, emphasize your . . .

Libra Jupiter Strengths	♎♃	*Be charming, your very own love story, a partner, idealistic, true to the letter of the law* (ipsissima verba), *beautifying and illuminating, tolerant, psychic, hopeful.*

But because Jupiter rules excess, watch out for . . .

Libra Jupiter Excesses	♎♃	*You can become lazy, overconfident, or conceited.*

Understanding the
Libra Energy and Influence

Libra begins the Autumnal Equinox. You were born when day and night hours are perfectly equal, ideally, a time of complete balance on our planet. In the old days this was the time of year when the scales *ruled*. We weighed crops and grains, grown through our long spring and summer labors, on the village scales after the harvest. In Libra time, we enjoy the fruits of our labors. Libra loves everyone to enjoy.

Besides love and beauty, Libra rules the law and books. In olden times, the harvest was the month in which the laws were of supreme importance: The laws and written customs governed the portion of the harvest to which each family was entitled. Libra Jupiters like to stick to the letter of the law; you respect laws and expect them to make sense. Your luck brings you a real old-fashioned sense of justice, going back to the time of Solomon, a power to see all the issues, weigh things appropriately, and apportion them wisely.

The Libra symbol is Themis, Goddess of Justice, holding the golden scales (an identical symbol represents Justice in every court of law today). Those golden scales go up and down, weighing the different options, until they right themselves when they are perfectly balanced. Great luck will come to you if you make decisions in the same way, by weighing and balancing, coming out, eventually, on the side of what's right.

To attract the luck and power of Jupiter, you must also be like the figure of Justice—blindfolded. Don't let passion or impulse sway your balance and objectivity. Stay close to both your intuition and your rationality. Follow logic in making your decisions, yet at the same time confirm those decisions through your powerful inner sense of what's right.

LIBRA SYMBOL
The Scales

The Kidneys, the Lower Back, and the Skin

⊿Ω⊿Ω⊿Ω⊿Ω⊿Ω⊿Ω⊿Ω⊿Ω⊿Ω⊿Ω⊿Ω⊿Ω⊿Ω⊿Ω⊿Ω⊿Ω⊿Ω⊿Ω⊿Ω⊿Ω⊿Ω⊿Ω⊿Ω⊿Ω⊿Ω⊿Ω⊿Ω⊿Ω

Your Jupiter sign governs the skin, the kidneys, and the lower back. The skin is like a Libra thermometer: You will find that imbalances in the body show up on the skin and that every emotion and mood you experience is reflected in your skin's condition. Libras generally have perfect skin, so when skin disturbances appear, it's a sign for you to look at what's bothering you, what's wrong in your environment, and to correct it. Kidneys are the filters, or go-betweens, of the internal organs; in life, you act as a filter or go-between with people and in different environments. To keep your kidneys healthy, you need to drink lots of water to prevent kidney stones. The lower back is directly affected by the state of the kidneys, as well as by pressure (including emotional pressure). When your back hurts, ask yourself where the pressure in your life is coming from.

⊿Ω⊿Ω⊿Ω⊿Ω⊿Ω⊿Ω⊿Ω⊿Ω⊿Ω⊿Ω⊿Ω⊿Ω⊿Ω⊿Ω⊿Ω⊿Ω⊿Ω⊿Ω⊿Ω⊿Ω⊿Ω⊿Ω⊿Ω⊿Ω⊿Ω⊿Ω⊿Ω⊿Ω

Lavender and Cornflower Blue

These are rich, soft, romantic colors. More than any other sign, Libras are affected directly by color—not only in their emotional state but also in their physical health. You are so sensitive that the wrong colors will actually make you physically ill, nauseous, tired or sleepy. Pastels are considered the best for Libras, because they are soft, unobtrusive, and pleasing to the eye. But for exceptional luck, wear and surround yourself with lavender or cornflower blue. Use these shades in your home—in accents, towels, and sheets. But if this does not suit your dress sense—if, for example, you're a fiery Sun sign—remember that as long as the color's energy touches the skin, it does the trick, so you can wear it under your normal clothing.

Libra Jupiter ♎♃ Strengths

And now your Libra Jupiter strengths:

Be Charming

As Raymond Chandler said of the darling of the Jazz Age, Libra Jupiter F. Scott Fitzgerald (born a double Libra on September 24, 1896): "He had one of the rarest qualities . . . the word is charm. . . . It's a kind of subdued magic, controlled and exquisite."

To draw on the power of your lucky Jove, highlight *your* exquisite magic and charm, do your best to be kind, sympathetic, and gracious to all. Stretch your butterfly wings and enjoy flitting from one social event to another. Your luck, remember, comes most from people's favorable reaction to you (no matter your sexual orientation, you benefit most from charming the opposite sex). Jay Gatsby has become synonymous with glamour, enchantment, and charm. Encourage the Gatsbyesque in your own life. "Life is a romantic matter. In the best sense one stays young."—Libra Jupiter F. Scott Fitzgerald.

Libra Jupiter charm is not always, by any means, related to the cosmetic and superficial. No book is more charming than the classic *Heidi*, by Libra Jupiter Johanna Spyri. The antithesis of Libra charm is revealed in Spyri's quote: "Anger makes us all stupid." To bring luck into your life, work on transforming areas of frustration into illuminating experiences. "If you want a better body, you must exercise more. If you want a better mind, you must read more" (remember: Libra rules books!); this is sage advice from a Libra Jupiter—Marilyn Voss Savant, born in 1946, with the highest IQ ever recorded (228).

Be Your Very Own Love Story

You're always wishing for a romantic existence; by holding on to your image of truth and beauty, you can hope to make it a reality. To be lucky, you should think in terms of bringing beauty to the world, whether as a painter or a beautician, by growing organic vegetables, working for world peace, or raising happy children. In the case of the famously attractive Libra Jupiter lovers John Reed and Louise Bryant, their vision of a more perfect world led them—as writers, painters, and actors—to be radical activists, first in Greenwich Village in the early part of the twentieth century and then in Russia, where they witnessed the Russian Revolution. Their glamorous, romantic, dramatic lifestyles were depicted in the movie *Reds*, which was also born Libra Jupiter in 1981! Even their tragic ends are glorified and romanticized. John Reed, who died the day before his thirty-third birthday, was interred in the Kremlin, and his journalist wife, Louise, wrote at his death: "He was never delirious the way most typhus patients are. He always knew me, and his mind was full of stories and poems and beautiful thoughts. He would tell me that the water he drank was full of little songs." John Reed believed in a world that was "richer, braver, freer, more beautiful." He was one of history's great romantics.

Romance is a Libra Jupiter province; Libra Jupiters practically invented the word "romance." Not just the flowers and moonlight scenario—though that's a lovely start—but romance in the sense of

(continued)
painting *The Birth of Venus*; another is the Greek statue known as the Venus de Milo.

great heroism and love until death. Libra Jupiter, never lose your romantic view of the world.

Be a Partner

Libra Jupiters are always lucky in partnerships—marriage, career, finances. Even if the partnership itself isn't ideal, it will benefit a Libra Jupiter in some way: expansion of self, professional rewards, greater financial opportunities. In fact, more than any other Jupiter sign, you benefit from marriage. Libra, ruling the seventh house of marriage and partnership, explains why Libra Jupiter natives will often gain whole new careers—and perhaps develop new personas—from their marriage and romantic relationships.

Famous examples include Libra Jupiters Brigitte Bardot, Sophia Loren, and Diane Keaton, whose mates furthered their acting careers. Bardot, who said that she had only wanted to be a wife and raise a family, was presented to the world by her husband, Roger Vadim, as the ultimate sex kitten in *And God Created Woman*. That film made her a legend. Sophia Loren was discovered by producer—and eventual husband—Carlo Ponti. Diane Keaton won both her Oscars in movies made by her then-partners, Woody Allen (*Annie Hall*) and Warren Beatty (*Reds*).

The number of Libra Jupiters who join professional forces with their mates is extraordinary: Cher (Sonny Bono), Catherine Zeta-Jones (Michael Douglas), Annette Bening (Warren Beatty), Jennifer Aniston (Brad Pitt), David Lynch (Isabella Rossellini), Diego Rivera (Frida Kahlo), Candice Bergen (Louis Malle), Geena Davis (Jeff Goldblum), Jennifer Lopez (Sean "Puffy" Combs and Ben Affleck), Elizabeth Hurley (Hugh Grant), Madonna (Guy Ritchie), even Bill Clinton (Hillary Rodham). Gertrude Stein's companion, Alice B. Toklas, not only cooked the meals but typed all of Stein's works.

Other successful Libra Jupiter professional pairings include Judy Garland and daughter Liza Minnelli (both Libra Jupiters); Donny Osmond with sister, Marie; and Libra Jupiter Jim Lehrer with Robert MacNeil.

Be Idealistic

The constant awareness of something better, a more perfect world, is at the root of Libra idealism. The Boy Scouts, guided by the chivalrous ideals of service to others and being that best that one can be, are Libra Jupiter, as is the Emancipation Proclamation of 1863, which brought freedom to the slaves. And so is the father of American democracy: "Labor to keep alive in your breast that little spark of celestial fire, called conscience," once advised Libra Jupiter George Washington. Washington believed in true Libra fashion that the United States was founded upon such high ideals as "the stupendous fabric of Freedom" and on "protecting the rights of human nature" in order to "establish an asylum for the poor and oppressed of all nations and religions." And, as a true Libra Jupiter, Washington stuck to his ideals: He risked his life by leading his troops into battle, uncommon at that time, and he refused the title of king that others wanted to bestow upon him.

Whatever field you are in, dear Libra Jupiter, you can embody the principal of idealism. Throughout her life, Mother Teresa gave tireless, cheerful, and loving care to those in extreme poverty, those who had no other advocate or hope of help. A combination of Libra Jupiter idealism with her Virgo self-sacrifice, she is now in the process of elevation to sainthood. Libra Jupiter Lynn Cox, the only person to have swum the Bering Strait, between the United States and Russia, said: "It made me realize that a swim could be more than an athletic event. It was a way of bringing people together."

Be True to the Letter of the Law (Ipsissima Verba)

Libra Jupiter Martin Luther, the great instigator of religious reform, began the Reformation when he nailed his Ninety-five Theses to the door of the Wittenberg Church. He called for the Church to conform to its original principles, and he protested against papal abuses and the corruption of church officials. Though his campaign ended in his excommunication and the start of the Reformation, Luther stated that he would have happily yielded every point of dispute to the pope, if only the pope had affirmed the Gospel (that is, God's law). Revealing the contrast between Luther's warrior image (more typical of a Fire sign) and his actual identity is his quote: "Music is discipline, and a mistress of order (law) and good manners, she makes the people milder and gentler, more moral and reasoning."

Like Martin Luther, you, Libra Jupiter, should try to stick to the letter of the law. Once you know the principles and ideals you want to hold high, stick to them, and don't bend any rules. You should be true to yourself and others, to tell the whole truth and nothing but the truth. Be authentic and sincere; be artless and guileless. Drawing upon your strength as a Libra Jupiter, what can you reaffirm in *your* life "as gospel"?

Be Beautifying and Illuminating

Did you know child psychologists say that if a child listens to the music of Wolfgang Amadeus Mozart (a Libra Jupiter, of course), the child's IQ is elevated? It's

❋ Stay even keeled and composed to get what you want; trust that everything will work out.

❋ Trust your instinct above all else.

❋ Conduct your business through partnerships and close relationships.

❋ Relate to the public; share, cooperate, and encourage others.

❋ Give calm direction and assurance.

❋ Grow and improve through a balanced and objective attitude, fairmindedness, and always strive for a diplomatic approach.

❋ Wear and surround yourself with lavender or cornflower blue—clothes, towels, sheets, and home decor.

❋ Hang out with more Geminis, Aquarians, and Libras.

almost as if Libra Jupiter Mozart had secret, amazing knowledge of the nature of our world and that listening to his music brings children in touch with the harmony of the universe. Libra Jupiters do have the special ability to sense nature's underlying order, the pattern that lies behind what looks to most of us like chaos, life's secret book of rules. As Libra Jupiter Johann Sebastian Bach, born March 21, 1685, said, "There's nothing to it. You only have to hit the right notes at the right time and the instrument plays itself."

Use your sensory powers to discover the true nature of things: Abd Al-Rahman Al Sufi was a Libra Jupiter astronomer born A.D. December 7, 903. It was in the palace of Baghdad, the fairy city of *The Arabian Nights* and the foremost center of culture during medieval times, that he discovered the Andromeda Galaxy, the first galaxy outside of our own, and romantically (Libra Jupiter!) named it "Little Cloud." Libra Jupiter Gustav Holst was born 971 years later, and he, too, found his inspiration in the heavens ("The heavenly spheres make music for us"): He composed the magnificent symphony *The Planets*. Listen to his awesome "Jupiter" movement; it will honestly inspire you!

Use your creative spirit to embrace truth and beauty. As poet and Libra Jupiter Percy Bysshe Shelley said, "Poetry is a mirror which makes beautiful that which is distorted." And if your creative powers aren't in the same league with Shelley, Mozart, and Bach, you do get luckiest by association with the arts, so make sure you buy your season tickets to the symphony, attend an art opening, read one poem a day, and fill your senses with music, color, and vision.

Be Tolerant

"If nobody spoke unless he had something to say, the human race would very soon lose the use of speech," stated Libra Jupiter Somerset Maugham. So let yourself, Libra Jupiter, enjoy the art—and it is an art—of conversation. Lighten up and be tolerant: You gain all your best associations from playful, *fun* social contacts. As Libra Jupiter Sir Winston Churchill said, "A fanatic is one who can't change his mind, and won't change the subject." Indulge in light repartee. You don't need to be noble and profound all the time! Despite his serious responsibilities, Churchill was a master of banter. Just consider this one: At a dinner party, a female political opponent remarked, "Mr. Churchill, if you were my husband, I would put poison in your coffee." To which Churchill replied: "Madam, if I were your husband, I would drink that coffee."

Libra Jupiters are all about having a good time—think Libra

Jupiters MTV, Woodstock, *The Brady Bunch*, and *Sesame Street*. So, relax, Libra Jupiter, and let the good times roll!

Be Psychic

Your psychic sense, Libra Jupiter, is the ability to sense the natural law of the universe, to know on a profound, individual level what's really right and wrong. Joan of Arc, born a peasant, on January 6, 1412, convinced the Dauphin and all of France to follow her in battle against the English. She had heard the voices of Saints Catherine and Michael in the fields of her native Domremy, telling her that the English should be driven from France and that she would raise the siege of Orleans. Wishing to trick her, the French court had another man sit on the throne of the Dauphin, but Joan correctly identified the real Dauphin, hiding in the crowd, and convinced him of her heaven-sent mission. The information from her voices proved to be accurate, and leading the army of France, she won the battle!

To tap into your psychic luck, try not to have any presupposed ideas; be open to what is out there. Libra Jupiter Carl Jung introduced the idea of the collective unconscious, the ultimate interrelatedness of all things, people, times, and places. In order to be in touch with this knowledge, Jung stated that one must also go within: "Who looks outside dreams; who looks inside wakes." Like Libra Jupiter Helen Duncan, one of the greatest psychics of the twentieth century, or Libra Jupiter M. Night Shyamalan, whose films *The Sixth Sense* and *Signs* explore psychic abilities to see the true order of things, use your Libra Jupiter powers to embrace the real nature of the world. And, like Holst, composer of "The Planets," don't be afraid to explore and proclaim your interest in the powers of the planets and the bounty of Jupiter!

Be Hopeful

"Men wanted; for hazardous journey, small wages, bitter cold, long months of complete darkness, constant danger, safe return doubtful. Honor and recognition in case of success." Sir Ernest Shackleton, the Antarctic explorer, put this ad in a London newspaper, hoping to attract recruits for his 1914 expedition to the South Pole. Believe it or not, this Libra Jupiter hope paid off, and he was flooded with applicants. He stated, "Optimism is true moral courage."

Most great optimists are Libra Jupiter. George Gershwin, perhaps the greatest composer of popular music in the twentieth century, wrote songs that kept Americans humming and whistling for generations, including " 'S Wonderful," "I Got Rhythm," and "They Can't Take That Away From Me." Who can forget these two playful, hopeful, optimistic images: the skyline of New York opening Woody Allen's *Manhattan*, with the swell of Gershwin's "Rhapsody in Blue" rising over it; or the scene of Libra Jupiter Cher (in her Oscar-winning performance) playfully kicking a can down the street in front of the same skyline in *Moonstruck*?

What Makes You Lucky in Love

You benefit more than any other Jupiter sign from being in love. Choose the one who gives you joy! Share your ideas, hopes and dreams.

For your relationship to work, there must be a strong mental and spiritual bond, as well as a physical one. As important as good looks are to you in choosing a partner, remember that Libra is too astute and sensitive not to desire the inner as well as the outer beauty. Because Libra Jupiters tend to live so completely through mates, it is crucial that you choose someone with whom you're happy to spend every moment—as they are with you. "Husbands are like fires; they go out when they're left unattended," once said Libra Jupiter Cher. However, a word of warning: As much as you think you have to be in constant attendance, do be aware that this can drive the person you love away. Use your measuring Libran scales to ascertain when they need distance, when they want closeness. The exchange between you and your significant other is so constant that harmony is a necessity. Arguments with their mates are particularly upsetting to Libra Jupiters.

If you want to find the perfect partner for a lasting relationship, be faithful and balanced, and don't act on impulse! Even if your Sun sign is in impatient Aries or image-conscious Leo, you will find success at love if you pick your partners carefully and practically. Think of the ways you are compatible, the plusses and the minuses. Don't fall head over heels with the "perfect image."

The myth of Venus is that she was so enchanting that flowers grew wherever she walked. The romantic idealist Libra Jupiter tends to project this image onto their love object in the initial stages of infatuation. When this fantasy starts to fade, the disillusion can be shattering, unless you look for the partner who on balance seems *right for you,* imperfections and all.

Your Best Career Moves

Diane Sawyer, born a Libra Jupiter on December 22, 1945, gives the perfect advice for your Libra lucky choice of career: "Take the road ahead and love the ride." There are three questions she says you must ask yourself: What is it you love? Where is the most satisfying (adventurous) place you could do it? Are you certain it will serve other people?

Libra Jupiter, your luck comes through other people. You probably got your current position, project, or venue through the tip or word of a friend or someone you know. Luck comes to you best through others no matter how hot your reputation already is or how many résumés you send out (don't even bother going through these channels). Friends talk you up and promote you even in other countries, even when they haven't seen you for years! The phrase "It's who you know" must have been coined by a Libra Jupiter. Don't ever start a business on your own. You are most successful in partnerships and collaborations. Host parties, dinners, any type of social event; you will get most of your professional contacts and help this way.

Whatever your Sun sign, to achieve success at work borrow from the Libra Scales and try to put everything in a balanced perspective. Be objective and don't form close ties with any one competing group at the office. Be fair, and, especially, be diplomatic. You would make a great second-in-command, and if you exercise your diplomatic skills, you might find yourself quickly promoted to first-in-command!

Particularly in rough times, it's important to draw upon your Libra Jupiter ability to look on the right side. It worked for Libra Jupiter Katie Couric, whose first appearance on camera prompted the president of CNN to call the assignment desk and say, "I never want to see her on the air again." But as Katie herself has said, "Even the most critical words can bring about the most positive results." You can't get any more Libra Jupiter than that. Look what it did for Katie!

How to Reach Optimum Health

A paradox: While Libra Jupiter benefits most from moderation and balance, Librans must also respect the ups and downs of their energy levels. So while you must make sure you take proper care of yourself by getting enough rest, it's lucky for you to work very hard (or play very hard) for days, perhaps weeks on end, and then suddenly switch to nothing but rest, and sleep, reading in bed, stumbling around your home, fixing yourself snacks as you watch TV for hours on end. It's your recharging time and what you need to stay balanced and healthy, both physically and emotionally. But while your energy can go up and down, in terms of your diet, your intake of alcohol and caffeine, and your exercise plan, you must stick strictly to daily moderation.

You must have balance for health: balanced diet, balance of work and play, balanced emotions, balance with the important people in your life. No extremes in environment. You need mild exercise—evenings are the best time for this—especially back exercises, to strengthen a weak lower back. Work on good posture or you will feel vague aches and pains in the back. You need to drink *tons* of water to keep your system flushed out and free of toxins. Your sensitive skin shows at once the effect of lack of sleep, rich food, caffeine, smoking, and alcohol. Dark circles under puffy eyes are an instant giveaway that you're not taking care of yourself.

Your Lucky Foods, Element, and Spices

Libra's cell salt is sodium phosphate, which equalizes the balance of acids and alkalis (Librans always need the proper balance!) and rids the body of waste material. The skin is the litmus test: Sallow skin shows that you need more sodium phosphate in your diet. Foods that are especially high in this mineral are strawberries, apples, raisins, almonds, asparagus, peas, corn, carrots, spinach, beets, radishes, tomatoes, brown rice, and oatmeal. You thrive on a high-protein diet; forgo bread and sugar. Broiled fish, seafood and poultry, low-fat cheeses, yogurt, plenty of fresh fruits and vegetables, salad greens—all are lucky for your health. Avoid alcohol and carbonated drinks, which are bad for the kidneys and will show up on the skin immediately. Your especially lucky foods: kidney beans (these are even lucky talismans, carried by hand) and anything made with tomatoes—called the "love apple," blessed by Venus, and very lucky for you!

Your Best Environment

Your health and day-to-day functioning are ruled completely by your environment. This is why the choice of environment is more important for the Libra Jupiter than for any other Jupiter sign. It's crucial! Your environment must be calm, beautiful, serene, and inspiring in its loveliness. You cannot live with disharmony or strident noises. Chaos is extremely unlucky for you; it will make you physically ill. Pay attention to the people you surround yourself with and make sure your relationships with them are honest and harmonious. You do best in a moist climate, with plenty of fresh air.

Your Best Look

You shine, Libra Jupiter, when you emphasize your voluptuous Venusian shape and form. Soft or rich colors and fabrics, not sharp black-and-whites or crisp business neutrals, are the way to go for you. You can accessorize with flowing scarves and enchanting touches—nothing is too magical for the love goddess. Curves are always in, from Libra Jupiters Jennifer Lopez to Dolly Parton to Brigitte Bardot. Even masculine Libra Jupiters are hardly ever angular—Libra Jupiter Diego Rivera being a prime example. The Venusian influence softens the features of the men, producing pleasingly round visages, such as Libra Jupiters Brendan Fraser, Ray Romano, and Tommy Lee Jones.

The truth is that most Libra Jupiters are such stunners—or if not actually classically beautiful, like Libra Jupiters Michelle Pfeiffer or Natalie Portman, have such an aura that you believe they are—that they can wear anything, or nothing.

Your Best Times

The luckiest times for you are when the Moon is in Libra, which happens every month for two to three days. Find these special days in your astrological calendar. During these days, schedule crucial meetings, tell that special person you love them, ask for a promotion or raise, push the envelope, and plan to shine. In your everyday, the best time for you to exercise is early evening. The luckiest times during the year for you are September 23–October 22 (when the Sun is in Libra); January 20–February 19 (when the Sun is in Aquarius); and May 21–June 20 (when the Sun is in Gemini).

The very luckiest time of your life—as in Lotto lucky—is when Jupiter reenters the sign it was in when you were born (Libra), which happens every twelve years and lasts anywhere from a few months to just over a year. This is called your Jupiter Return, and it launches your luckiest streak. You can count on the *extraordinary* to happen the very day your Jupiter Return begins. Amazing events can begin happening up to six weeks beforehand, so plan ahead!

How to Be Lucky Financially

Be logical. The Scales are nothing if not objective, weighing, and considering. Look at all the possibilities, take your time, then go firmly with your decision. You'll get there in the end. It can just take a while. If you try stock and real estate markets, adopt a careful, balanced, open approach. Do lots of research and use your own judgment.

Don't let anyone—bank, spouse, kids, friends, accountant, investment adviser—push you before you're ready. You're the one who has to live with your decision. You may find it easier to make money through ideas, invention, and creativity rather than through investing and financial planning—try to make money betting on yourself as opposed to betting on the markets. Careers in writing, graphic design, corporate communications, broadcasting, public relations, and strategic planning will be lucrative for you. For Libra Jupiter, work in the arts is particularly lucky, as are careers in the healing arts—Libra Jupiters make fine massage therapists.

Whatever your field, remember: The more you communicate, the more money you make. A word of warning: Beware of your love of overabundance and tendency toward extravagance. Learn early how to save and be responsible, to avoid going without later in life.

Success in Family and Friendship

You gain luck with family and friends by, first and foremost, being tolerant. People feel that they can really relax with you and be themselves, and that you value them as such. Yours is a soothing presence. At the same time, you are number one on everyone's party-invite list. The more you appreciate the differences—the negatives as well as the positives—in family and friends, the more you're fulfilling your Libra Jupiter role.

Tune in to your lucky intuition to sense what your loved ones need. Last but not least, do not hesitate to mix business and pleasure— usually a no-no—by working with, or going into business with family and friends. You are loved for your matchmaking skills, both in love and assembling either a working ensemble group or a dinner party, and you'll find that the favor will be returned. There's someone hoping to be introduced to you; family and friends are the means of introduction.

Actually, you hardly perceive a difference between family and friends: while family roles may not be traditional, you do well to count your family—particularly your parents—among your closest friends. The only drawback to this making-the-world-your-friend is that the mailman becomes as important as your dad, the new acquaintance as special as the friend of many years standing. Toning down your need to make an impression can lessen the resentment that this attitude might cause. You are actually a loyal and devoted supporter of all you love.

The perfect Libra Jupiter cautionary tale is that of Dr. Faustus, who exchanged his soul for the promise of youth, success, fame, riches, and knowledge. He followed the Libra Jupiter inclination toward partnership—but his partnership was an unholy alliance with the Devil. His unwise choice was motivated by his grandiose illusions, conceit, and supreme laziness. He believed that he was above mortal consequences. He preferred to rely on assistance, not from his own efforts, but on the Powers of Darkness. Such a shortcut cost him his soul.

The prime Libra Jupiter excess is confidence without responsibility. This will bring you nothing but disaster and leave you sore and disappointed for years. You may have to reteach yourself true confidence if it was taken from you as a child. Enjoy your skill and superiority, and be sensitive to the burdens such gifts give you.

Excess #1 — *You Can Become Lazy*

Your challenge is to chill out but not check out! Your excessive charm and public appeal can make it easy to get lazy and let slide your ethical responsibilities. If you are too lazy to make your own decisions, you can become too dependent on advisers; learn to trust your instincts.

Libra truly rewards substance, objectivity, balance, and work. Those charmers who do not do their homework are headed nowhere fast. Perhaps the most extreme example of Libra Jupiter laziness is the Libra Jupiter date of March 9, 1981, when the U.S. Department of Agriculture declared ketchup a vegetable. On the flip side, Paris Hilton, born Libra Jupiter in the same year, made her "easy life" a number one TV series by *working* it.

Excess #2 — *You Can Become Overconfident*

You can hurt a lot of people's feelings by being overconfident. Be careful! They may retaliate. Confidence is a Libra Jupiter lucky plus,

DON'TS

✳ Become the next Unibomber and go live in a shack, never washing, never saying hi.

✳ Have so many advisers that you stop listening to yourself.

✳ Get flustered, panic during a crisis, or become forceful and angry with others.

✳ Get overconfident and lose balance.

✳ Embrace escapism in any form.

✳ Live for yourself alone.

✳ Give up on your ideals and word.

but overconfidence means the Libra Scales become out of balance and topple over. Libra Jupiter Bill Clinton's fall came when he believed he would not be called to task for lying. The beautiful charmers Marie Antoinette and Mary, Queen of Scots, were both Libra Jupiters, but both were overconfident and believed too much in their own romantic illusions. Marie Antoinette had a rustic farm constructed on the grounds of the palace of Versailles, where she and her ladies-in-waiting played at being costumed "milkmaids," out of touch with the starvation of the real peasants. Mary, Queen of Scots, believed that she was the rightful queen of England and that she could overthrow her cousin Elizabeth I. Both Marie and Mary were beheaded.

History will tell if Libra Jupiter George W. Bush is overconfident in his right to wield the military might of the United States against the consensus of the United Nations and international protest.

Excess #3 *You Can Become Conceited*

"Solitude gives birth to the original in us, to beauty unfamiliar and perilous—to poetry. But also, it gives birth to the opposite: to the perverse, the illicit, the absurd." These words from *Death in Venice* were written by Libra Jupiter Thomas Mann—the author also of *Dr. Faustus*—and could be used to describe Libra Jupiter Michael Jackson, who has this to say about himself: "I have been the artist with the longest career and I am so proud and honored to be chosen from heaven to be invincible."

Perhaps the greatest problem for Libra Jupiters is the fact that media rewards excessive confidence, so that Libra Jupiter Madonna can say "Basically, I want to rule the world" and no one blinks an eye. Remember, Justice is blindfolded, so *look within,* dear Libra Jupiter. Don't let your conceitedness take you places you don't want to go.

Sun Sign-Jupiter Sign Combinations

You know what you want and you want it now, so lower your Ram's horns and *charge!* Whoa!—not if your Jupiter is in deliberating, balancing Libra. All those fiery, me-first, go-for-it attributes come naturally to you, but if you want a little luck and help from the Stars, look to your Jupiter. It's riding high in Libra, asking you to stop, consider, use the Libra Scales to weigh the options—maybe even use a little Libra charm and diplomacy (look it up, Aries, if it's not in your vocabulary) to handle the situation.

Aries-Libras Gloria Steinem and Paul Robeson both acted from their groundbreaking, pioneering Aries force, but they succeeded so well through their Libra Jupiter charisma and charm. Your Libra Jupiter reminds you that a leader has to have followers; you'll succeed when you work *with others*—and when you use Libra objectivity to choose your direction.

The Bull's hooves are firmly on the ground, but you'll find your rewards—whether the love of your life or the comfort of luxurious surroundings—by lifting your head and, like Ferdinand the Bull in the children's story, smelling the flowers. Balance your Taurus Earth with fresh Libra Air; widen your view with expansive Jupiter.

Libra Jupiter gives you the charm to draw others to you, with Taurus practicality—a potent combination for the creation of wealth. But Libra Jupiter fortune is built by sharing, by working with others: the Libra Scales require balance and justice for all. Working with partners or family is lucky for you: Taurus-Libras Michelle Pfeiffer and Candice Bergen worked with their husbands to attain great artistic and financial success.

It's no wonder that you can enjoy being a social butterfly, entertained and entertaining, With your Jupiter in Libra, your charisma is emphasized; you can stand in the spotlight and shimmer, to the delight of your numerous acquaintances—or audience. As Libra Jupiter Anna Kournikova said, "I'm beautiful, famous, and gorgeous." But when the applause becomes monotonous—or, worse, the spotlight fades—you won't be left alone in the dark if you've developed your deeper Libra Jupiter qualities of balance and purpose, relating to others not just as audience but with genuine care and fairness.

Use your Gemini-Libra combo sparkle for a greater good, whether it's writing a book of witty insights, entertaining the masses—or the neighborhood kids—or becoming a spokesperson for a cause. Fail to do so, and, like Gemini-Libra the Artist Formerly Known as Prince, fading from the spotlight can be painful, indeed.

♃♎

The Moon of Cancer and the Venus of Libra Jupiter create beautiful dreams and beautiful people. From Woody Guthrie to Gilda Radner, we have always listened to your kind. Remember: You must *follow* your Libra Jupiter voice to get what you want. *Dreaming* of a romantic mate won't bring him to you door, and if you scuttle away and hide, he won't see the sensitivity beneath the brittle exterior.

For real success, join with others and work in group and ensemble situations. Your boss doesn't appreciate how much you do; could it be because you hide behind the ferns you've put on your desk to make it your private space? Use Libra Jupiter objectivity to find out *why* you're moody and Libra Scales to balance. Libra Jupiter gives you confidence to be charming, and you'll find yourself appreciated for your other qualities, too.

♃♎

Leo, the showman of the zodiac, with the wow-power of Libra Jupiter! Famous Leo-Libras include luminaries of all the ages, from artist Marcel Duchamp to actress Angela Bassett. But will it be a momentary showy flash or a brilliant, steadily burning sun? The temptation, especially if success comes early—which it will with this combination—will be to believe one's own publicity, accept adulation as your due, and see how far you can go, how much you can get away with. But remember: With power comes responsibility.

Libra Jupiter urges you to create substance beneath the show: Take classes, develop your talents, *grow.* You find satisfaction in knowing that what you give is true value, not sham; that's fair-minded Libra Jupiter.

Use your Libra Jupiter to not stand back and observe. Like Virgo-Libras Beyoncé Knowles and Gloria Estefan, get in there and work it! Ever considered becoming a director—of a film, a library, a clinic? As a Virgo, you spot—and often point out—how things could be done better, more efficiently, if attention were paid to details. Use your Libra Jupiter to work with people and to charm them while you criticize. You're hardest on yourself, of course, but with Libra Jupiter, the experience takes on a feeling of creating something great *together*.

Follow your Libra Jupiter guide and you'll have what you seek, whether a near-perfect (nothing's ever *quite* perfect, for a Virgo) piece of work or a lasting romantic relationship. Now, relax and enjoy it!

Venus, the goddess of love, charm and voluptuousness, rules! You delight and inspire others by your very being. Remember the double-Libra TV series *The Brady Bunch*? You guys are fun! Then the need for Libra balance appears, and you start weighing the other side. The scales tip; all the pleasure in your "you-ness" must be wrong; there must be another side. So you doubt, you run away, you look for answers—perhaps from the wrong sources, the wrong people. Your Libra concern for others, for balance and justice, can become so all-absorbing that it's turned on its head and becomes selfishness.

Trust your Libra Jupiter; know that your charisma can be a powerful force for good. Trust *yourself*. Go forward by looking at the present and future; stop worrying about the past. People root for you and cheer you on like crazy; think of double-Libra ice-skating champ Nancy Kerrigan, who had everyone on her side.

The passion of Scorpio channeled into charismatic Libra Jupiter produces incredible power. Note that I said "charismatic," not "charming." Scorpio-Libra is too intense to be merely charming. From devastatingly sexy Owen Wilson to challenge-the-world surreal-ist René Magritte, Scorpio Libras make us see life differently. You can change the world: with Libra Jupiter, you can draw others to your cause, but balance the Scales of Justice to be sure it's a cause that bene-fits them, too.

You don't intentionally deceive but may deceive yourself by the intensity of your feelings. This is particularly true in love. Use your

Libra Jupiter objectivity to choose a mate worthy of your respect; otherwise, love becomes an obsession, not the deep and rewarding relationship you need.

♎♃♎

Venus and the Horseman of the Skies—a mythical romance. This combination creates a lovable, restless, life-of-the-party creature, but with a perceptive mind, a sharp wit, and a fiery devotion to causes under the attractive exterior. Just think of Sagittarius-Libras Bette Midler and Lucy Liu. The danger is that this jolly mix will rely on charm and spur-of-the-moment action.

Your charm and energy put you in the limelight, which you enjoy, but can also make you a target for resentment from those who misinterpret your motives. The secret—with the love of your life or on the job—is to maintain your inner balance, to weigh your actions more carefully. Be practical and *consistent* in love, career, and money matters. You'll enjoy success in all three by following your Libra Jupiter guide.

♎♃♎

Capricorn will reach the top of the mountain—the goal of every Goat—but the path is less difficult with a Libra Jupiter. Earthy Capricorn moves with steady determination, choosing each step warily, while Libra Jupiter *inspires* Capricorn and tells you to lighten up, to enjoy companionship and the moment. You know that hard work, a will of bendable but unbreakable steel, and a practical approach are assets.

With Libra Jupiter, you balance these qualities with a broader outlook, a more humanitarian goal. You succeed *with* and *for* others, as well as for yourself. With Libra Jupiter, you can also have it all. Capricorn-Libra race car driver Michael Schumaker, who's made millions winning many races, is an example.

♎♃♎

New ideas are your reason for being, Aquarius, but with Libra Jupiter, detached curiosity finds a closer, warmer connection with the rest of humanity. Iconoclasts Bertolt Brecht and Sergei Eisenstein introduced totally new forms in theater and film but balanced this passion with activism and concern for humanity. Aquarians are sympathetic, helpful friends, but they're more interested in concepts—preferably their own, though anything original will do. You'll have greater luck tapping into the balance and charm of Libra Jupiter, weighing your original ideas and concepts against deeper concern for others. Embrace your Aquarian originality and offbeatness, but recognize the consequences your actions may have for others. Use your Libra Jupiter, and you'll delight the world with your novel ideas and shining ideals.

♒♎♒♎♒♎♒♎♒♎♒♎♒♎♒♎♒♎♒♎♒♎♒♎♒♎♒♎♒♎♒♎♒♎♒♎♒

Pisces imagination, grace, and wisdom, with Libra Jupiter charm, can create one of the most fascinating creatures in the zodiac— artistic, immensely appealing, and sensitive. Pisces compassion with the Libra sense of justice creates a desire to right society's wrongs, but without the Libra Jupiter *air* (intellect), watery, emotional Pisces can despair, rather than take action. Libra Jupiter helps you find the delicate balance between ideals and reality.

You *can* win an Oscar, found a much-needed clinic, express your creativity on canvas, save the whales—if you use disciplined effort to add substance to your dreams. Find your truth, work for it, and let Libra Jupiter charisma draw others to your cause and into your life. Pisces-Libra Jack Kerouac certainly did this—"on the road" and everywhere else!

Relationships with Libra Jupiters

Finally, for those of you who aren't Libra Jupiter yourselves but who have a spouse or significant other, boss, child, or parent who is, here are some great tips to help you learn to deal! This is how the luck of someone else's Jupiter touches *your* life.

Your Significant Other

No doubt, when you first met your Libra Jupiter, you were completely enchanted. The question now is, how much of them can you take? The words "Libra" and "love" are the same. They are said to have been the ones who invented romance, and it could be true. The thing is, they never want it to stop. Your challenge will be the off-days when your mate's scales have dipped low and you just need to get through day-to-day living. Your Libra Jupiter always wants it to be special and may be disappointed if you simply need to be on your own and simply just exist as yourself. They idealize you, you see. They want to give you everything, including all of themselves, and want you to do the same for them. This usually extends beyond the romance of the bedroom and into career, partying, even washing the dishes and taking out the garbage. However, don't be resistant to going into business or creative partnership with them; they are fantastically good at this and will help you succeed, possibly beyond your wildest dreams. From a Libra Jupiter mate you will gain the luck of being able to take much advice (and either listening to it and following it, or not, as you think fit).

Sexy as existence with them may be, learn how to tell them, without hurting their feelings, when you need a breath of fresh air. Remember: You are their heaven and earth, and they expect to be yours.

Your Boss

Sometimes your Libra Jupiter boss will sit silently and not speak for periods of time, but at other times, she will talk your ears off. She will want your suggestions, listen to each one, ask for more, and then discard them all and do what she said she was going to do in the first place. Sometimes it makes you feel as if you don't even have to be there, or you're just a sounding board. No offense, but the latter is often the truth. Libra Jupiter bosses know what they want, they just may not know now, or at the time they're *supposed* to know. As a result, you can sometimes get blamed for something not going right or not working out.

My dentist is a Libra Jupiter boss, and whenever I go to have my teeth cleaned, I feel as if I've been invited to a tea party or—when it's more rushed—to a full-blown social convention. Not one, but three technicians and assistants will seat me, ask me how I am, tell me how they are; the questions come flying, and even when my mouth is full of instruments, the talk never ceases. It's also hard to tell—even when you know—who's really the boss. Everyone's chattering, laughing, giving advice, and so on.

It would be surprising if you haven't been invited to your Libra boss's home yet. They gain luck by entertaining their employees at home. If not at their home, then at a restaurant or venue that has a pleasant atmosphere and tasteful décor. Social skills are important for them. Libra Jupiter Steve Case, who fell from grace as darling of AOL Time Warner, failed to sufficiently use his Libra Jupiter social skills. One colleague stated that sitting next to him for six hours on an airplane was the longest six hours he'd ever experienced. So if you want to help the company's success, encourage your boss's conversational and public relations skills. You'll have a more pleasant time on the job, too.

Your Child

Some of the most compassionate insights into the world of childhood come from the moving stories of Libra Jupiter Hans Christian Andersen. He aroused the sensibilities of children to the needs, sufferings, and joys of others. From the encouragement of potential in "The Ugly Duckling" to the destructive obsession of "The Red Shoes," you, dear parent, can learn how to help your Libra Jupiter child from his wisdom.

To give your Libra Jupiter child the best opportunity for growing into a happy, stable, responsible human being, follow the formula: Do not rush him or her. Ever. Structure—one step at a time. Libra Jupiter children have incredible daydreams, desires, and ideals. There's a pure innocence to all of them. Encourage their creativity and artistic talents, and make sure they have access to a library. Don't yell. They need peace, harmony, order, beauty, quiet. Truth seeking and kind hearted, they fear making mistakes or misjudging, hence they hesitate and need time for decisions. Show by example. Suggest a solution, gently, several times.

Your Parent

An amazing bit of astrological wisdom—or lore—here, but children of Libra Jupiter parents are said to be usually exceptionally bright! So if your parent is Libra Jupiter count yourself lucky. You were born bright and you will remain that way. You probably grew up with plenty of socializing, discussion at the dinner table, and a beautiful home. Your Libra Jupiter parent wanted you to be intelligent, interested, fair, neat, look nice, have responsible, interesting friends, and achieve good to excellent grades. They're not pushy like the Fire signs, nor despairing like a Water sign might be if you failed a grade (or relationship), but, oh my, they have high *ideals* for you, and in those they place all their hope.

The hardest part about being a child of a Libra Jupiter parent is the fear that you would disappoint them. Either that or to have to listen to their advice, help, and encouragement for hours (and hours) on end. I am sure you listened to them a lot: They love to talk. If someone did something bad to you, instead of your Libra parent instantly defending you they might have asked, well, what did *you* do? It's not a lack of support, it's simply the Libra need to see all sides of the issue and to assess impartially. From this you may gain the luck of not taking life so personally that you can't be smart and keep going forward. You may have learned early to stop and think, consider, and be as objective as possible. That can save one years of drama and unnecessary emotional ruts later on in life.

That, in the end, is what astrology is really all about. Appreciating how different we all are, and loving the differences! Nature gives an antidote for every poison. Isn't that the miracle of life? Problems appear, but always, hopefully, there's a wonderful solution that will make everyone happy, and Libra Jupiter is guaranteed to find it. And you will make sure it is most fair and beautiful. *Remember, Libra Jupiter: You are the fair and just charmer of the zodiac.*

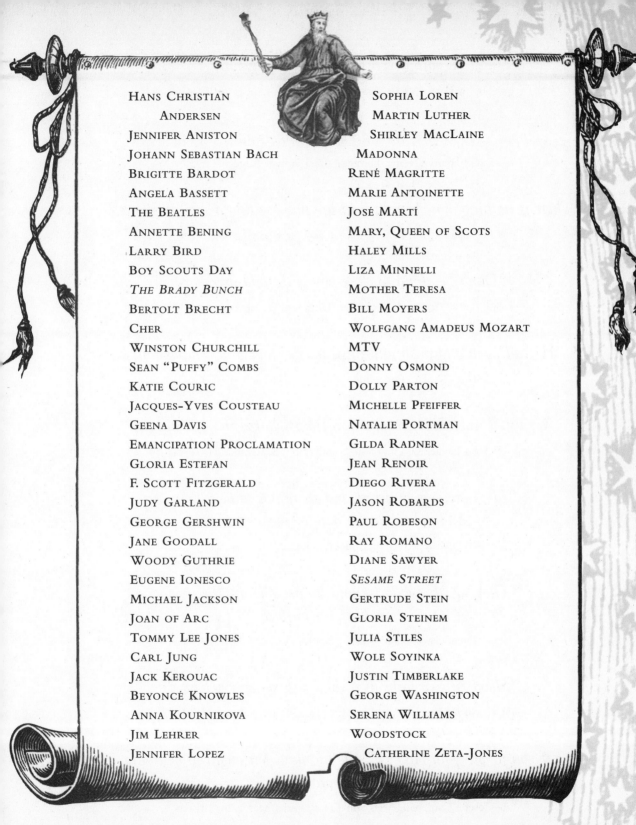

HANS CHRISTIAN
ANDERSEN
JENNIFER ANISTON
JOHANN SEBASTIAN BACH
BRIGITTE BARDOT
ANGELA BASSETT
THE BEATLES
ANNETTE BENING
LARRY BIRD
BOY SCOUTS DAY
THE BRADY BUNCH
BERTOLT BRECHT
CHER
WINSTON CHURCHILL
SEAN "PUFFY" COMBS
KATIE COURIC
JACQUES-YVES COUSTEAU
GEENA DAVIS
EMANCIPATION PROCLAMATION
GLORIA ESTEFAN
F. SCOTT FITZGERALD
JUDY GARLAND
GEORGE GERSHWIN
JANE GOODALL
WOODY GUTHRIE
EUGENE IONESCO
MICHAEL JACKSON
JOAN OF ARC
TOMMY LEE JONES
CARL JUNG
JACK KEROUAC
BEYONCÉ KNOWLES
ANNA KOURNIKOVA
JIM LEHRER
JENNIFER LOPEZ

SOPHIA LOREN
MARTIN LUTHER
SHIRLEY MACLAINE
MADONNA
RENÉ MAGRITTE
MARIE ANTOINETTE
JOSÉ MARTÍ
MARY, QUEEN OF SCOTS
HALEY MILLS
LIZA MINNELLI
MOTHER TERESA
BILL MOYERS
WOLFGANG AMADEUS MOZART
MTV
DONNY OSMOND
DOLLY PARTON
MICHELLE PFEIFFER
NATALIE PORTMAN
GILDA RADNER
JEAN RENOIR
DIEGO RIVERA
JASON ROBARDS
PAUL ROBESON
RAY ROMANO
DIANE SAWYER
SESAME STREET
GERTRUDE STEIN
GLORIA STEINEM
JULIA STILES
WOLE SOYINKA
JUSTIN TIMBERLAKE
GEORGE WASHINGTON
SERENA WILLIAMS
WOODSTOCK
CATHERINE ZETA-JONES

"Everything you see I owe to spaghetti."

—Sophia Loren, Virgo Sun-Libra Jupiter. September 20, 1934

"Happiness is so rare, the mind does somersaults to protect it."

—David Henry Hwang, Leo Sun-Libra Jupiter. August 8, 1957

"There is nothing more beautiful in life than getting a second chance."

—Ron Kovic, Cancer Sun-Libra Jupiter. July 4, 1946

"Don't be afraid to close your eyes and dream."

—Sean "Puffy" Combs, Scorpio Sun-Libra Jupiter. November 4, 1969

"We can push and push and push toward love and compassion."

—Jane Goodall, Pisces Sun-Libra Jupiter. March 4, 1934

"Genius is the ability to put into effect what is in your mind."

—F. Scott Fitzgerald, Libra Sun-Libra Jupiter. September 24, 1896

*"Ideas are a great arrow, but there has to be a bow.
And politics is the bow of idealism."*

—Bill Moyers, Gemini Sun-Libra Jupiter. June 5, 1934

*"They say that exercise and proper diet are the keys
to a longer life . . . oh well."*

—Drew Carey, Gemini Sun-Libra Jupiter. May 23, 1958

"Good girls go to heaven, bad girls go everywhere."

—Helen Gurley Brown, Aquarius Sun-Libra Jupiter. February 18, 1922

*"If you follow your bliss . . . the life that you ought to be living
is the one you are living."*

—Joseph Campbell, Aries Sun-Libra Jupiter. March 26, 1910

SCORPIO JUPITER

♏♃

Dig Within

From Marcus Aurelius, born A.D., April 26, 121, who urged, "Within is the foundation of good, and it will ever bubble up, if you will ever dig," to Susan Sarandon, born October 4, 1946, who declared, "It is not difficult to be successful. But it is difficult to remain human," these Scorpio Jupiters knew, and so should you, that your greatest power and choices lie within—and *it is all about your power and choices.*

The world is surely wide enough to walk without fear.

— SCORPIO JUPITER JEANETTE WINTERSON

The sign of the Scorpion is where we find the deep ones, the sexy ones, the psychics, the undersea explorers, the ones who walk the world without fear, digging up nasty truths. Just think of Scorpio Jupiter reporter Christiane Amanpour, cheerfully investigating the difficult truth on a daily basis. This Jupiter sign often brings a lashing of genius; it won't pose any challenge to be brilliant! The real challenge is to avoid the self-destruction and self-neglect that often accompanies genius. You must take care, also, not to seduce simply for the sake of seduction.

Your luck will come in extremes, will be born, die, and then be reborn again. In fact, as Scorpio rules death and rebirth, transformation in all its forms, you are often even "seen" after death. There are still countless sightings of Elvis, and Scorp Jupe Aleister Crowley's ghost is still said to be plaguing various houses in England. Scorpio Jupiters are also guaranteed the success of the Great Comeback. Elvis did it, Napoleon Bonaparte did it, even Scorpio Jupiter Susan Lucci finally got her award.

Scorpio Jupiter is all about breaking with the old, bringing in the new, transforming, and changing. Whatever you guys do, you seize the unconscious—whether it's as psychic Uri Geller or Georgia O'Keeffe's sexual stamens—and you're never forgotten. Transformation is key to your success. You're much stronger than you think.

Take a moment, look back long and deep over your life, recount as many events as you can, and ask yourself what has made you lucky or unlucky in your life. You'll notice that whenever you act like a Scorpio, *you create miracles in your life.* No matter what your Sun sign is, *you get lucky when you live out the Scorpio legacy gifted to you by the benevolent planet Jupiter.*

To take advantage of your luck, emphasize your . . .

Scorpio Jupiter Strengths ♏♃	*Be transformative, regenerative, in touch with the power of your unconscious, magnetic, probing and shrewdly observant, devoted to the challenge, open to change, dangerously imaginative, and dedicated.*

But because Jupiter rules excess, watch out for . . .

Scorpio Jupiter Excesses ♏♃	*You can become calculating, addicted to power, or destructive.*

Understanding the
Scorpio Energy and Influence

Scorpio is the sign of transformation, of death and rebirth. You are born as the night is waxing stronger than the day and the Earth is being pulled down into the sleep of autumn. You are born with the falling of the leaves; like a snake shedding its skin, nature is already moving on to the next chapter. As Scorpio is drawn down to a profound depth of being, you plumb your unconscious, from which you get your strengths. Scorpio is more aligned to the elemental forces, including sex and death, than any other sign. What others see of Scorpio is usually only *the tip of the iceberg*. Scorpio's power is compelling, unyielding, and undying. Still waters run deep.

Scorpio has such access to power that its natives may choose from three different ways of expressing themselves. In fact, Scorpio needs to be superaware of the implications of the choice, for each one packs a wallop of its own. The Scorpion is the most familiar symbol, the most basic, and the first step on the evolutionary ladder of psychic development. In nature, the scorpion is a survivor: Its direct ancestors extend back 200 million years before the first dinosaur appeared. It is fiercely self-protective, waiting for the right moment to strike, and its powerful sting is legendary.

The Phoenix is a mythical bird with incredible magical powers and the ability to rebirth itself through great challenge. It dies in flames and then rises from its own ashes.

The Eagle is the highest, wisest, and most evolved choice to which a Scorpio can aspire. The eagle soars high and proud, protecting its loved ones. America uses the eagle as its symbol of might; Native Americans also consider the eagle to be *their* strongest symbol—both for its power and as a healer of the spirit.

SCORPIO SYMBOLS
The Scorpion, the Phoenix, the Eagle

SCORPIO PARTS OF
BODY
The Sexual and Reproductive Organs

Scorpio is all about birth and rebirth, so it is logical that this sign rules our reproductive organs. (Without these, we couldn't be born.) You need to be especially discriminating not only in your sexual practices and partners but also by making sure you get regular and complete physical exams. Scorpio's intense sexuality and connection to the life force means that you go relentlessly toward your objective—no beating about the bush here.

Because Scorpio governs the unconscious, Scorpio Jupiters must be careful to allow their unconscious to grow through dreams. Be especially careful to get enough sleep, Scorpio Jupiter—your mind requires hours of sleep each night so that you can dream.

SCORPIO COLORS
Black and Russet

These are the colors of passion and power: strongly defined, deep, and intense, the colors also of the dying year. Black is, in Western cultures, the color associated with death and also with witchcraft or magic; originally, black was worn in mourning, or draped across doors and windows where someone had died because it was believed to ward off evil spirits or spirits of the dead. Black is not one color but *all* colors; that is, it *takes in* all hues, not reflecting but *absorbing* and *holding* them. If you look into your unconscious—that which is hidden and in the dark—all the answers (like all the colors) are there. Russet is deepened red, the color of aged wine—or blood. It is also the color of autumnal turning leaves. These are sexy, compelling, classy colors. Black accentuates your formal dignity; it is a statement of your serious intent, your controlled power and ultimate containment. Russet is grounding, the warmth of a banked fire, ancient lifeblood of wisdom.

Scorpio Jupiter Strengths ♏♃

And now your Scorpio Jupiter strengths:

Be Transformative

You become luckiest transforming yourself and others, forcing yourself and others to look at, and experience, things in a totally new way. Examples include such opposites as the Buddhist Dalai Lama—whose message of peace and tolerance is a strong influence in the world today—and Aleister Crowley, high priest of the occult. The Dalai Lama (meaning "Ocean of Wisdom") is considered by Tibetans to be the reincarnation *(rebirth)* of the Bodhisattva of Compassion, a holy figure who chooses to reincarnate to serve the people. In contrast, Crowley's message to his followers was, "Do what thou wilt shall be the whole of the law." When Crowley—who became known as a master of the dark arts and who created the Thoth Tarot deck—died, he was reviled by the English press as the wickedest man on Earth. The Dalai Lama is, today, regarded by many as the holiest.

Scorpio Jupiter transformation is revolutionary and extends to all aspects of life: the physical—Chuck Yeager breaking the sound barrier on October 14, 1947—to the sociopolitical—Jackie Robinson breaking the color barrier in major league baseball on April 10, 1947. The planet Uranus, which rules revolution, upheaval, and electricity (all that shocks), was discovered during another Scorpio Jupiter passage in 1781. Remember: You were also born during a Scorpio Jupiter passage. What must you do to fulfill this power of transformational energy in your life? It needn't be inventing the wheel; it can be creating a new art form, breaking through barriers of a rigid belief system (your own?), inventing a new marketing strategy, developing a new product, championing a controversial political or social action, or coming out on television in front of 46 million viewers as Ellen DeGeneres did in her own show, *Ellen*. Scorpio Jupiter Ellen bared her soul and shocked a nation: "I said the words I was most terrified to say throughout my whole life. 'I'm gay,' I admitted, 'and I'm afraid for people to know I am.' I started crying. Those words changed my life. Many people have told me that they came out after watching that show with their parents."

Be Regenerative

To the Sioux—and to many other American Indian Nations—the birth of a white buffalo calf is a promise of spiritual and physical resurgence and renewal. On August 20, 1994, Miracle, the sacred white female buffalo calf, was born Scorpio Jupiter on the Heider farm in Wisconsin. (The first white buffalo calf born in more than sixty years.) A sacred symbol and fulfillment of Native American prophecies and beliefs, her birth was also heralded as a sign of hope and harmony for all humanity.

Miracle isn't the only example of Scorpio Jupiter people, animals, and events who have helped the world refresh its power and regenerate.

THE RULING PLANET OF SCORPIO

Proud Pluto

Pluto is cold and dark, so distant it takes 248 years to orbit the sun. It is the only planet an Earth spacecraft has not yet explored. It affects earthquakes and volcanoes, endings and beginnings, and our unconscious. It was given its name by a young British girl; her letter was the first to arrive at the Lowell Observatory, where the planet was discovered in 1930. Before this, it was nameless and mysterious; it was called Planet X.

YOUR HISTORY
The Romans called you Pluto; the Greeks named you Hades. King of the Underworld, ruler of the dead, you lived there alone for years until you fell in love with Persephone, stole her, and took her down below. Her mother, Demeter, Goddess of Fertility, was so sad at losing her daughter that the whole Earth went into mourning. Nothing grew and

(continued)
everyone started to die. Finally, it was agreed that Persephone could live above ground for half a year and must return to the Underworld for the other half. This is why for half the year the Earth is barren (autumn and winter) and half the year it is fertile (spring and summer), and why Pluto is only happy *half* the time.

The Marshall Plan, Secretary of State George Marshall's plan to revitalize and regenerate war-torn Europe, is Scorpio Jupiter. After the plan went into effect, new, free countries emerged, Phoenix-like, from the ashes of Europe. Aslan, the lion, the great symbol of death and regeneration in the classic *Chronicles of Narnia* books, was created by Scorpio Jupiter C. S. Lewis. In the last of the *Narnia* books, Aslan dies to save the world and is reborn, transformed.

Dear Scorpio Jupiter, what do you need to renew in your life? Apply this lucky power of regeneration to a relationship that has become stale, a physical makeover, a spring cleaning, revitalization of your sex life, reestablishment of professional contacts to revive your career, renovation of your home or work space, involvement in a neighborhood recycling program, or a deep look at your aims, beliefs, and goals to redirect yourself toward success.

Be in Touch with the Power of Your Unconscious

Your power, like the 90 percent of our brain activity that is unconscious, is very deep. What others see of you only scratches the surface. Scorpio's reputation for secretiveness grants you the power to involve yourself with the hidden inner workings of the psyche—and achieve positive results. The unconscious renews you and increases your confidence. This is why it is so important, Scorpio Jupiter, for you to get a lot of sleep! The more you sleep, the more contact you have, through dreams, with your unconscious. Sex, too, is a renewal: Orgasm has been known for centuries as the *petit morte,* the "little death."

From the ability to control physical objects through the power of the unconscious, as demonstrated by Israeli Uri Geller, to Oglala Sioux shaman Black Elk—who followed his vision of being taken to the center of the Earth to acquire the wisdom of the Six Grandfathers— Scorpio Jupiters naturally access the unconscious. For those of you who think that this is a male-dominated province, don't forget Lyra, Scorpio Jupiter Phillip Pullman's child-heroine in *The Amber Spyglass,* who saved the world with her daemon, the animal manifestation of her unconscious.

Be Magnetic (Especially with Your Torso!)

Primal instincts, the enticing forbidden, and interest in the dark underbelly bring success to you, Scorpio Jupiter! Speaking of bellies, you Scorpio Jupiters use yours to great effect, such as Scorpio Jupiters

Sade, Britney Spears, and the King of Pelvis, Elvis Presley. Ever think of being a belly dancer? (The world's sexiest spy, Scorpio Jupiter Mata Hari, was one.) Although it's hardly believable now, Elvis was incredibly controversial with his gyrating torso being famously cut from *The Ed Sullivan Show*.

Scorpio is always associated with sex, and many of you are sexy, either in the traditional sense, or on the cutting edge, as in "sexy" meaning "great, new, more than cool." Scorpio Jupiter Alfred Hitchcock's films are sexy. They're also scary.

Some of your kind are known only for being scary and little else. Scorpio Jupiter Linda Blair peaked in *The Exorcist* (she was nominated for an Oscar at age twelve), and Glenn Close, though known for her long list of brilliant film portrayals, is particularly renowned for her role as the maniacal thwarted woman in *Fatal Attraction*. The same goes for Scorpio Jupiters Boris Karloff, Vincent Price, and Sharon Stone as temptress-murderer in *Basic Instinct*, and Miranda Richardson's genius portrayal of real-life prostitute Ruth Ellis, the last woman to be executed in England, in *Dance with a Stranger*. Think of these other Scorpio Jupiter dangerously sexy talents: Gary Oldman, River Phoenix, Anna Paquin, and Patti Smith. Scorpio Jupiter Uma Thurman is at her best when she, too, walks on the wild—and darker—side. (Both the men she married are Scorpio Jupiters as well: Gary Oldman and Ethan Hawke!)

So compelling are you, Scorpio Jupiter, that you can let your magnetism speak for yourself. Just look at the great master of mime, Scorpio Jupiter Marcel Marceau.

Be Probing and Shrewdly Observant

Born shrewd, Scorpio Jupiter grants you great powers of observation. You can use them to gather information that will further your ambition. Your observational skills, together with a sense of brevity and containment, make being a reporter—the ultimate observer—your forte, rather than a celebration of the self. You benefit most from your brilliant, analytical mind, from your pride but not your ego, and from your subtle persuasion rather than aggressive control. Be a worker, not a personality.

Scorpio Jupiter Ernest Hemingway was praised for his terse, incisive writing (indicative of his Scorpio Jupiter) but ultimately rejected for his macho posturing (a product of his Leo Sun). Novelist Kurt Vonnegut, reporter Christiane Amanpour ("My early experience at CNN taught me to have absolute clarity of vision"), social critic Jacob Holdt, scientist-author Carl Sagan, filmmaker Paul Thomas Anderson, playwrights David

* Explore the deep end in every area of your life.

* Transmute your desire into your work, and constantly challenge yourself to bring out your best.

* Take risks when the time is right and wait out the dry spells.

* Shrewdly observe, gather information, analyze, and deduct.

* Tap into powerful transformational energy and constantly rebirth yourself.

* Surround yourself with the colors of russet and black—clothing, decor, linens.

* Be a gourmet as well as a gourmand—use leeks, ginger, allspice, chili, mint, and red peppers.

* Hang out with more Cancers, Pisces, and Scorpios.

Hare and Dennis Potter, talk-show host David Letterman—all have taken a keen observation of the universe and a shrewd investigative ability and changed the way we look at the world around us.

Be Devoted to the Challenge

Challenge brings out the best in you, Scorpio Jupiter. Fellow native Napoleon Bonaparte—a brilliant military tactician with unswerving ambition—was Corsican born rather than French, started his military career as a lowly corporal with no aristocratic preferment (considered necessary at that time), rose to commander of the French army, waged tireless campaigns against the great European powers of the eighteenth and nineteenth centuries, and eventually crowned himself Emperor of France. However, his defeats are as famous as his victories. Twice exiled, he returned from the first exile to resume military command only to be defeated at Waterloo; and only death, possibly by poison, ended his relentless march to fate. His return from exile mirrors the challenge faced by fellow Scorpio Jupiter Salman Rushdie, author of *The Satanic Verses*. Despite *fatwa,* the Islamic death threat that killed his publisher and his assistant, Rushdie emerged from hiding with new novels and, eventually, became highly visible in public again.

Let's assume that most of you Scorpio Jupiters don't plan to conquer the world *literally*. How will you conquer your own world? By seeking out, rather than shying away from, the challenges presented to you: a job that involves learning new skills and exploring new fields, tackling the ad account—or client—that no one else wants to touch, resculpting your body through exercise, investigating and exposing a hidden business or political agenda, or seeking environmentally sound methods in modern agriculture. Think of fellow Scorpio Jupiter George Washington Carver, born of slave parents but able to earn a master's degree in agriculture at Iowa State University. Carver went on to revitalize Southern agriculture by developing new uses for soybeans, sweet potatoes, peanuts, and other crops. Nothing could be a clearer statement of Scorpio Jupiter's mission than Carver's words: "When you do the common things in life in an uncommon way, you will command the attention of the world."

Be Open to Change

Extreme change is often disastrous for people of other Jupiter placements, but it brings you tremendous luck, Scorpio Jupiter! Allow yourself to accept and fly with total changes—in job, home, finances.

Even if all is going well, have the courage to risk change; it always works out well for Scorpio Jupiter. Consider the actor with a successful Hollywood career, who switched to politics, changed political parties, and accepted a new position as president of the United States—Scorpio Jupiter Ronald Reagan. Or ponder the member of the Black Panther Party and writer of two books—*Soul on Fire* and *Soul on Ice* (written while the author was in jail)—who became a born-again Republican, Scorpio Jupiter Eldridge Cleaver. Actress Sally Field used her Scorpio Jupiter to escape her saccharine TV role in *The Flying Nun* and her trivial image as a beach bunny, playing the gritty union activist in *Norma Rae*, which won her an Oscar and launched her highly respected acting-producing career. Whether you make the choices, as these people did, or whether fate dumps it in your lap, you come out on top by going with, rather than resisting, change.

Be Dangerously Imaginative

You have more success by expressing the unseen, the side of life that most people ignore. Look at the grotesquely terrifying fifteenth-century paintings of Scorpio Jupiter Hieronymous Bosch; Diane Arbus's grotesque twentieth-century photographic images, which made visible those who were, at that time, "invisible"—dwarves, transvestites, nudists; Georgia O'Keeffe's erotic images of flowers, suggesting genitalia; Charles Schultz's humorous cartoons of children as they had not been seen before: insecure misfit Charlie Brown, bossy, five-cent psychiatrist Lucy, filthy Pig-Pen. Read the surrealist modern prose of Jorge Luis Borges; the horror thrillers of Stephen King; and the strange fantasies of novelist Mervyn Peake.

As Persephone was brought back from Hades, the world of the dead presided over by Pluto, Scorpio's ruler, so were the fifty thousand war dead symbolically brought back to the living in the captivating Vietnam Veterans Memorial in Washington, D.C. Out of fourteen hundred designs submitted for the memorial, the one chosen was that of a Scorpio Jupiter, twenty-one-year-old architecture student Maya Lin. Her design has been credited with healing wounds and transforming the country.

Take the unexplored path, Scorpio Jupiter. Whether you're a website designer, an insurance salesperson, or a stockbroker, don't be afraid to take on the riskier ventures, even those with ostensibly dark undertones. As Diane Arbus said, "My favorite thing is to go where I've never been." Don't follow the crowd. Scorpio Jupiters Lorne Michaels, creator and producer of *Saturday Night Live*, and Nicholas

Ray, director of *Rebel Without a Cause*, challenged public taste and succeeded wildly. As Ray said, "There is no formula for success. But there is a formula for failure, and that is to try to please everybody."

Be Dedicated

Scorpio Jupiter Dr. Albert Schweitzer, philosopher, humanitarian, and medical missionary, established first a hospital, then—with his thirty-three-thousand-dollar Nobel Prize award—a leper colony, in French equatorial Africa. He based his personal philosophy on a "reverence for life" to the extent that he refused to kill even a fly or a mosquito. His statement, "We see no power in a drop of water but let it get into a crack in the rock and be turned to ice and it splits the rock. The power of idealism is incalculable," is typical of Scorpio's slow-burning, undying passion.

Your own passion need not be an immediate fireball; it can start small—a "drop of water"—such as volunteering at a home for the aged or at a day-care center, becoming a teacher's aid, selling a few collectibles on eBay. It can grow, over time, to a sustaining and satisfying hobby or vocation. Be proud of your dedication, whether it be regularly donating to the blood bank, collecting oral histories of your area, quilting, or training for your annual scuba-diving holiday.

What Makes You Lucky in Love

Your fellow Scorpio Jupiter, poet Theophile Gautier, had this to say about love: "To love is to admire with the heart; to admire is to love with the mind."

Even if your Sun sign encourages you to indulge momentary passions, you will succeed more in love if you follow your shrewd instincts and deep intuitions. To be successful, don't play it cool, don't be coy or reticent or play games; be direct, and be the one to pursue. Your first impression is always right about someone. No matter how much your reasoning mind will try to dissuade you later, reconnect with that palpable electric sensation you felt on first sight. Be aware that you will experience urges to distance yourself from, and negate, the bond; this is only Scorpio's need for self-protection. (Do not project this onto your mate!) Look for deep, inner connections and ignore superficial differences. You are best with partners who like to compromise. As a Scorpio Jupiter, you also get lucky listening to your mate and responding with your valuable insights and advice. Remember how good you are at this.

Your Best Career Moves

Success for you, Scorpio Jupiter, comes when you follow your profound intuition. Trust your immediate take on future trends, creative ideas, unrecognized talent, and new markets in business. You become successful by seeing what others miss, either pulling the best-seller out of the slush pile, recognizing the next diet trend, or creating the right product or business to fill a need in the market. Just look at how lucky following their Scorpio Jupiter made Conrad Hilton, Steve Forbes, and Marshal McLuhan. McLuhan, famous for "The medium is the message," predicted, in the 1960s, that electronic media would link the world in a global village.

You could be the next wizard financial forecaster, fashion designer, advertising guru, or sports talent scout. Or—talk about seeing things others miss—the next champion whistleblower, like Scorpio Jupiter Sherron Watkins, who exposed the more-than-creative bookkeeping at Enron. (Whether you save your boss extra expenses—or jail time!—you can be a valuable asset in the office.)

You create your luck by concentrating on life's deepest issues and questions. Find roles within your company that allow you to explore the big picture and think about the meaning of the company, and how it can grow to match the future. Trust your gut feeling and follow your shrewd hunches. You are often very psychic and need to listen to what your ESP tells you. You are also an expert at transformation, and your Scorpio Jupiter means that you will find success leading organizations that are in the transformation process. You can revivify seemingly dead situations, products, and companies. It also means that moving and changing jobs and positions are often quite advantageous for you.

How to Reach Optimum Health

Because the part of the anatomy Scorpio rules—the reproductive organs—are symbolic of life-giving force and renewal, your energy is usually constant as well as constantly renewing itself. A tendency to brood, however, and finding it hard to relax and rest completely often leaves you feeling exhausted. It is imperative to not only eat healthily but *regularly,* as you can skip meals and substitute feel-good alternatives (alcohol, drugs, sweets), which will make you feel more tired than ever as well as bring you down into depression. Eat many small meals throughout the day rather than one or two large ones—keep your evening meal light—and drink bottled, rather than tap, water. Keep to high protein with fresh fruit and vegetables and whole-grain breads. Above all, you need a peaceful environment in which to play and to rest.

Your Lucky Foods, Element, and Spices

Scorpio's cell salt is calcium sulphate. Deficiency of this crucial mineral brings on skin eruptions, infertility, and colds and viruses that linger forever. Foods rich in calcium sulphate are tomatoes, onions, asparagus, figs, radishes, black cherries, prunes, kale, parsnips, cauliflower, watercress, radishes, and coconuts. Make calcium a priority in your diet with plenty of milk, cheese, yogurt, and cottage cheese, or calcium alternatives and supplements. Make sure you eat five of the following foods every day: leeks, lentils, artichokes, green salads, fish and seafood, escarole, beets, romaine, brussels sprouts, almonds, walnuts, wheat germ, citrus fruits, apples, berries, and bananas. Leeks protect the reproductive system even when dried, hung up in your home, or used as a talisman! Extraordinarily lucky for you also are ginger, allspice, mint, horseradish, chile, and fresh and dried red peppers.

Your Best Environment

The best environment for you is on or near the water, preferably the ocean. It must be a warm, even a hot, climate; you are happiest as a tropical creature. A Scorpio Jupiter friend of mine lived, for a time, in a houseboat and claimed that the repetitive lapping of the waves and the boat's rocking simulated the protection of the womb and healed all her ills. It is always good for you to have fresh flowers—preferably chrysanthemum and rhododendron. Countries ruled by Scorpio include Norway, Tahiti, and Morocco; Scorpio-ruled cities include New Orleans; Washington, D.C.; Liverpool; and Newcastle. You will increase your luck by visiting or spending time in them.

Your Best Look

Form-fitting is best—no frills. Your strong physical presence is enhanced by baring the neck and collarbone. Dark, dramatic, and streamlined is compelling and magnetic on you and gives you confidence and that sexy aura—clothes that show your body and reveal your sinuous moves. You can play the femme fatale or Casanova like no one else. Black is, of course, a winner. You can add russet and other deep, red-brown tones for richness; avoid pastels; hats are a lucky accessory. Jewelry should be striking rather than ornate; stick with silver. For greatest impact, cultivate the mysterious, deliberate grace of a panther. Or for an alternative, you get lucky with the 1930s detective look created by the film characters of Scorpio Jupiter writer Raymond Chandler.

Your Best Times

The luckiest times for you are when the Moon is in Scorpio, which happens every month for two to three days. Find these special days in your astrological calendar. During these days, schedule crucial meetings, tell that special person you love them, ask for a promotion or raise, push the envelope, and plan to shine. In your everyday, the best time for you to exercise is in the evening. The luckiest times during the year for you are October 23–November 21 (when the Sun is in Scorpio); February 20–March 20 (when the Sun is in Pisces); and June 21–July 22 (when the Sun is in Cancer).

The very luckiest time of your life—as in Lotto lucky—is when Jupiter reenters the sign it was in when you were born (Scorpio), which happens every twelve years and lasts anywhere from a few months to just over a year. This is called your Jupiter Return, and it launches your luckiest streak. You can count on the *extraordinary* to happen the very day your Jupiter Return begins. Amazing events can start happening up to six weeks beforehand, so plan ahead!

How to Be Lucky Financially

Scorpio Jupiter's sense of money-making opportunities is phenomenal. You can sniff out profit potential in seconds flat. You instinctively gravitate toward powerful, rich, well-known people; often, unknowingly, you find yourself conversing with Mr.—or Ms.—Big, someone who has access to the gold. Finance is an area where you really can trust your instincts. You have natural brilliant timing; your advantage is being able to wait until exactly the right time to make your move. You have an uncanny knack for spotting investments that will pay off down the road. In the stock market and in real estate, look for distressed or struggling companies that are in need of transformation. Some will be losers but others, you will find, will reinvent themselves and bring you wealth as they rise again. In today's stock markets, you might want to carefully invest in industries undergoing transformation, or companies that are transforming others. If you look for a house or apartment that seems rundown but could be converted into something marvelous, you will reap returns on your investment that seem unbelievable now.

You will find success when you enter a career that allows you to explore deep regions of yourself and to express yourself: you can be an artist, an entrepreneur, or a deep thinker. Passionate emotional expression or an emotional tension in your creative work brings good luck and money. More successful writers boast a Scorpio Jupiter than any other Jupiter placement, from Charlotte Brontë, Thomas Hardy, and Vladimir Nabokov to Ernest Hemingway, Norman Mailer, Joan Didion, and Dave Eggers.

Success in Family and Friendship

You have the ability to see the core of truth or value in a person, thing, or situation, which most other people have completely given up on, and therefore, you can help people heal or reform themselves, even if the condition seems desperate. Friends and family count on you to be the level head in a crisis and to come through when needed, but don't sacrifice your own deep needs, or you'll have nothing to give to anyone else. For success, don't waste your time with superficial acquaintances or general social requirements. For you, every relationship must have a purpose, hopefully connected to what is deep inside you. From both family and friends, you expect confidences to be kept, as you keep theirs. Intensely connected but fiercely independent, until you live with the one you love, it is better to live alone and without roommates. Even when you have a family of your own, make sure you have a space to go to in order to be alone. Professional and financial opportunities often come through friends; make sure you pick their brains once in a while. For success, work with friends, as you need to work with people you can trust—but work with them as equals, not as employees, or you'll lose the friendship.

Scorpio Jupiter Excesses ♏♃	*When you go too far!*

The perfect cautionary tale for Scorpio Jupiter is *An American Tragedy*, by Theodore Dreiser, which was made into the film *A Place in the Sun*, starring Montgomery Clift, Elizabeth Taylor, and Shelley Winters. It is the story of a young man from a poor background who is motivated by his ambition for social status, wealth, and success. He takes a job in his rich uncle's factory and, using his family connections, determinedly starts his advance toward a position of importance. He becomes involved with a female factory worker, and at the same time, begins pursuing his beautiful, wealthy cousin. When his worker-girlfriend becomes pregnant, his pride and fear cause him to see this as the death of his dreams of advancement, and he murders her.

Be careful that you do not allow the pleasure of power to warp your fine, analytical Scorpio Jupiter mind into becoming a calculating tool. This could cause you to remove yourself from the rest of humanity, believing you have the right to judge others as less worthy, and therefore losing your own humanness. Warning: Do not become a machine.

Excess #1 — You Can Become Calculating

Your superior powers of analysis can naturally lead you to be calculating, not only in situations where this is called for but also where compassion and vulnerability are essential. You tend to be the recipient of other people's confidences while not revealing your own. While this can make you a great success as a spy—remember Scorpio Jupiter Mata Hari!—this will ultimately leave you alone and lonely. Though this may give you an assurance of invincibility, safely plotting every move, the truth is you are not loving yourself. You feel unworthy to have a relationship without all the extra machinations and work on your part.

Excess #2 — You Can Become Addicted to Power

Power is not only your aphrodisiac, it is also a familiar form of self-protection for you. With Jupiter's excess of Scorpio comes an excess of sex. Be careful not to confuse sex and power. For Scorpio Jupiter women, their sexuality has such a strong effect on people that it becomes a form of power and can be misused. For Scorpio Jupiter men, it translates more into direct control and physical force. The most extreme example is Charles Manson. With total power over his cult harem, he could—and did—direct them to carry out anything he wanted, including murder. He achieved this power through the primary areas ruled by Scorpio sex, imposition of his extreme belief system, challenge of survival, and isolation outside of society's rules and protections. It is no surprise that Manson, born November 12, 1934, has not only his Jupiter in Scorpio but also his Sun, making him a double Scorpio.

DON'TS

* Give away too much of yourself or chatter.

* Let fear have the upper hand.

* Encourage suspicion or forget the edict "Innocent until proven guilty."

* Allow your power and mental control to be taken over or defined by another.

* Ignore your first impressions, intuitions, and gut feelings.

* Live in a cold climate, forgo massage, or miss sleep.

* Trivialize your sexuality.

Excess #3 You Can Become Destructive

Born on the same day in 1970, and also a double Scorpio, Tonya Harding acted out her destructive impulses by having her ice-skating competitor's knee assaulted in a brutal attack. Here the sense of invincibility leads to complete abdication of human responsibility and accountability to morality and law. Other examples of this are Scorpio Jupiters O. J. Simpson and Ted Bundy. Sexual manipulation was also a motivation in the actions of all three.

Fortunately, very few Scorpio Jupiter destructive tendencies reach these extremes, but the challenge is to not let detachment from humanity and an overblown sense of power lead to even small acts of mindless destruction. This is certainly the saint-or-sinner Jupiter placement and must be respected as such.

Sun Sign-Jupiter Sign Combinations

Scorpio Jupiter adds intensity and tenacity to Aries' direct, active force. It gives the Ram staying power and intuition, continually refueling Aries fire from the depths of the subconscious. Your natural tendency is toward passionate involvement and quick action, but it is luckier for you to stand back, assess, and strike when the time is right. Using your genius intuition, not your Aries impatience (and sometimes temper), brings you greater success. Scorpio Jupiter reminds you to act from the essence of *yourself.* Connect to your inner self, and as Shakespeare said, "Thou canst not then be false to any man."

With Scorpio Jupiter insight, you avoid the do-they-*like*-me? doubts and disappointments of putting energy into unworthy pursuits—or people. Aries-Scorpio David Letterman is watched by millions every night; his Scorpio is in total control, as producer, and allows him this satisfaction. Scorpio Jupiter helps you meet these challenges with the power of continuous self-renewal, infinite patience, and the unwavering purpose necessary to attain final victory.

No one needs to tell a Taurus to be persistent. (Ever tried to turn aside a bull?) But the Bull can be fooled into lowering his head and charging the matador's cape. Use the deep wisdom, shrewd judgment, and infinite patience of Scorpio Jupiter to size up the situation—your next job offer, an investment plan, or your attractive new neighbor—to see if they're right for *you.* Not just the surface you, who wants the comforts of the good life and earthy, sensual pleasures—Scorpio Jupiter can supply these, particularly the sensual—but it tells you to go deeper. Sexuality becomes Eros, the primal force, a physical, mental, emotional blend of your total self. Choose your significant other for a lifetime.

As any farmer knows, in life and business you'll have good years and bad, drought and plenty. Plants may die, but those with the deepest roots survive. With Scorpio Jupiter, you accept change as renewal and growth. Taurus-Scorpio Andre Agassi, champion tennis player, wins using these strengths—a wizard at returns and rebounds, too!

Gemini
(May 21–June 20)

Nothing could be further apart than the airy skies of Gemini and the subterranean river of Scorpio Jupiter. But as a Twin, you're used to dualities, and a combination of opposites is often beneficial. Scorpio Jupiter urges you to connect your quicksilver mind to your emotions. You often take refuge in cleverness and wit, but with Scorpio Jupiter, you use your bright intellect to examine and express the dark heart of the matter—perhaps as a writer, psychologist, or actor.

Scorpio Jupiter gives you the power to transform yourself and others through creativity or political action. Geminis are natural politicians; with Scorpio Jupiter, your successful campaigns deal with deep issues—perhaps those others are afraid to touch. Gemini-Scorpio Henry Kissinger, constantly juggling diplomatic timebombs, is a prime example. Scorpio Jupiter urges you to trust your gut reactions (feelings, psychic intuition) in choosing career, locale, romantic involvements. Geminis are life's eternal tourists, moving on to the next interest, the next relationship. With Scorpio Jupiter, you acknowledge hidden needs and desires. You stop floating in shallows and find courage to dive deep into the life-renewing primal sea.

♏♃

Cancer
(June 21–July 22)

Emotions are emphasized by the association of these two Water signs but expressed in very different ways. While the Cancer Crab is at home in tide pools, finding security under rocks, Scorpio Jupiter calls from the unknown depths. Cancer moods are affected by the changeable Moon, by reaction to the tides. Scorpio Jupiter tells you to *connect* with the tide, the deep, powerful currents. See the difference? Cancer is moved—and vulnerable; Scorpio Jupiter urges you to forget fear, be one with the movement, and reminds you that renewal is constant. Are you still hanging on, with crablike tenacity, to the thrift-store chair you scrounged for your first apartment, your collection of teddy bears—or the not-too-exciting guy you've dated for four years? You're sabotaging your Scorpio Jupiter luck. Don't throw out your old loves; transform them. Cancer-Scorpio Danny Glover, as a fine actor, is able to grow and shed his character shells at will.

Brilliant Sun on the surface, dark depths below—Scorpio Jupiter gives the Lion the passion and the deep *need* for self-expression that makes this natural showman more than an entertainer. You have a message for the world and are determined it will be heard. Be careful that your need isn't so strong that you *impose* yourself on the world, or on friends and lovers. With Scorpio Jupiter, you have purpose in achieving power and glory, a deep connection to basic human desires (Scorpio Jupiter sexuality) and the tides of history, the moving forces of your time. Whether in the media, the political arena, or on a personal level, your influence will be felt, especially if you judge people and events shrewdly. You'll experience extreme ups and downs of popularity but will never lose your intense motivation, psychic intuition, and powers of personal and public rebirth.

Leo-Scorpio Magic Johnson is not only one of the greatest basketball players—and court showmen—of all time, but also uses his fame to aid charitable causes.

᛭♏︎

Virgo's intellect, precision, and critical abilities are given force and depth by Jupiter Scorpio. You see the writing on the wall and discern what you feel needs to be done.

Your insistence on virtually making a map for change and improvement will at times anger people, but steely dedication and your steady, sure way of handling whatever you do makes you appreciated by coworkers and friends, even if they find your capability daunting. For the sake of relationships with your mate, coworkers, or the public, you need to avoid withdrawing into your secret world. You have too much to share with the rest of us! Virgo-Scorpio Leo Tolstoy gave his genius to the world of great literature. In keeping with the luck of complete ends and beginnings that Scorpio Jupiter brings, think on this: Tolstoy left his only copy of *War and Peace* on a train! He never recovered it and he had to write the entire book *from scratch*. But you know what? He said the one he had to do all over again was *much better*! Thus, your greatest masterpieces are—literally—*reborn*.

IF YOUR SUN SIGN IS:

Libra

(September 23–October 22)

Sweet, soothing Libra charm is given an infusion of erotic intensity by Scorpio Jupiter and becomes more than just a pretty face. You are more determinedly—even ruthlessly—ambitious, and more obsessed with sex and sexual conquests.

The talent of Luciano Pavarotti, arguably the greatest operatic tenor of today, is magnified ninefold by his Scorpio Jupiter charisma, magnetism, passion, and sexuality. He did not hesitate to break with his past, against public opinion, and it brought him happiness. Use the power of your Jove in your decision-making process; always trust your first instinct and make a firm, final decision based on that. Act resolutely, not caring if you're pleasing anyone, and don't be afraid to stand alone. It actually will be a great relief to be able to dispense with the superficial niceties and the sometimes codependent diplomacy of your Sun sign. Finally, remember that your luck lies more in silence and dignity than in debate.

⚖♏♃♏⚖♏♃♏⚖♏♃♏⚖♏♃♏⚖♏♃♏⚖♏♃♏⚖♏♃♏⚖♏♃♏⚖♏♃♏⚖♏♃♏⚖♏♃♏⚖♏♃♏⚖♏♃♏⚖♏♃♏⚖♏♃♏⚖♏♃♏⚖♏♃♏⚖♏

IF YOUR SUN SIGN IS:

Scorpio

(this makes you double)

(October 23–November 21)

"Double, double, toil and trouble"—were Macbeth's witches double Scorpios? Could be; they appear from out of the dark, stirring a bubbling cauldron of spells and ambitions, and offering tempting but dangerous prophecies. Are they manifestations from the Underworld or from the mind? Your double Scorpio Jupiter, like these crones, offers you immense power, intense personal visions—often with a skewed and totally original viewpoint, as in the writings of double Scorpio Kurt Vonnegut—and loads you with the responsibility of making radical choices. You, more than anyone, are in touch with the primal in your nature and in humankind as a whole, and your eroticism is inextricably mixed with this primitive power, giving you a magnetism for individuals and the public.

To find your luck with this double dose, your challenge is to not lose control; the danger here is not in giving it away to others, but in abdicating your responsibility to yourself and to society.

Sadge rides again! but in a more purposeful, serious manner, rather than the usual happy-go-lucky gallop to wherever. Scorpio Jupiter may be confusing to the Horseman: You have conflicting urges, both emotional—Fire and Water. (You see the conflict in *that*.) Do you follow your deeper, life-is-serious Scorpio feeling and intuition? But it's much more fun to play—and travel, and. . . . If only you could shrug off the feeling that there's something *more*. Scorpio Jupiter gives you incisive insights that can be expressed with brilliant wit, the forte of Sagittarius-Scorpio playwright–songwriter Noël Coward. ("Only mad dogs and Englishmen go out in the noonday sun.") You are vulnerable to erratic changes of mood and direction, both day to day and over the years. There's more than a hint of strong Scorpio sexuality, but what you let the rest of us see is often only the naughty flirt, while you hide and protect your more deeply erotic nature.

Persistence is no news to the Capricorn, and with Scorpio Jupiter, you have the most controlled and enduring power in the zodiac. This is great for your family, friends, and company, but can be hard to live with, especially if you're *you*. Your combination is made up of the two most solitary signs, so your tendency toward self-sufficiency is greatly exaggerated. And dear Capricorn Scorpio, it is good—in fact, *essential*—for you to lighten up among friends. Your humor is subtle, wicked, and often goes over people's heads. Thus you are discriminating in your choice of companions. With such a reserved Sun sign, for luck, add Scorpio sex and passion. You benefit from more actual immersion in water, whether it be long, sensuous baths or daily swimming.

You will, in your lifetime, build an empire, whether it be like Capricorn-Scorpio hotel magnate Conrad Hilton, or by building a strong family or a network of Internet correspondents. Your Capricorn would keep you building forever; let your Scorpio renew your spirit and transform your task to joyous passion.

IF YOUR SUN SIGN IS:

Aquarius

(January 20–February 19)

Scorpio Jupiter gives direction to the explorations of the iconoclastic Aquarius mind. To access Jupiter's luck, open the floodgates and let Scorpio passion and elemental energy flow through. This will float your boat further and faster, and will bring a richer variety of people and experiences—and shoot-the-rapids adventures—into your life. You may feel torn at times, because Scorpio and Aquarius radically oppose each other in areas such as commitment and achievement. Don't follow your natural impulse to run away. Last but hardly least, sex: Make more of it. Jules Verne, author of *Twenty Thousand Leagues Under the Sea* and many other works combining revolutionary scientific ideas with visions of the future, is the epitome of the Aquarius-Scorpio.

IF YOUR SUN SIGN IS:

Pisces

(February 20–March 20)

This is the most magnetic of all Scorpio Jupiter combinations. From the enchanting Queen Latifah, to sex-siren Jean Harlow, Pisces-Scorpios rule the world with their charisma and red-hot talent. Double Water not only enhances your sensuality but also makes you extremely psychic. For greatest success, do not indulge in your receptive sensitivity, but profit from it, as Scorpio would. Trust and stick to your first instincts and be careful of letting your Pisces empathy have full sway. The luckiest route for you is not the soft, vulnerable Pisces side, but the determined, shrewd, self-protective Scorpio side. This is a very artistic combination, and you might be throwing away potential for profit if you do not nurture your creativity. You will have times of severe disappointment and recoveries, close and binding relationships, and the ability to spot brilliance far ahead of its time. You make the best friend in the world.

Relationships with Scorpio Jupiters

Finally, for those of you who aren't Scorpio Jupiter yourselves, but have a Scorpio Jupiter spouse or significant other, boss, child, or parent, here are some tips to help you learn to deal! This is how the luck of someone else's Jupiter touches *your* life.

Your Significant Other

In love, Scorpio Jupiters want to be silent—silent and deep. It's very powerful, down there in the primal sea, too powerful for words, so let your unconscious do the communicating. If you're with a Scorpio Jupiter, sex will be very important in your relationship, more than with any other Jupiter sign. Day to day may not always be smooth, but is it worth it? Yeah.

As Scorpio Jupiter Britney Spears said, "When you're comfortable with someone you love, the silence is the best." In matters of the heart, Scorpio Jupiters like to be silent with their mates. They're more comfortable talking to everyone else about their love life, but not to their mate. So much communication with your Scorpio Jupiter lover will be between the sheets, and psychically as well. You will dream of each other. It's a very intense connection. Make sure you're up for it before you dive in. Take long breaks at the beginning of the relationship. It's almost like you're entering another frequency when you start to seriously date a Scorpio Jupiter. It's like going into space—and all the training that must precede that. Scorpio Jupiter Neil Young was speaking every Scorpio Jupiter's secret fear when he sang, "Only love can break your heart." Once they get past that fear and decide to take the plunge, be aware that they are probably feeling what Scorpio Jupiter Rainer Maria Rilke wrote, "Love is only the beginning of a terror we can barely endure." They believe it, too. They feel it. It's terrifying. They are so terrified of letting go of control. So if a Scorpio Jupiter is hanging around you, they like you, they really, really like you! They wouldn't be wasting their time—and precious resources—otherwise.

Your Boss

Never try to fool your Scorpio Jupiter boss; he or she will see straight through whatever excuses or pretenses you offer. Scorpio Jupiter bosses not only have eyes in the back of their head but are psychic to boot. They can give their workers, assistants, and colleagues a hard time. They expect you to take your job seriously and to be devoted to it and to the success of the business.

When you've passed the tests, you will be part of the inner circle of success (celebrities may be an extra perk) and reap the rewards, but it will be less in ego gratification and more for the sake of the company or cause. Your Scorpio Jupiter boss is tough, some might even say a slave driver, not for the sake of wielding power but because they give every drop of their blood and expect you to as well. A word of warning for the fainthearted: They do best with challenges, so there may be a crisis a day, even simulated for the sake of drawing out the best work from everyone. But you will gain confidence in you own abilities, a strong sense of your own boundaries, and the ability to be professional rather than take things personally. You will be fortunate to observe and learn from a mind whose first impressions are astute and whose ability to forecast future trends will guarantee success. If you can hang in there, the rewards are immeasurable: You'll be proud of what you've helped to create.

Your Child

Like the choice of symbols for Scorpio, this child has the greatest number of life choices of any other Jupiter placement. A Scorpio Jupiter can, with good discipline, take the high road of loyalty, integrity, and achievement; or with neglect, take the low road of self-destruction. They're born old, so this direction begins at an early age. Of particular danger to the Scorpio Jupiter child, as he or she grows older, is the matter of experimentation and drugs. It is their natural desire to experience all deeply and often at expense of health and safety. They are intensely loyal, loving, and devoted, fiercely protective of their siblings (even if they fight with them at home). They need their privacy to be respected; their room will probably have a huge Do Not Disturb sign on the door, and inside, will probably be surprisingly bare. Don't be concerned if they hide things away; that's just their nature—it gives them a sense of control of their lives. In school, they're utterly brilliant, but again, it comes down to discipline and choice: They almost always will be either the best or the worst in their class.

Your Parent

With a Scorpio Jupiter parent, you will be fiercely loved and cared for and very protected. You'll get all the essentials and what they deem to be important and crucial for your growth, but little in terms of frills or extravagances. They'll be honest and straight with you, and you'll know exactly what they expect at all times. They will always ask the best from you, push you toward achievement, and be proud of you. You won't often receive praise, but it will heartfelt when it comes. There will be strict codes of behavior, stern discipline; they require respect. Unlike kids with parents in other signs, you won't be able to get away with things. They can be extremely generous but on their own terms, and if they think it's good for you. Their code for you is extremely moral and practical, and if you disappoint hem, they can hold on to it for years and not let you forget it. As part of them, what you do reflects on them and their pride. Your childhood may not be as much fun as some, but while your friends may feel, years later, let down by their parents, you have the supreme fortune of never feeling that way.

From your Scorpio Jupiter, you gain the strength, fortitude, and confidence to handle your life, which in the end means a world more than fun and games. As demanding as it may be, you'll never doubt the depth of their love.

This is what astrology is all about, and, I'd venture, a civilized life. Appreciating how different we all are, and loving these differences! Nature gives an antidote for every poison. Isn't that the miracle of life? Problems appear, but there's always a solution, and if anyone can locate and resource it, Scorpio Jupiter can. *Remember, Scorpio Jupiter: You are the magnetic transformer of the zodiac.*

Madalyn's Movie Choice for Scorpio Jupiter

Raging Bull
This gritty black-and-white film about boxing champion Jake LaMotta's brutal extremes, which cost him his wife, brother, and career, gives us a ringside view of Scorpio Jupiter's desire to win, pain, and physical challenge. Martin Scorsese directs this tale of fury and loss and its Scorpionic choices and consequences with hard-edged compassion.

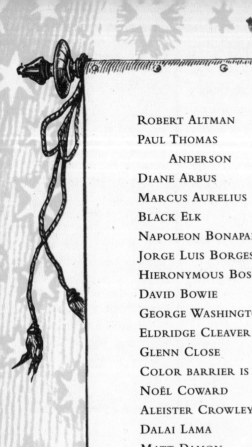

ROBERT ALTMAN

PAUL THOMAS
  ANDERSON

DIANE ARBUS

MARCUS AURELIUS

BLACK ELK

NAPOLEON BONAPARTE

JORGE LUIS BORGES

HIERONYMOUS BOSCH

DAVID BOWIE

GEORGE WASHINGTON CARVER

ELDRIDGE CLEAVER

GLENN CLOSE

COLOR BARRIER IS BROKEN

NOËL COWARD

ALEISTER CROWLEY

DALAI LAMA

MATT DAMON

DOROTHY DANDRIDGE

ELLEN DeGENERES

DAME JUDY DENCH

JOAN DIDION

DAVE EGGERS

SALLY FIELD

LUCIAN FREUD

URI GELLER

DANNY GLOVER

DAVID HARE

JEAN HARLOW

ERNEST HEMINGWAY

ALFRED HITCHCOCK

MAGIC JOHNSON

BORIS KARLOFF

KEN KESEY

STEPHEN KING

DAVID LETTERMAN

C. S. LEWIS

NORMAN MAILER

MARCEL MARCEAU

THE MARSHALL PLAN

MATA HARI

VLADIMIR NABOKOV

GEORGIA O'KEEFFE

GARY OLDMAN

LUCIANO PAVAROTTI

ANNA PAQUIN

MERVYN PEAKE

RIVER PHOENIX

JOSÉ GUADALUPE POSADA

ELVIS PRESLEY

MIRANDA RICHARDSON

RAINER MARIA RILKE

SADE

CARL SAGAN

SUSAN SARANDON

CHARLES SCHULZ

ALBERT SCHWEITZER

SOUND BARRIER IS BROKEN

KEVIN SPACEY

BRITNEY SPEARS

ALFRED STIEGLITZ

SHARON STONE

GLORIA SWANSON

LEO TOLSTOY

URANUS DISCOVERED

KURT VONNEGUT

THE WHITE BUFFALO CALF

TENNESSEE WILLIAMS

"What do I ask of a painting? I ask it to astonish, disturb, seduce, convince."

—Lucian Freud, Sagittarius Sun-Scorpio Jupiter. December 8, 1922

"Do not the most moving moments of our lives find us without words?"

—Marcel Marceau, Aries Sun-Scorpio Jupiter. March 22, 1923

"The monster was the best friend I ever had."

—Boris Karloff, Sagittarius Sun-Scorpio Jupiter. November 23, 1887

"If my devils are to leave me, I am afraid my angels will take flight as well."

—Rainer Maria Rilke, Sagittarius Sun-Scorpio Jupiter. December 4, 1875

"I always had a repulsive need to be something more than human."

—David Bowie, Capricorn Sun-Scorpio Jupiter. January 8, 1947

"Acting is like a Halloween mask that you put on."

—River Phoenix, Leo Sun-Scorpio Jupiter. August 23, 1970

"If I were white, I could capture the world."

—Dorothy Dandridge, Scorpio Sun-Scorpio Jupiter. November 9, 1922

"Life itself is a quotation."

—Jorge Luis Borges, Leo Sun-Scorpio Jupiter. August 24, 1899

"Filmmaking is a chance to live many lifetimes."

—Robert Altman, Pisces Sun-Scorpio Jupiter. February 20, 1925

"Inspiration is a guest who does not like to visit lazy people."

—Pyotr Tchaikovsky, Taurus Sun-Scorpio Jupiter. May 7, 1840

Believe That You Can

Find the desire, the ability, and the gumption to press on—you can win!

— SAGITTARIUS JUPITER

ERIN BROCKOVICH

From Geoffrey Chaucer, born in 1342, whose *Canterbury Tales* exuberantly urged "Lat every felawe telle his tale aboute," to Chief Joseph, born in 1840—"Hear me, my chiefs!"—to James Baldwin, born August 2, 1924, who reminded us: "For while the tale of how we suffer, and how we are delighted, and how we may triumph is never new, it always must be heard," these Sagittarius Jupiters believed, and so must you, in the power and luck of exuberant self-expression. Tell your tale, Sagittarius Jupiter! And the more colorful, the more rags-to-riches (or riches-to-rags), and the more triumphant it is, the more luck and success will come to you.

The original father of medicine, Hippocrates, born a Sagittarius Jupiter in 460 B.C., wrote: "Prayer indeed is good, but, while calling on the gods, a man should himself lend a hand." And there you have your success, dear Sadge Jupiter—it's *you*! As Jupiter is Sagittarius' natural ruler, you get a *double* whammy of Jove's qualities: *jovial,* lucky, optimistic, expansive, vigorous, fun, risk taking, generous, philosophical, liberal minded, and you do it big or bust—which you must do, if you are to fulfill your potential. In astrology, this placement is considered to be dignified, allowing Jupiter's nature to be expressed in pure form. Express it, dear Sagittarius Jupiter, by shouting it out into the world!

Paul Revere was a Sagittarius Jupiter. The classic image of him galloping on his horse, yelling "The British are coming! The British are coming!" is the actual motif of the Sagittarius symbol—half man, half horse, galloping forward. You should always be in motion to be lucky, whether physically or symbolically, always propelling yourself forward.

You also do best embracing the journey, in love with life and all parts of it. No matter how hard a life you've had, no matter how reserved, or melancholy, or cautious the rest of your astrological chart may make you, this is how you get lucky. No other way. Push yourself to be open, unprejudiced, embracing, life affirming, exuberant, goofy, open, sunny, effervescent, and just plain glad to be alive.

Take a moment, look back long and deep over your life, recount as many events as you can, and ask yourself what has made you lucky or unlucky in your life. You'll notice that whenever you act like a Sagittarius, *you create miracles in your life.* No matter what your Sun sign is, *you get lucky when you live out the Sagittarius legacy gifted to you by the benevolent planet Jupiter.*

To take advantage of your luck, emphasize your . . .

Be a reacher for the moon, life affirming, an enthusiast, a believer in miracles, your own drummer, on the move, eager to learn and to teach, a gutsy risk taker, and unaffected.	♐♃ **Sagittarius Jupiter Strengths**

But because Jupiter rules excess, watch out for . . .

You can become reckless, wasteful, or senselessly heedless to danger.	♐♃ **Sagittarius Jupiter Excesses**

Understanding the
Sagittarius Energy and Influence

Sagittarius is born at the beginning of holiday time! Sadge is coming to town, bearing excitement and gifts, and getting everyone psyched for Thanksgiving, Hanukkah, Christmas, and Kwanzaa—all the major gift-giving days. Sadge wants to be best friends with everyone, and this is such a great time of year, full of hope and joy, fun and laughter, horsing around (Sagittarius is ruled partly by the horse), playing with costumes and partying, telling stories, being funny, getting to act like a *kid* again. Life is too short to be a pretentious, stuffy-boring *grownup!* Innocent honesty is at the heart of Sagittarius vision. Deeply philosophical, fiercely ambitious, seeing the big (and bigger) picture, and doing it *their* way with tons of wow-enthusiasm and creative energy is the way of the Sadge. Being a Fire sign, Sagittarius is impulsive and adventurous, and this clumsy colt often puts its foot in its mouth, blurting out truths and embarrassing others. Sadge always bounces back, however, as its ruling planet is Jupiter and its guiding force is incredible, unbelievable *luck.*

SAGITTARIUS SYMBOL
The Archer or Centaur

Sagittarius is the Centaur, a magical beast who is half human and half horse—specifically, Chiron, the kindest centaur. Chiron, the Archer, was a wise teacher and healer; his student, Hercules, accidentally shot him in the thigh with an arrow during a hunting lesson. Chiron chose to give up his immortality to Jupiter, in order to free Prometheus, the giver of fire. In return, for giving fire back to the world, Jupiter rewarded the centaur by placing him among the stars, and he became the constellation Sagittarius. Remember that your luck comes through a *Fire* sign, and honor yourself as the giver of fire by acting with vigor, passion, and inspiration. Live as though you are immortal. Run wild and free; remember that half of you is the horse. You gain the wisdom of Chiron not through introspection, but from action and experience, shooting your arrows ambitiously at your target and galloping after them, wherever they may lead.

Remember Chiron's wound in the thigh? It's the legacy of Hercules that Sun sign Sagittarians need to protect their thighs. But as a Sagittarius Jupiter, it is luckiest for you to use the hips and thighs as much as possible—run, walk, bicycle, dance. The hips and thighs propel your courage forward into adventurous risk taking in your life. Your sign also governs the liver, which you must protect. The liver symbolically represents courage, and you must exhibit plenty of it. Damage of the liver by excess of alcohol or by toxins can lead to hepatitis or cirrhosis. You're encouraged to be an energetic party person, but because your ruling planet, Jupiter, governs excess, watch out for too much of a good thing!

SAGITTARIUS
PARTS OF BODY
The Liver, the Hips, and the Thighs

⚹⚹⚹

Purple, the color of royalty, is dramatic, artistic, flamboyant. Silver is shimmering, instant magic, and it is recorded that its gleam in nature attracts animals (including humans!). These are not subdued, practical, everyday office colors; wearing them requires courage and adds fun and flare and a touch of mysticism—and suits you perfectly. To lift your spirits and escape the ordinary, add these colors to your wardrobe and surround yourself with them—linens, furnishings, wall accents, dishes. If you're required to follow a conventional dress code at work, remember that the colors' energy against your skin is even more powerful—so how about getting yourself sexy silver, or lacy purple, lingerie. Scarves—think blowing in the wind—are the luckiest accessory for Sagittarius. And, of course, silver jewelry.

SAGITTARIUS COLORS
Purple and Silver

And now your Sagittarius Jupiter strengths: ♐ ♃ **Sagittarius Jupiter Strengths**

Be a Reacher for the Moon

Sagittarius Jupiters succeed when they accentuate the positive, trust in their gut, dream the impossible dream, and act on pure gumption. Aim for the moon with your ideals intact, like Sagittarius Jupiter Antoine de Saint-Exupéry's Little Prince. In astrology, the Moon rules the unconscious and inner emotions, so it is essential that you go from your heart. As the creator of *The Little Prince*

Giant, Generous Jupiter

Jupiter is one thousand three hundred times the size of our earth and weighs twice as much as all the other planets *combined*. It is a miniature sun, with its own solar system, boasting more satellites than any other planet—fifty-eight, at last count—and emitting its own energy. Our solar system's largest planet has an atmosphere made mostly of *laughing gas* (helium). No wonder Jupiter is such a fun and happy planet—and such a gas!

YOUR HISTORY

Jupiter was first named Zeus by the Greeks, then called Jupiter, or Jove, by the Romans. This is where we get the word "jovial." Jupiter was ruler of all the gods, the supreme law-giver and spiritual chief of Rome, and believed to be just and merciful.

wrote: "It is only with the heart that one can see rightly; what is essential is invisible to the eye." So, dear Sagittarius Jupiter, not everyone will be able to see what you are reaching for, or indeed, even believe that it exists. You will need to have the faith to keep on, despite the dissuaders, the cynics, the critics, or those who just can't be bothered. You must aim high, past any indifference or doubt that you yourself may have absorbed from your past or from those around you. Just think of Sadge Jupiters Erin Brockovich, who refused to take no for an answer, or Elena Kagan, the first female dean in the history of Harvard Law School. Reaching for the moon also means reaching inside yourself and trusting in your greatest dreams and, like the Archer, finding your target in the sky.

Be Life Affirming

If your Sun sign—say, Virgo or Capricorn—makes you naturally cautious or a skeptical realist, you will have to try especially hard: commit to daily affirmations, fill your life with joy, positive people, wagging dogs, bright colors, and happy music and films. On a positive spiritual note, think (and listen to) Sagittarius Jupiter Aaron Copeland's symphony *Appalachian Spring*. And of course, " 'Tis a gift to be simple, 'tis a gift to be free, 'tis a gift to be how we want to be"—being gifted is what Jupiter is all about! is the main hymn of the Shakers, a small unaffected community who engage in "shaking" (the ultimate dance of joy and being) as their shared religious experience. They also were the first community at the time to have equal relations between men and women, and believed that all should be treated equally.

Think also of Sagittarius Jupiter Grace Kelly dancing and singing the irrepressibly happy "I'm Singin' in the Rain" ("I'm singin' in the rain. What a glorious feeling—I'm happy again!"), and Sagittarius Jupiters the Flintstones (Yaba daba doo!), and Doris Day with her peppy, upbeat numbers, and Cameron Diaz in *There's Something About Mary* (and that hair gel, too!), and Harpo Marx's hysterical antics—who can help but *love them*?

They make us laugh and they make us feel good. They make us glad to be alive. This is what you were put on Earth to do, Sadge Jupe, and for yourself as much as us!

Be an Enthusiast

Draw on the fire of your Jupiter sign to fuel you up and ignite your energy. What is most important isn't success, money, or other people's expectations; it's that you are excited about your life and whatever it is that you're doing. But to be excited about something, it has to interest you—you can't fake it. You, Sagittarius Jupiter, won't be happy or lucky if you choose your job because of the pension plan, the convenience of locale, or because your dad or mom worked there for thirty years. Likewise, don't waste your time on a "suitable" date, a blind date your best friend *knows* is "just right for you," or someone with whom you're merely comfortable. "Appropriate" is the antithesis of your idea of real life. Don't let anything stop you from liking what you like!

Consider your fellow Sadge Jupiter Jim Henson and his zany Muppets! Or ever-perky Barbie. When Barbie was born into the world on March 9, 1959 (Sun sign Pisces—a *dream* girl), at the New York Toy Show, she started with only a few basic outfits—stewardess uniform, cocktail attire, secretary wear—and like a true Sadge Jupiter, expanded into a range that just keeps *going!* Like a true Archer gal, she never seemed to need Ken all that much and just gets more ambitious as she moves forward with the times. As with many a Sadge Jupiter, there's a certain amount of chuckling and incredulity, but it's tinged with affection. Millions of girls love Barbie—she's their *best friend* (Sadge's favorite expression), all the way from the sixties to the twenty-first century, and probably beyond.

One wildly inappropriate enthusiast was Sagittarius Jupiter Ed Wood, dubbed "Worst Director of All Time" after his death, a title which his wife said he would have loved. He was best known for his gender-bending cult film *Glen or Glenda*. His other films, ranging from science fiction to horror, featured senseless dialogue punctuated by wild and weird special effects with no relation to plot. He wrote best when partially drunk and wearing feminine clothing. He used the pen name Akdov Telmig ("vodka gimlet" spelled backward), and books poured out of him, sometimes in two or three days. As portrayed brilliantly by Johnny Depp (overlooked for an Oscar) in Tim Burton's film *Ed Wood*, Ed approached everything and everyone with undying enthusiasm. Enthusiasm cannot be ordered. Whatever seizes *your* imagination—excites *you*—is it.

Be a Believer in Miracles

Nothing's impossible for a Sadge Jupiter. "If the doors of perception were cleansed, everything would appear to man as it is—infinite," wrote Sagittarius Jupiter William Blake, English poet and painter, born November 28, 1757. Infinite equals not finite—endless, open, free. Time is not the same for you: When you're with someone you love or involved in a work project that excites you, time will cease to exist. This works to your advantage. And what is a miracle? Something out of the ordinary, something sudden and unexpected, bending the laws of cause and effect, positive, uplifting, and often healing. It creates spontaneous belief. You need to align yourself with "thinking outside of the box" in an expansive and hopeful way. Open your eyes to what is beautiful and amazing; allow yourself to see and experience life in positive superlatives. It's no accident that Dale Carnegie, proponent of *How to Stop Worrying and Start Living* ("Believe that you will succeed and you will") and author of *How to Win Friends and Influence People*, was a Sagittarius Jupiter. He inspired millions not to doubt themselves and to approach all situations positively, and, by so doing, to improve their lives. Today imaging—using positive mind images of physical action, achievement, and winning—is used successfully in all sports, and often (appropriately for Sadge) by runners!

Be Your Own Drummer

"If a man does not keep pace with his companions, perhaps it is because he hears a different drummer. Let him step to the music he hears, however measured or far away," wrote Henry David Thoreau, born a Sagittarius Jupiter on July 17, 1817. You, Sadge Jupiter, are not meant to have an ordinary life, but an extraordinary one. Remove yourself from a limited, circumscribed world and set of beliefs; push the envelope; expand your horizons; the mundane and conventional are not for you. Sagittarius Jupiter Nicholas Copernicus, born February 19, 1473, went against the prevailing trends of his time—and authority of the Church—by arguing his heliocentric theory: That we revolve around the Sun instead of the other way round. Like a true Chiron teacher, he made way for modern science. So what if the rhythm of your thinking, ideas, and life does not fit the norm? Allow yourself to follow your own path.

You may have success earlier than many of your contemporaries; then, at the time they are achieving, your success seems to have flown. Don't worry—it will come back. In the meantime, try not to compare or judge yourself by others' timing and standards.

Be on the Move

Sagittarius Jupiter Hippocrates wrote, "Walking is man's best medicine." You can take it further than that! Chief Joseph, born a Sagittarius Jupiter in 1840, was given the name Hin-mah-too-yah-lat-kekt, or "Thunder Rolling Down the Mountain," a typical movement for your sign! For luck, invest in any type of movement: walking, running, horses, or cars—and, yes, even in motorcycles. The Hell's Angels, born Sagittarius Jupiter on March 12, 1948, have always declared themselves "Born to Ride." Atop their Harley-Davidsons, they still argue today, when denied entry somewhere, that they are not a gang, but "a motorcycle enthusiast's group."

"The man who interviews America," Sagittarius Jupiter Studs Terkel, born May 16, 1912, has been drawing this lucky trait of constant travel and movement for seventy years, going around the country collecting oral histories. From his best-seller *Hard Times*, which documented Americans' experience of the Great Depression; to *Working*, which recounted Americans' work lives, from prostitutes to policemen, from farmers to fashion models; to *Race* and his Pulitzer Prize–winning *The Good War*, Terkel has given us, as he says, "history from the bottom up rather than history written by generals." Remember: You get lucky by embracing the "everyman." Your Jove motto is, "One for all, and all for one."

Be Eager to Learn and to Teach

Chiron the Centaur was a good centaur precisely because he was a great teacher and mentor (the other centaurs of the time were apparently off ravishing maidens).

Sagittarius Jupiter Sidney Poitier refused to listen to those who would put him down and became the first African American to win a Best Actor Oscar. When he started, he could barely read a word of English. He taught himself English from discarded newspapers in a restaurant, where he worked as a dishwasher. His most beloved role the world round is in *To Sir with Love*, where he plays a teacher in a working-class area in London, and through making the students excited to learn, changes their lives. The famous scene of Lulu as a student singing "To Sir with Love" to him is etched in everyone's minds forever. Lulu, from working-class Glasgow, is also a Sadge Jupiter!

"I owe the City University of New York for my daughter's education, I owe the streets of New York City for mine," Sidney Poitier said. Sadge Jupiters can start with nothing, aim for the Moon, and end up with everything. You must remember this if you're sitting with empty pockets reading this book!

Be a Gutsy Risk Taker

Think of Sagittarius Jupiter Daniel Boone, born November 2, 1734—the great frontiersman, hunter, explorer, and adventurer, who pushed all the boundaries, constantly moving westward through constant danger, seeking new horizons, places where it was much less crowded, opening new territories. One of the first into Kentucky, which in his day was still the far west! And remember, the South Pole was discovered on a Sagittarius Jupiter day, December 14, 1911.

Act on your gumption. You can start with nothing, and through sheer gutsy risk taking, accomplish miracles. By his own account, one day, a bright young man wandered onto a Universal studio lot, and without an office or a job, he began advertising his talents as a film-maker. He found an empty room, set up an office for himself, and set about learning all he could—by watching what was happening around him. No matter what people said about the risks involved, he refused to listen. He began with a small feature that was a success, and now he is one of the most famous people in the world. He is Sagittarius Jupiter Steven Spielberg.

Be Unaffected

Your own subtitle for being unaffected can also be: "Get Your Hands Dirty!"

The real Rosie the Riveter, Rose Will Monroe, was a Sagittarius Jupiter. She was the nation's premiere poster girl during the 1940s, and she mobilized a generation of women into joining the workforce during World War II. Rosie remained a tireless worker after the war, driving a cab, operating a beauty shop, and founding Rose Builders, a construction company that specialized in luxury homes.

Another Sadge Jupiter has also become famous for working hard with his hands, in this case building homes for the needy. Despite becoming president of the United States, and winning a Nobel Peace Prize, Jimmy Carter is most famous for hard work, dedication, and unaffected simplicity.

Even romance and sex should be unaffected and without illusions to grandeur. As the great classical actress of the nineteenth century Mrs. Patrick Campbell said, "It doesn't matter what you do in the bedroom as long as you don't do it in the street and frighten the horses." Of course, she was born a Sagittarius Jupiter.

What Makes You Lucky in Love

Whatever you do, try to resist getting caught up in tangled emotional dramas and long, drawn-out arguments. Remember what Sadge Jupiter Hermann Hesse said: "Love must not entreat or demand." The luck of Sagittarius Jupiter truly is reflected in this sentiment: give tolerant, generous, accepting, unconditional love. Love your mate for who they are, *all* of who they are, faults and all—or let them go. Seriously. Don't waste your—or their—time trying to "change" them, or waiting for them to somehow change themselves.

Even if your Sun sign is geared toward detail and not missing a thing, it is better for you to see the whole forest rather than the individual twigs or leaves. When you can see the whole picture at the beginning of a relationship, for example, it will save you a lot of time in the long run. Negatives that you catch early on are your warning sign not to continue. Sagittarius Jupiter wants you to keep moving forward with hope, and if the road is too rocky to allow that, you need to accept reality and the fact that even your optimism cannot change another person.

First and foremost, your date or mate must give you a feeling of joy, fun, adventure, and make you laugh. As Sadge Jupiter Woody Allen said: "Love is the answer, but while you're waiting for the answer, sex raises some pretty good questions"; and about sex: "That was the most fun I ever had without laughing." What was the most fun *you* ever had without laughing? Think about with whom—and that's your mate!

Your Best Career Moves

Whatever your field, for you to gain the luck of your Jupiter, you must take risks and follow the Archer's arrow as far as it will go. Travel is extraordinarily lucky for this Jupiter placement, so powerful that it is a force in itself, and can propel you forward to success. This translates figuratively into ideas spreading outward—which makes you the best teacher of all the Jupiter signs—and literally, in terms of physical travel. It can be making business trips or taking yourself off to a different locale to work on a project—roving journalist, traveling salesman, on-site engineer, or chronicler of a town's and its people's history. Luck comes not just through change of locale but through learning—new cultures, customs, languages, and ideas—and through work that requires you to learn new job skills. The beginning of any new endeavor, that first burst of fiery vigor, is the most exciting part for you; your enthusiasm inspires others to join your cause. Big gestures, not playing it small, also work well for you. Remember Sagittarius Jupiter Abbie Hoffman's dramatic demonstration on Wall Street: He dropped dollar bills into the air and stopped trading on the floor for several minutes as people scrambled for the bills!

You are an excellent team player, but bear in mind that your spirit of independence can clash with rigid rules and others' methods of operation. Accent the positive in the workplace and try to be as encouraging as possible. Others will soon see you as a great booster and a resource for them, too! You must trust your inspiration and your heart, dear Sagittarius Jupiter. You work best in spurts, with dry spells and fertile spells, and can handle a freelance work life very well. You are attracted to the immediate and fun solution; what's harder is keeping on when it's no fun at all. When you know something is wrong or should be fixed, speak out until people take action. If others torpedo deeply laid plans, ignore them and bound ahead, regardless.

How to Reach Optimum Health

As the Sagittarian-ruled thighs and hips represent volition and locomotion, it makes sense that you need to be physically active in the great outdoors, with plenty of fresh air and sunshine. The best sports for you are running, walking, horseriding, and archery. You can literally become ill if you don't exercise enough; your thinking can become cloudy and foggy. Sagittarius Jupiter can get away with more excesses than the Sadge Sun sign, but you do need to avoid overuse of alcohol because it will damage your liver. For optimum health, also avoid boredom! You need to be constantly mentally stimulated to keep your spirits uplifted and your life involvement strong. You're lucky—with this Jupiter placement, you generally have the boon of robust health and longevity.

Your Lucky Foods, Element, and Spices

Sagittarius' cell salt is silica, which keeps the connective brain tissue healthy; stops numbness in legs, feet, and fingers; and fortifies the nervous system. You will know if you're lacking in this mineral if you have problem gums, sallow skin, or limp hair. Silica is found in these foods: green peppers, strawberries, apples, prunes and figs, the skins of fruits and vegetables, potatoes, raw salads, oats, whole-grain cereals, and eggs. The best diet, generally, is high protein with plenty of fish, poultry, eggs, yogurt, and skim milk. Make sure you eat two of these food every day: cherries, asparagus, brussels sprouts, lemons, beets, plums, oranges, brown rice, and tomatoes. Your luckiest seasoning of all is sage, and foods thought to bring prosperity and financial energy are peanuts and maple sugar. As Sadge Jupiter gets lucky gambling, particularly on horses or cars, why not try?

Your Best Environment

You, Sagittarius Jupiter, need fresh air, mountains, and a sunny climate. If you live in a city—and you'll enjoy all the new people to interact with if you do—choose one with parks, open spaces, and riding, walking, and jogging trails nearby, surrounded by mountain peaks, if possible. Denver or San Francisco, for example; or L.A.—newer, more liberal-minded cities. Group—and party— situations suit you; don't isolate yourself. Silver and purple hues will make you happy, whether in a glorious sunset across the shimmering water or in Persian wall hangings.

Your Best Look

Sagittarius Jupiters are lucky here, too, as they have the most choices of anybody! You can go wild and dramatic, sporty, offbeat chic, vintage, period costume, serious or sober—as long as it's *your* own expression, *your* individual take. Probably a fashionista's nightmare precisely because, for your success, you won't follow their rules! Follow your own drummer when it comes to what you're going to wear and how you're going to look. *Vogue* summed up Sagittarius Jupiter Cameron Diaz's fashion in their May 2003 issue as "joyful dressing . . . no one can throw a dress over jeans like Cameron." *Vanity Fair* simply said on its cover "Cameron Diaz Just Wants to Have Burping Contests." Even if you're wearing the most conventional business attire, for surefire success add an individual touch of your own, however subtle—a scarf, a wacky pen, a way-out tie!

Your Best Times

The luckiest times for you are when the Moon is in Sagittarius, which happens every month for two to three days. Find these special days in your astrological calendar. During these days, schedule crucial meetings, tell that special person you love them, ask for a promotion or raise, push the envelope, and plan to shine. In your everyday, the best time for you to exercise is *anytime it suits you.* The luckiest times during the year for you are November 22–December 21 (when the Sun is in Sagittarius); March 21–April 19 (when the Sun is in Aries); and July 23–August 23 (when the Sun is in Leo).

The very luckiest time of your life—as in Lotto lucky—is when Jupiter reenters the sign it was in when you were born (Sagittarius), which happens every twelve years and lasts anywhere from a few months to just over a year. This is called your Jupiter Return, and it launches your luckiest streak. You can count on the *extraordinary* to happen the very day your Jupiter Return begins. Amazing events can begin happening up to six weeks beforehand, so plan ahead!

How to Be Lucky Financially

Sagittarius Jupiters have the greatest numbers of tales of rags to riches—*and* riches to rags. You can start with nothing and end up with everything, or vice versa! Constant movement, remember? Never still and steady. Even though constant movement will make you luckiest, common sense would say to be careful. Many of you begin a new project or company with great enthusiasm—but with completely the wrong person or people. Be careful of teaming up with, and sharing all your resources with, people you've just met or don't know very well. Trusted friends help you out and give you jobs, so ask them if in doubt. If you do what you love, you will make money however—and how many people can say that? Don't be afraid to publicize yourself—the good word about you spreads easily. You can even change jobs and partners quickly, you lucky dog!

You should focus your investing on companies that make people happy and make the world a better place. Invest in people and projects that work on education and justice. Your lucky focus on learning should make it easier to invest in educational publishers, and your interest in travel should help you make money in international companies. Stay away from complex industries that are hard to understand or harmful to the planet, and be careful of gambling. You are luckier than most, however, when it comes to horses and cars, both ruled by Sagittarius—or chariots, as Sagittarius Jupiter Charlton Heston found to be the case.

Success in Family and Friendship

Sagittarius Jupiter Zelda Fitzgerald wrote, "Both of us are very splashy, vivid pictures, those kind with the details left out, but I know our colors will blend, and I think we'll look very well hanging beside each other in the gallery of life."

Zelda wrote these words to the love of her life, F. Scott Fitzgerald, but for Sadge Jupiters, they apply to their friends as well. Have you ever seen photos of friends and family in a Sadge Jupiter's home? They all look like they're following Zelda's model! Everyone's clowning around or grinning from ear to ear; rarely will you see soberly posed photos taken by a portrait photographer.

The perfect Sagittarius Jupiter cautionary tale is Voltaire's *Candide:* " 'Alas,' said Candide, 'it is a mania for saying things are well when one is in hell.' " As you will notice if you look at other Jupiter signs' cautionary tales, yours is hardly even cautionary!

Candide's name comes from a French adjective with essentially the same meaning as the English "candid." "He combined an honest mind with great simplicity of heart," wrote Voltaire. Candide is perpetually the optimist and is rarely able to make his own judgments; instead, he leans on the teachings of his dear Dr. Pangloss. His blind devotion to Pangloss's optimism renders him comical, especially in the wake of any disaster.

You will say silly things in your lifetime (it is inevitable) that you may later regret. But again, you bounce back—always. From Prince Charles's "To get the best results you must talk to your vegetables" to George Bush Senior's "No, I don't know that atheists should be considered as citizens, nor should they be considered as patriots. This is one nation under God," these Sagittarius Jupiters do not seem to have suffered any ill effects. Sadge Jupiter histories are easily forgotten. For instance, Bush Senior's specialty in the CIA was *election fraud,* a detail that escaped mainstream media coverage during his son's questionable winning of the presidential election in 2000. (It was reported only on NPR.)

Although, as you will see from the excesses below, because the very nature of your ruling planet, Jupiter, is excess, you guys get away with most things—more than anyone—and you often bounce back wonderfully well. So your real goal should be not to ignore, throw away, or waste your natural good luck!

Excess #1 You Can Become Reckless

Usually for Fire sign Jupiters recklessness is not a negative, but for Sadge Jupiters it leads so often to being publicly humiliated that it's never worth it! Perhaps it is because the Sagittarius Jupiter spirit is just so expressive, so irrepressible, that their actions cannot be hidden. The funny part is that their actions are not necessarily worse than most other people's—they just do it in such a reckless, clumsy way that they're inevitably caught. Just look at this list of publicly embarrassed Sadge Jupiters: Hugh Grant arrested with prostitute Divine Brown in his car on Sunset Boulevard; Winona Ryder for her shoplifting spree; Woody Allen for naked photos of and affair with his adopted seventeen-year-old daughter; Prince Charles's infamous "I wish I were a tampon" conversation; Chevy Chase's publicly asking one of *SNL*'s writers for oral sex; Lauren Bacall's moving account on the *Today* show of her close friend's life being saved by a certain medication (the *New York Times* revealed that a pharmaceutical company had paid her ahead of time to tell this story!); John McEnroe's temper tantrums on the tennis courts of Wimbledon; and on and on. The public reaction is more incredulous than horrified—as in, *How stupid can you get?* (Even Jay Leno said to Hugh Grant on his show, "Man, what were you *thinking* of?") But, as you can see, generous Sadge Jupiter is forgiven by a generous public—none of their careers were ruined by these events.

Excess #2 You Can Become Wasteful

Look at the original "poor little rich girl," Woolworth heiress Betty Hutton, born during the passage of a Sagittarius Jupiter in 1912, who, despite her millions, died alone and penniless, having given most of her money away foolishly to bad-risk investments, lovers, and untrustworthy people who deliberately took advantage of her.

When you are stuck in situations that are dull and going nowhere, these are the times that you're most in danger of "breaking out," becoming reckless, spending irresponsibly, committing horrible gaffes, or indulging in vulgar antics. Then you are wasting your own talents—and other people's time.

Excess #3 You Can Become Senselessly Heedless to Danger

The sinking of the *Titanic* could have so easily been avoided. Not by caution, just by *sense*. The very worst of the Sadge cautionary tales—and there aren't many—but the ultimate one happened during a Sagittarius Jupiter passage on April 14, 1912—the sinking of the *Titanic*, in which thousands of lives were lost. The ship was declared "unsinkable," so much so that it had been equipped with only enough lifeboats for one third of its passengers. During the riskiest part of the crossing, the ship's speed was *increased,* urged on by the ship's designer so it could "make the papers." As was quoted in the film *Titanic*, when the ship's sinking became an inevitability, the captain said to this ship's designer, "You will have your papers."

Sun Sign–Jupiter Sign Combinations

Aries

(March 21–April 19)

Two Fire signs—the Ram and the Horseman—what a team! Keep yourself fueled up on sunny hope and your fantastic energy. You accomplish so much, whether it's starting your own company (your energy inspires others), starting a fund drive for the volunteer fire department (you'll probably fight the fires, too), or starting a storytelling hour for kids at the library. And *you* are the one acting out *Cinderella*—any tale with a happy ending!

So much of what you do is the enthusiastic "starting" that ignites from double Fire, but remember that you need to finish your projects too—or turn them over to others, without hanging on to details, which aren't your thing. Even in love, you can find luck in moving on, like the galloping Archer. Like fellow Aries-Sadge Emma Thompson, who has kept going forward in her life and career, taking on new and different roles and remaining friends with an ex-husband (Kenneth Branagh) as only Sagittarius can do. Turn your energy away from how people make you feel (your ego) and into the world (your expanded consciousness) and you will find everything you need there, dear Aries-Sadge!

Taurus

(April 20–May 20)

With this combination of determined optimism, you can successfully write your own ticket. The Bull's steadiness of purpose supports the Horseman's fiery go-forth attitude, but you find most luck with Sadge fun enterprises. For long-term success, use your Taurus patience and wisdom to accept that you may need to act contrary to your nature. When the time is right, throw Taurus caution to the wind. Overcome your skepticism to *believe* in dreams, lucky numbers, finding that four-leaf clover, and Pollyanna's eternal optimism. You get lucky with changing scenarios and locations, with movement and with giving. Think of donating to your favorite charity. The reward of Sadge Jupiter giving is that it comes back to you twofold!

Dear Taurus, more than any other sign, you will stay in situations, places, and relationships that aren't good for you, simply for the sake of security. Think of the great Oscar-winning actor Taurus-Sadge Spencer Tracy, who did not follow his Sadge Jupiter luck to marry the love of his life, Katharine Hepburn (he stayed in a sad marriage instead). Let yourself plan what you want, and then leap forward to grab it with both hands!

Geminis are brilliant thinkers and analysts, but you must know that if your Jupiter is in Sagittarius, you need to dive in, wholeheartedly commit to going as far as the Archer's arrow takes you, with sunny, just-glad-to-be-alive optimism. This is the end of distancing yourself from emotional involvement, action, and commitment.

Both Gemini and Sagittarius have ideas galore, but for carrying them out, call on your successful fire-giving action and physical energy to propel you forward. This also is a lifesaver with decisions, narrowing the endless options, and making your dreams a reality. Both parts of you love your freedom, so commitment may come later in life. Gemini-Sagittarius William Butler Yeats wrote in a letter to Olivia Shakespeare, whom he had not seen for forty years: "I came upon two early photographs of you yesterday while going through my file. You know how one looks back to one's youth as to a cup that a madman dying of thirst left untasted?" Don't leave your life untasted, dear Gemini-Sagittarius.

Cancer is cautious and home oriented, devoted to family and security, and Sagittarius is risk taking, travel oriented, devoted to a wide circle of friends and new horizons. These two opposites will frequently tear at you, but Jove is the answer! Let's say you want to buy a home: Well, take a chance on the old fixer-upper in an up-and-coming area, instead of waiting (perhaps forever) for the perfect house in an established neighborhood. Rather than go for the safe-and-familiar relationship, trust yourself to approach the new person who really lights your fire. Rather than sticking to the same-old, same-old as far as career, guarantee yourself—ironically—more security (and money!) by seeking a higher position with more to challenge and excite you.

In other words, dear Cancer, break the mold. Cancer-Sagittarius Kathy Bates did just that, refusing to play it safe by relegating herself to being a character actress, someone's aunt or weird sister, but instead, carrying movies on her own, winning an Oscar, and at the age of fifty-four, climbing naked into a hot tub with Jack Nicholson in *About Schmidt*! Consider areas in your own life that could do with renovations in the adventure department, and breathe some lucky fire on them. Watch your success sizzle!

⤔⤔⤔⤔⤔⤔⤔⤔⤔⤔⤔⤔⤔⤔⤔⤔⤔⤔⤔⤔⤔⤔⤔⤔⤔⤔⤔⤔⤔⤔

This fabulous combination adds regal fire to leader fire. Become so involved in any activity—from career moves to social gatherings, home decor to new love interest—that you forget maintaining the image. Don't sit on your royal throne but leap on your horse and *go!*

You'll succeed by encouraging others and by always being open to new friendships and the next adventure. Don't think about designer clothes, which wine to serve, or an impressive presentation of your ideas in the boardroom. Instead, create your own style, invent a new cuisine (ever thought of being an inventive, kooky cooking guru?), and get everyone enthused by your sheer exuberance. Look at Leo-Sagittarius Sean Penn, and Antonio Banderas, and Julia Child!

⤔⤔⤔⤔⤔⤔⤔⤔⤔⤔⤔⤔⤔⤔⤔⤔⤔⤔⤔⤔⤔⤔⤔⤔⤔⤔⤔⤔⤔⤔

You want life to be perfect in every detail—and help is at hand from Jove. As a Virgo you can stop worrying "Will it be right?" and ask yourself, as a Sadge, "Will it be *fun?*" Both your Jove and Sun (Virgo) urge you to help others often through frank criticism, but stop before you backhandedly criticize your best friend's haircut or her new boyfriend. Let your Sadge Jupiter voice blurt out your honest opinions but spare the subtle digs. The latter may not offend so much, but it is not so easily forgotten.

Think of Sadge Jupiter Maurice Chevalier, the unforgettable French film star with flawless timing and delivery, but with such pure pleasure in his own singing that millions were wooed and won. Or Queen Elizabeth I, whose sharp perceptions were presented with dash and the "common" touch. She succeeded as a great and regal leader; she was loved as the "Good Queen Bess."

⤔⤔⤔⤔⤔⤔⤔⤔⤔⤔⤔⤔⤔⤔⤔⤔⤔⤔⤔⤔⤔⤔⤔⤔⤔⤔⤔⤔⤔⤔

Sagittarius fire can blaze in the air of Libra, but to find your luckiest, most successful path, *toss away those scales and seize the moment.* You find your harmony in going with the flow. If you follow your Sagittarius Jupiter, you'll trust yourself—and the world—more. Never be afraid to forgo taste for fun and flare. It worked for Sadge Jupiter Sarah "Fergie" Ferguson, the Duchess of York, whose American Weight Watchers job (for big bucks) was a royal embarrassment but personally gave her fame, fortune, and popularity. It also succeeded for classic heartthrob leading man Marcello Mastroianni, when he dropped glamour roles for those like the caricatured macho husband in *Divorce Italian Style*. Throw your hands up in the air, pretend you're on a roller-coaster—don't worry, you'll never lose your capacity for brilliant thought—and you'll do better than *ever*!

⚞⚟

So you're a Scorpio—deep, mysterious, and masked—but even the Lone Ranger had his friend Tonto! As a Sadge Jupiter, stop riding alone; you'll have more luck if you open up, communicate, and *ride*! Think of galloping hooves in *The Lone Ranger* theme music—which was really *The William Tell Overture*, composed by Rossini in 1829 under a Sagittarius Jupiter passage. The Lone Ranger's horse, the metal in his lucky horseshoes, and his bullets were all the magical Sagittarius color of *silver*.

Dear Scorpio-Sadge, go all the way toward success with your very own Lone Ranger creed, in business and relationships: That to have a friend, one must be a friend; that by working together we can create a better world; to be prepared physically, mentally, and morally to cheer for what is right; to pay for what you've taken; and "that God put the firewood there but that every man must gather and light it himself." Hi Yo, Silver, Away!

⚞⚟

You, double Sadge, love people, and you want them to love you. This applies both to the love of your life and to your public. With this extravagant Sadge energy, you often give yourself to the world as performer or spokesperson. Just take care where you place your trust and remember to cherish yourself as well as those around you; otherwise, you can literally burn yourself out. Double Sadge diva Maria Callas gave the world the soaring beauty of her voice, but she gave so much (including her heart to Aristotle Onassis) that she had nothing left, not even her glorious voice, and died before her time.

A sense of humor can help; as double Sadge actress Christina Applegate said about her mother: "She always says she couldn't afford a babysitter, which is why she put me on the stage." Give your free spirit the healing touch of nature and the joy of moving with neither constraint nor purpose. Give yourself the gift of shooting your goals and arrows into the air for the sheer joy of it, as high and as far as they will go.

IF YOUR SUN SIGN IS:

Capricorn

(December 22–January 19)

When the Goat follows his or her lucky Sagittarius Jupiter, the climb is so much more fun all the way, as well as even more successful. Take your cautious, self-doubting approach and throw it to the wind. Lift your head—and heart—look about and see how lucky you really are. Share your multiple talents with the world rather than going it alone, as you're apt to do. Just look at Capricorn-Sagittarius singer Ricky Martin, the Latin heartthrob who took the world by storm with the sexy "Livin' La Vida Loca." With each outrageous swivel of his hips, the Sadge-Jupiter out-there persona is expanded. Take a page from this flamboyant Sadge attitude and apply it to your own performance, whether it be in love and or at work. Create energy and draw people to your side with your enthusiasm and excitement, and see how you get your work ethic across more forcefully—and teamwork done more quickly. Being on the top will be more enjoyable than you thought possible.

IF YOUR SUN SIGN IS:

Aquarius

(January 20–February 19)

You get most successful when you use the fire of Sadge to warm the coolly analytical Aquarian mind, and share those flashes of brilliance with all and sundry. Remember that Aquarius-Sagittarius actor Burt Reynolds gave his career a giant leap ahead by posing for the first male centerfold.

You'll not only benefit by replacing your sense of aloneness—and sometimes loneliness—with the pleasures of gregarious existence, but your into-the-future intellectual ventures will become *fun* and much more playful. You can be the mapmaker for us all, Aquarius-Sadge. With your shared visions you'll do well in any situation where you devote yourself wholeheartedly. This can be in a relationship, an architectural firm, a community-planning session—anywhere. For greatest success, go for the sheer *fun* of it.

IF YOUR SUN SIGN IS:

Pisces

(February 20–March 20)

Like Pisces-Sagittarius Spanish film director Luis Buñuel, you succeed by bringing your bizarre and beautiful Pisces dreams before the public. Use your lucky Jove to come out of yourself, to leave any melancholy introspections behind, and to lead the charge in your desired direction. You sometimes need to throw some good old Sadge fun into your relationships, especially when it becomes a tangled love knot.

Pisces and Sagittarius are the two kindest signs of the zodiac, with great compassion, caring, and sympathy. Time to be kinder to yourself. Put a big smile on your face as you ask for that promotion, raise, or more reasonable working conditions. Both Pisces and Sadge are imaginative and creative; use your Jove to express those talents more forcefully. The world is waiting for you to shine like a Sadge, dear Pisces.

Relationships with Sagittarius Jupiters

Finally, for those of you who aren't Sagittarius Jupiter yourselves, but have a Sagittarius Jupiter spouse or significant other, boss, child, or parent, here are some tips to help you learn to deal! This is how the luck of someone else's Jupiter touches *your* life.

Your Significant Other

You're out of luck if you're a Puritan, Blue-stocking, no-fun Nancy, but if you are, why for heaven's sake did you ever start dating a Sadge? Sagittarius Jupiter adores romance, travel, fun, and delight; they are a very sexual sign, and they love all of the fripperies—love letters, Valentine's Day, lace, and lingerie. It can be a fun ride if you want to go along. But know they enjoy the secrecy and passion of early love affairs: as Sagittarius Jupiter Aphra Behn said, "Love ceases to be a pleasure when it ceases to be a secret."

Your Sagittarian Jupiter mate gets lucky when they teach and when they learn—thus these activities can also be lucky for you. Find out what they enjoy—whether it's golf or Italian or needlework or building projects—and ask them to teach you how it all works. Make sure you don't take it very seriously (they tend to be happy-go-lucky themselves), and you'll have a fine old time learning alongside them.

Be prepared for your partner to make sudden changes, almost on a whim, to travel nonstop and to feel spontaneous enthusiasms. So long as you know these things are temporary, you'll be able to adjust, relax, and have fun. It can be a shock if you're a homebody Cancer or a critical Virgo, but you'll do best if you let go and see what life brings you.

Yet, even though your Sadge Jupiter loves movement and change, know this: They are very emotional and they can't stand being left alone. They change their minds over little things (Sadge Jupiter Paul Williams said, "Lady Luck is fickle—but a lady's allowed to change her mind") but once they've chosen you, they're hooked for life and can be deeply hurt if the relationship ends. Don't play with them: with this force, don't sign on the dotted line unless you're sure they're the one.

Your Boss

Your Sadge Jupiter boss is always moving here and there too much for you to ever sit him or her down. This boss, who can't stay in one place for long enough, is delightful and infuriating by turns. My advice to those who work for a

Sadge Jupiter is to hang on to the virtues and try to get over or ignore the rest.

There are many advantages to having a Sagittarius Jupiter boss. These Jupiters are generous, happy-go-lucky, optimistic, cheerful, and fair about sick leave and vacations. But they will be painfully honest about your faults and the parts of your life that need improvement. These bosses are great morale boosters, sympathetic, frank, never phony, and they can shock more sensitive natures. Sometimes they go from being publicly critical (which can feel terrible) to gushing with very sincere compliments. They will give no notice at all about certain things that crop up out of the blue. It's kind of like working for a child.

Sadge Jupiter bosses are kindhearted, but they aren't afraid to step on toes. These ambitious Jupiters can throw orders about regally—but they're not *really* trying to put you down. They are genuine, very intuitive, and they can spot the phonies a mile off. They can talk your ear off and will launch into things, and are very warmhearted, like a good friend. They want to trust you with everything.

From your Sagittarius Jupiter boss you can gain the luck of finding out how much better it is to be honest than not, and how, in one famous Sadge Jupiter's words, "The best way to know God is to love many things" (Van Gogh). And when things screw up, how best to say to *yourself*, "Nothing can be soul or whole that has not been rent" (Sadge Jupiter William Butler Yeats).

Your Child

"One is the loneliest number that you'll ever know . . ."—Sagittarius Jupiter Aimee Mann is really singing for all Sadge Jupiter children, as loneliness is their deepest fear. They never really grow up. These children are brutally honest, friendly, enthusiastic, and they want warmth and friendliness in return. But they also want everything else! As Sadge Jupe Snoop Dogg said, "I'm back to take my throne back from all these suckers in the game." Frank, curious, investigative, Sagittarius Jupiter children ask constant questions, and most of their questions are aimed at puncturing adults' answers for the sake of convenience, hypocrisy, and deception. They don't give a fig about social mores (don't even think of threatening with embarrassment—they don't care!). But they need to learn about money and allowances very, very early—or they will spend with no thought otherwise. (Be very firm about this.)

Sagittarius Jupiter children hate routine, rules, and orders, and they do best in an open, progressive education. They are idealistic and deeply honorable, and fighting for causes develops their strength. Rambunctious and clumsy, they need affection and demonstrative cuddles like waggy puppy dogs. Never be indifferent—that's the cruelest for them. And always watch out for them. Sagittarius Jupiters are innocent and trusting, and you must ensure they learn to trust the right people or they can fall under dangerous influences.

Your Parent

Your Sadge Jupe parent is first and foremost a friend, even though, sometimes, you may have wanted them to be more like a "parent." Fun, joking around, always sharing, they respect you as a little human being—though they will intrude sometimes, wanting to know your secrets, stepping on your privacy (there isn't a single child of a Sagittarius parent who hasn't been embarrassed on at least one occasion). They will not take lying, they don't care about pretentiousness or social airs other parents might get into to measure your success. Most people with Sagittarius Jupiter parents remember them as fun above anything—and loyal. At times, they may have even preferred the company of their parent (particularly the mother) more than other kids. (Unless you're a Sagittarius Sun sign, and then you probably wanted more freedom and independence.) They could have overlooked some essentials, though, and given away your confidences and embarrassed the hell out of you. But there was nothing mean about it, no ill intent. And that's the best thing.

This is what astrology is all about, and, I'd venture, a civilized life. Appreciating how different we all are and loving these differences! Nature gives an antidote for every poison. Isn't that the miracle of life? Problems appear, but there's always a solution, and if anyone can locate and resource it, Sagittarius Jupiter can. *Remember: You are the life-affirming mover and shaker of the zodiac.*

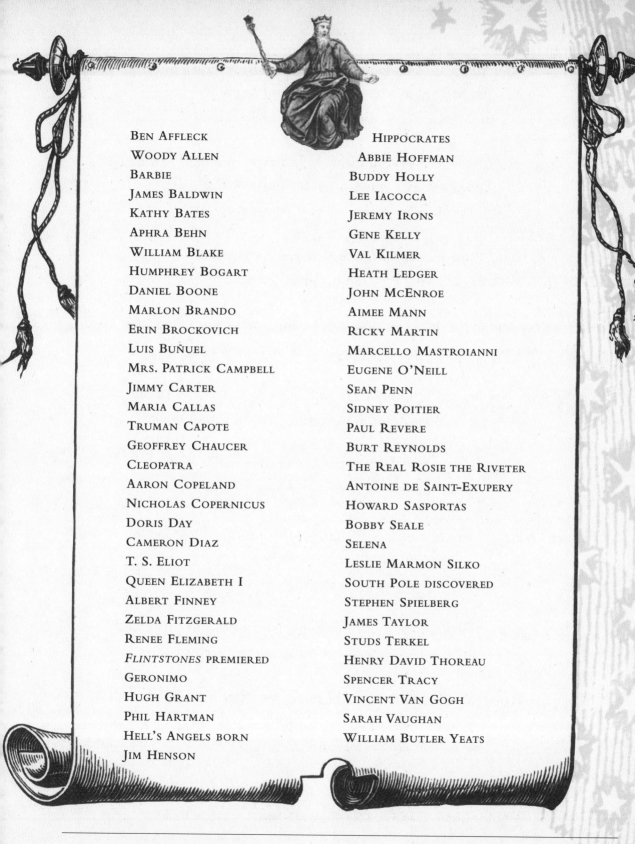

Ben Affleck
Woody Allen
Barbie
James Baldwin
Kathy Bates
Aphra Behn
William Blake
Humphrey Bogart
Daniel Boone
Marlon Brando
Erin Brockovich
Luis Buñuel
Mrs. Patrick Campbell
Jimmy Carter
Maria Callas
Truman Capote
Geoffrey Chaucer
Cleopatra
Aaron Copeland
Nicholas Copernicus
Doris Day
Cameron Diaz
T. S. Eliot
Queen Elizabeth I
Albert Finney
Zelda Fitzgerald
Renee Fleming
Flintstones premiered
Geronimo
Hugh Grant
Phil Hartman
Hell's Angels born
Jim Henson

Hippocrates
Abbie Hoffman
Buddy Holly
Lee Iacocca
Jeremy Irons
Gene Kelly
Val Kilmer
Heath Ledger
John McEnroe
Aimee Mann
Ricky Martin
Marcello Mastroianni
Eugene O'Neill
Sean Penn
Sidney Poitier
Paul Revere
Burt Reynolds
The Real Rosie the Riveter
Antoine de Saint-Exupery
Howard Sasportas
Bobby Seale
Selena
Leslie Marmon Silko
South Pole discovered
Stephen Spielberg
James Taylor
Studs Terkel
Henry David Thoreau
Spencer Tracy
Vincent Van Gogh
Sarah Vaughan
William Butler Yeats

"There'll be nobody like you ever again. Make the most of every molecule you've got as long as you've got a second to go."

—David Brower, Cancer Sun–Sagittarius Jupiter. July 1, 1912

"For the pattern is new in every moment / And every moment is a new and shocking / Valuation of all we have been."

—T. S. Eliot, Libra Sun–Sagittarius Jupiter. September 26, 1888

"Oh boy, I want the world to see that you were meant for me."

—Buddy Holly, Virgo Sun–Sagittarius Jupiter. September 7, 1936

"I won't sell my soul to the devil, but I do want success and I don't think that's bad."

—Jada Pinkett Smith, Virgo Sun–Sagittarius Jupiter. September 18, 1971

"I put my heart and my soul into my work, and have lost my mind in the process."

—Vincent Van Gogh, Aries Sun–Sagittarius Jupiter. March 30, 1853

"I belong to the people."

—Selena, Aries Sun–Sagittarius Jupiter. April 19, 1971

"The only reason to make a million dollars in this business is to be able to tell some fat producer to go to hell."

—Humphrey Bogart, Capricorn Sun–Sagittarius Jupiter. December 27, 1899

"An actor's a guy who, if you ain't talking about him, ain't listening."

—Marlon Brando, Aries Sun–Sagittarius Jupiter. April 3, 1924

"People pay for what they do, and still more, for what they have allowed themselves to become. And they pay for it simply: by the lives they lead."

—James Baldwin, Leo Sun–Sagittarius Jupiter. August 2, 1924

"I will tell you something about stories, they aren't just entertainment."

—Leslie Marmon Silko, Pisces Sun–Sagittarius Jupiter. March 5, 1948

♑♃

Going That Extra Mile

From Saint Thomas Aquinas, born in 1225, who wrote, "A man has free choice to the extent that he is rational," to Oscar Wilde, born 629 years later in 1854, who stated, "There is no sin except stupidity," these Capricorn Jupiters knew, and so must you, that rational wisdom is key. Rejoice. *You have more than enough to be a staggering success.*

> *I am not afraid.*
> *I will tell the truth wherever I please.*
>
> — CAPRICORN JUPITER MOTHER JONES

Traditionally, Jupiter is considered to be in its "fall" in Capricorn, meaning its qualities are not so readily available. This isn't bad news, Capricorn Jupiter! Capricorn is the sign of the Mer-Goat, and if your Jupiter is in Capricorn, you will generate Jupiter luck by being tenacious and hardworking like the goat, and by believing in the impossible, like the mermaid part of you. The fact that Jupiter's bounty doesn't come easy is actually good: Things that come easy *are unlucky* for you, and things that are scarce and hard to find come with Jupiter's blessing. You get lucky when you go that extra mile, work harder than anyone, triumph over the greatest number of obstacles, and cultivate patience. You must be mature and self-disciplined. But keep your sense of mer-magic and your sense of humor! Otherwise you run the risk of feeling sorry for yourself (well, who wouldn't?; your path *is* extra challenging—but it also *rewards* you so much extra!), overdoing it to the point of self-deprivation, and rigidity. Do beware, dear Capricorn Jupiter, as these characteristics drive your luck away.

Capricorn Jupiters who achieved great success usually had to work years for it. But sometimes they made it even harder for themselves than they needed to—the Goat likes that extra challenge. Watch out,

Capricorn Jupiter, as this can cause Jupiter's energy to desert you, and, indeed, many of your kin have ended their lives exiled, imprisoned, or ostracized from society. Charlie Chaplin was exiled from the United States; Karl Marx had only seven people turn up to his funeral; Lenny Bruce died alone; poor and isolated, Richard Nixon resigned from the U.S. presidency in disgrace; and even Capricorn Jupiter Margaret Thatcher was ousted from power when she became more extreme in her politics, turning her party against her.

Take a moment, look back long and deep over your life, recount as many events as you can, and ask yourself what has made you lucky or unlucky in your life. You'll notice that whenever you act like a Capricorn, *you create miracles in your life.* No matter what your Sun sign is, *you get lucky when you live out the Capricorn legacy gifted to you by the benevolent planet Jupiter.*

To take advantage of your luck, emphasize your . . .

Capricorn Jupiter Strengths ♑♃	*Be unbreakable, a looker outside the box, philosophical, a pragmatic idealist, a tragicomic master, tenacious, nobly enduring, brilliantly focused, and in it for the long haul.*

But because Jupiter rules excess, watch out for . . .

Capricorn Jupiter Excesses ♑♃	*You can become divorced from feeling, cut off from life and its joys, and power hungry.*

Understanding the Capricorn Energy and Influence

Capricorn is born during the most challenging time of the year. Your time is the cold, dark month that runs from the Winter Solstice to the middle of January. The trees and the ground are bare (or covered in ice and snow!), and it is time to squirrel away and save for the future. In the ancient times, this was a season when nature's bounty was slim indeed, and every root, every frozen berry, was worth its weight in gold. Those with plentiful, rich stores, organized after the harvest, were truly lucky. Remember this, Capricorn: Preparation, organization, and discipline are lucky for you.

Despite ancient fears about survival, there is birth and life even in the dead of winter. Capricorn, above all, is realistic, striving, climbing upward to come out on top. Sticking to your goals, staying on track, no matter what life or winter throws at you is lucky for you. You will even embrace, enjoy, and become thrilled by the challenge. Capricorn is the Winter Winner!

♃♑

Over six thousand years ago, there was a magnificent Sea Goat named Ea. He was a famous teacher and brought civilization to the people of Mesopotamia. He lived in the oceans of the Earth and came out every day to give his wisdom to the people, and every night he returned to his sea. Thousands of years passed, and as he adapted more and more to the Earth, *where he was needed,* he evolved into the sexual Pan, the god of nature, who played the most beautiful and alluring music in the world on his reed pipes. Thus, Capricorn has many different sides: teacher, creator, survivor.

Now that the goat has learned to live solely on the land, it is survivor *extraordinaire!* It maneuvers precarious paths and ledges that would kill other animals, and makes leaps of thirty feet to small ledges that are barely wide enough to stand on. The goat knows how to stretch and reach for new heights and goals. It always lands on its feet!

CAPRICORN SYMBOL

The Sea Goat

CAPRICORN PARTS OF
BODY
The Bones, the Teeth, the Knees, and the Joints

Capricorn rules the body's foundation; that is, the very bare *bones*. It is master of the entire skeletal structure; in particular, the teeth, the knees, and the joints. It is obvious that without Capricorn, there would be no foundation in the broadest sense. So, quite literally, dear Capricorn, you need structure, a strong foundation, a sense of history, and to be *grounded* in reality and to feel your life firmly footed *on the ground*—and if you do not have these in your life, you must create them! It is interesting that Capricorns themselves have beautiful bone structure, but they always seem to have a blow-out with their teeth—caused, I believe, by too much chewing over of things, worry, grinding of the teeth. Here is where the mastery becomes control and obsession. Rely on your foundation; do not try to move the Earth.

ୱଓ4ୱଓ4ୱଓ4ୱଓ4ୱଓ4ୱଓ4ୱଓ4ୱଓ4ୱଓ4ୱଓ4ୱଓ4ୱଓ4ୱଓ4ୱଓ4ୱଓ4ୱଓ4ୱଓ4ୱଓ4ୱଓ4ୱଓ4ୱଓ4ୱଓ4ୱଓ4ୱ

CAPRICORN COLORS
Granite Gray and Dark Green

Granite gray, the color of stone and hearth, is lucky for you. Think of a gray like the core of the Earth, the mountains, or ancient, wise, molten rock. Embrace this color in clothes and fabrics— wear a light gray cardigan or charcoal scarf—and you'll draw Jupiter luck and energy. A second traditional, lucky hue is another earthy color—dark green. Think of classic evergreen shades, the hue of winter pines on the mountains, the rich, dark green of the juniper hedge or the holly's leaves. Use dark green accents, like dark jade jewelry or evergreen shoes, for their lucky power.

Capricorn Jupiter Strengths ♑

And now your Capricorn Jupiter strengths:

Be Unbreakable

Capricorn Jupiter Langston Hughes summed it up when he wrote, "I've been insulted, eliminated, locked out, and left holding the key. But I am still here." Remember this, Capricorn Jupiter: Stick to your guns, stay determined, don't let anything the world sends your way have the power to hurt you, and you'll win.

Rubin "Hurricane" Carter, born during a Capricorn Jupiter passage on May 6, 1937, symbolizes indefatigability to the nth degree, first by overcoming his troubles with the law as a teenager to become a famous prizefighter, knocking out legendary boxer Emile Griffith, to his successful fight to overturn his wrongful murder conviction. In 1967 Carter was convicted and sentenced to life imprisonment for the murder of three white men at a bar, despite shaky testimony. While incarcerated he was a model prisoner, spending most of his time reading and writing. In the early 1980s, Rubin successfully worked with young activist Lesra Martin, championed by Bob Dylan, to have his conviction overturned and his innocence vindicated. Rubin Carter is a symbol of what being a Capricorn Jupiter is all about!

The more challenge the better for Capricorn Jupiter—it's your basic movement and, in a sense, your nutrition; it motivates and fulfills you. Frederick Douglass, born Capricorn Jupiter in 1818, wrote, "It is not light that is needed, but fire; it is not the gentle shower, but thunder. We need the storm, the whirlwind and the earthquake!" *This is what you, Capricorn Jupiter, need to succeed the most.* It's not the challenge, per se, that turns your Jupiter energy on, but the chance to overcome, requiring the need for absolute discipline, almost monastic discipline, commitment. Like Capricorn Jupiter, what happens when you meet the challenge? Astonishing success, luck, and the power to change.

Be a Looker Outside the Box

The lucky power of the benevolent planet is drawn to you when you look around, inside, above, and over the surface of things. When you've seen the essential *beyond,* as Capricorn Jupiter theater director Peter Hall famously did in his pioneering productions *When You Break Through the Fourth Wall.*

Capricorn Jupiter Edgar Cayce, the Sleeping Prophet, had his own way of finding out the hidden truth. On each day for more than forty years of his life, Cayce would lie down on a couch with his hands folded over his stomach and allow himself to enter a sleep-induced sleep state. Then, provided with the name and location of an individual anywhere in the world, he would speak in a normal voice and give answers to any questions about that person that he was asked. These answers were taken down by a stenographer. Today on file at the Association for Research and Enlightenment in Virginia Beach are copies of more than fourteen thousand of Cayce's readings. Cayce was also a Sunday-school teacher, father of two children, and a photographer. He used his psychic skills and his Capricorn Jupiter to become the pragmatic psychic: His reports were known for their detailed medical diagnoses and anatomical observations.

Another great example of Capricorn Jupiter insight is playwright

✳ Draw on your strength and wisdom to triumph over the greatest number of obstacles.

✳ Be philosophical about life and people, looking through the long lens of experience.

✳ Act with a view toward long-term plans, with great maturity, discipline—and humor.

✳ Cultivate patience, self-discipline, self-control, responsibility.

✳ Increase the measure of your success by implementing *twelve-year* plans.

✳ Eat regular meals and schedule weekly relaxation exercises.

✳ Wear and surround yourself with charcoal and granite grays and dark greens.

✳ Hang out with more Capricorns, Virgos, Tauruses.

Oscar Wilde. Wilde was very much a self-created man: He made himself famous through years of determined hard work, a biting wit, and a brilliant ability to see the truth and to express it back to us, using paradoxes and jokes. And in his plays, like *An Ideal Husband* and *The Importance of Being Earnest,* he expressed deep truths about life and Victorian society. "Earnest," for instance, was Victorian slang for "gay"—and the play was Queen Victoria's favorite!

Other Capricorn Jupiters uncover secrets and hidden information through journalism and investigation. Capricorn Jupiter Seymour Hersh broke the story of the Mai Lai massacre and continues to report the inside scoop on the Pentagon to his readers. He was called the "closest thing American journalism has to a terrorist" by Gemini Jupiter Richard Perle after Hersh uncovered Perle's business dealings in Iraq. Pulitzer Prize–winning Capricorn Jupiter Peter Arnett reported from Iraq during both the 1991 and 2003 wars, only to be fired by CNN during the second Gulf War for daring to present the "other" side of the story. These journalists are the scariest for the status quo, perhaps because their reports are, like Capricorn Jupiter, the most unflinching, direct, and history minded of anyone's. Be like them—unflinching, direct, a powerful truth seeker—and luck will follow.

Be Philosophical

It's no accident that many brilliant philosophers are Capricorn Jupiter. They find luck through their dedication to seeking the truth and their hard work. One example is Saint Thomas Aquinas—the medieval philosopher-theologian who attempted to synthesize Aristotle with the doctrines of the Church and revolutionized Western thought and philosophy. Aquinas wrote that there was no conflict between developing one's understanding of God through reason and through faith: This revolutionary idea permitted Christian thinkers, for the first time, to use science and logic to investigate the world. This Dominican friar awoke the West's modern tradition of philosophy, reason, and thought from the Dark Ages—a Capricorn Jupiter accomplishment, indeed. It is also interesting to note that Aquinas's *a priori* reasoning brought him to the revelation that "the celestial bodies are the cause of all that takes place in the sublunar world"!

Another great Capricorn Jupiter philosopher, Karl Marx, focused his thinking in the material, the real, and the concrete to produce the most brilliant critique of capitalism and modern society ever written. Marx refused to believe anything that happened in political history happened for other than economic reasons. This focus on the gritty realities of life (as opposed to far-flung idealist fancies) is typically Capricorn Jupiter. It's no wonder that both Aquinas and Marx are among the few philosophers who genuinely continue to influence the world we live in today.

Be a Pragmatic Idealist

Think of Capricorn Jupiter Rosa Parks: Her refusal to back down in the face of opposition was the spark that started the Civil Rights Movement. She sat, literally, for the extra mile on the bus that declared it illegal for her to sit down, and then walked every day to work during an eighteen-month bus boycott, at the end of which the bus company had no choice but to end segregation on its buses. (And it wasn't because Montgomery got idealistic; the buses were losing too much money!) Rosa famously said that her refusal to give up her seat to a white man was because she was *tired*—tired of putting up with injustice, tired of being treated so badly, and just plain tired after a hard day's work. Your Jupiter sign is, above all else, an Earth sign, and so you must listen and respond to the physical. Your idealism is not the abstract ivory tower type, but real and here, today; responding to what needs to be done in the real world.

As German theologian Martin Buber, born a Capricorn Jupiter, said: "The idea of responsibility is to be brought back from the province of specialized ethics into that of lived life. Genuine responsibility exists only where there is real responding. Responding to what? To what happens to one, to what is to be seen and heard and felt."

Draw upon your power, dear Capricorn Jupiter, to improve "what happens to you": those conditions that you live with every day. Whether it's at work; your living conditions; how you are relating to your mate, family, or friends; or the state of your neighborhood. Trusting your responses and your sense of responsibility is key. Even such a *romantic* poet as John Keats could write: "Nothing ever becomes real until it is experienced—even a proverb is no proverb to you till your life has illustrated it." He was born on October 31, 1795, a Capricorn Jupiter.

Be a Tragicomic Master

You, Capricorn Jupiter, are the tragicomic master, the comedian who reduces us to hysterics one moment, tears the next. If you engage the power of Jupiter, you will have the power to make people laugh and the power to see the fundamental truths about people and their lives. This can make *you* a power yourself.

Virtually every great comedian whose wit and humor belie the sadness of the clown falls within this Jupiter sign. Charlie Chaplin, the wonderful, human little tramp; Peter Sellers, the hilarious yet touching Inspector Clouseau; Danny Kaye, the great comedian of early Hollywood; John Belushi, humor touched with self-destruction; Lenny Bruce, the comedian whose work was so important and so close to the grain that his routines were banned: All are Capricorn Jupiter. Why are these guys so great at comedy? We love laughing at their routines but there's a wisdom here, too, an experience that just cannot be found in other Jupiter signs that make the dry wit, the hysterical pathos—perhaps it is best described by another Capricorn Jupiter, Langston Hughes: "Humor is laughing at what you haven't got when you ought to have it." It is a Capricorn trait to make us laugh while showing us the truth. You can use this to your advantage. Humor can help you communicate effectively, and with Jupiter's aid you can use it to touch people at their innermost core. Think of Capricorn Jupiter Jack Nicholson in *One Flew Over the Cuckoo's Nest,* a movie that makes us laugh, gasp, and cry. Use your skills at performance and comedy, too, and you can touch the world as well.

Be Tenacious

Capricorn Jupiter, think of those determined mountain goats, known for their fierce ambition, that endlessly trudge up the mountain to reach the top. But note that while Capricorn Suns are known for their focus on image, career, public relations, and ambition, these concepts are not particularly lucky for you. It's actually luckier for you to be tenacious and determined and not to think about what others think of you at all.

Many great Capricorn Jupiters achieved success by not worrying in the least what others thought of them! Margaret Thatcher, born a Capricorn Jupiter on October 13, 1925, became the first (and so far only) woman prime minister of Great Britain in spite of the pervasive "old boys" network. Known for her incredible stamina and determination, Thatcher refused to let the early snubs of her rivals hamper her political ambitions. Another Capricorn Jupiter, Richard Nixon, suffered electoral defeat, but his determination and tenacity won him the presidential election of 1968. Following his Capricorn Jupiter excesses of pride, ambition, and paranoia, which led to his downfall, he then came back into the political arena as a private citizen to develop relations with China.

Be Nobly Enduring

Marian Anderson, born a Capricorn Jupiter on February 27, 1897, was the first African American to perform at the Metropolitan Opera. In 1939 Marian planned on giving a concert in Washington, D.C., but the Daughters of the American Revolution refused to allow her to perform at Constitution Hall because of her race. Incensed with this, Eleanor Roosevelt resigned from the group and helped sponsor Marian at a concert at the Lincoln Memorial where she sang for an audience of 75,000 while millions more listened on the radio. Marian Anderson performed with the Metropolitan Opera in Verdi's *Un Ballo in Maschera* as Ulrica. Her last concert was in 1965 at Carnegie Hall. She served as an alternate delegate to the United Nations and received numerous degrees and awards, including the Presidential Medal of Freedom in 1963 and the National Medal of Art in 1986. Anderson kept her dignity intact, worked hard and tenaciously, and became a role model to millions. Be like her, Capricorn Jupiter—be nobly enduring, and the luck and wealth of Jupiter will come to you!

Be Brilliantly Focused

Carlos Castaneda, born a Capricorn Jupiter on December 25, 1925, put it perfectly and succinctly: "To get to your impossible destination, you have to focus on only the immediate point in front of you. Forget everything else surrounding that point, and you will get where you're going." Like this Jupiter philosopher, sage, and student, you should concentrate on the essential and overlook the unnecessary. Castaneda's journey began when he studied with the Indian teacher Don Juan and was initiated into the knowledge of the ancients. His work teaches us all about looking for the secret truth and discarding distractions.

This journey, Capricorn Jupiter, will involve reading and learning—they are lucky for you. Cultivate brainy discernment, history, knowledge, and wisdom; learn the truth that will guide your choices and shape your priorities. The great philosopher of religion William James, of course a Capricorn Jupiter, said, "The art of being wise is the art of knowing what to overlook." Overlook what is not necessary in your own life, dear Capricorn Jupiter, and find the joy and riches that remain.

Be in It for the Long Haul

To get lucky, go that extra mile! Think Capricorn Jupiter Charles Lindburgh, who made the first transatlantic nonstop flight—the *longest* flight—going without sleep in a tiny plane.

Or Capricorn Jupiter *long-distance* runner Jesse Owens, who so brilliantly made medal-winning runs at the Munich Olympics despite protests from the Nazis. Like these two Capricorn Jupiters, luck will come to you if you work hard, plan for the ultimate goal, persevere, and stick to your guns.

Born in 1973, the famous Supreme Court case *Roe v. Wade* established a constitutional statute recognizing a woman's right to choose. Talk about a long haul! This took years to be brought into the legal arena! And, precisely because it is Capricorn Jupiter, it doubtless still has a tough road ahead of it.

It's no accident that *most* of the pioneers who fought for the rights of African Americans were Capricorn Jupiters. Or that the greatest number of successful African Americans today are—still—Capricorn Jupiters. From Marian Anderson, to Langston Hughes, from Louis Armstrong, Hattie McDaniel, Sammy Davis Jr., B. B. King, Malcolm X, Medgar Evers to Bill Cosby, Johnnie Cochran, RuPaul, Garrett Morris, and Colin Powell—all Capricorn Jupiters! All had to fight long and hard. As Capricorn Jupiter Yogi Berra said, "It ain't over till it's over."

What Makes You Lucky in Love

In your love life, try to achieve the impossible. If you know the sort of person who would make a perfect partner, look for them as you date and don't settle for less. Your partner might at first be a bit unwilling, but if you are a persistent wooer, you will bring them around. You need to be extremely patient in relationships; it will be rewarded in the end. Think of the future; *do not dwell on your partner's past.* As Capricorn Jupiter Wayne Gretzky, born January 26, 1961, advised: "Skate to where the puck is going and not to where it's been."

Remember your best love luck comes when you stick to the classics. Dress conservatively, in muted shades and toned-down fashions. Emphasize your classic good looks without overwhelming them. Be the same on your dates: Pick classic, traditional nights out. Go at first to places that aren't too loud, where you can talk to your mate—you'll want to talk all night, Capricorn Jupiter! Once you make that connection, enjoy all that nightlife has to offer.

You must have patience in your relationships and allow them to develop. Jupiter won't be generous if you are too impulsive. Instead, plan things carefully, hang back, and watch things develop slowly and surely. Don't run away if a prospective mate plays hard to get. Instead, see it as one of your blessed Jupiter challenges! This does *not* mean you should be self-righteous or controlling. Make sure you know the difference between playfulness and disinterest.

Your Best Career Moves

You can fly to the Moon and achieve the impossible, but you must be willing to do the legwork and wait years, if possible. Don't look for sudden-success careers. Instead, try to learn a skill that can take years—painting, writing, or changing the world—and put all your energy into becoming the best you can over the years. Self-improvement is your mantra. Expand and develop extensively beyond the current limits of one's chosen field of endeavor.

To succeed at work, set tasks for yourself that others feel are impossible. If there's a job no one has signed up for, you are the one who can do it, and give yourself the chance! Patience and determination will lead you to the top if you persevere (which can be very hard and feel unnatural if your Sun sign is an Air sign like Gemini, and you gravitate to constant change). Stick with the job for years if necessary and avoid change. Work hard at fostering self-discipline and long-term thinking. Capricorn Jupiter Shaquille O'Neal said: "You work hard now. You don't wait. If you're lazy or you sit back and you don't want to excel, you'll get nothing. If you work hard enough, you'll be given what you deserve."

Or what about fellow Capricorn Jupiter Ludwig van Beethoven? One of the hardest-working and most prolific composers, Beethoven wrote so many piano sonatas that explored so many variations in the form that his contemporaries believed he had exhausted almost all the sonata's possibilities! Very few were written after Beethoven. Even deafness could not stop his ambition and his drive to compose. He won his success through patience, determination, and hard work, and he became famous, wealthy, and powerful in his lifetime by harnessing the bounty of his Jupiter sign. Be the Beethoven of your career, Capricorn Jupiter, and success will come to you.

How to Reach Optimum Health

While warmth is lucky for you, Capricorn Jupiter, extra exposure to the sun is not. Be especially careful to protect yourself by using a good sunscreen and staying out of the tanning salon. Take good care of your skin, which can be prone to dryness, by using sesame and almond oil, both of which carry special Jupiter energy for you. It is good for you to bathe to balance your Jupiter energy; add wintergreen oil or wintergreen extract—just a few drops—to recover your tenacity and determination. Wrap yourself well and warmly in cold or damp weather. And don't forget to eat regularly! It's easier for you than for most to go without. Capricorn Jupiter Julianne Moore was not only speaking for her profession but for all Capricorn Jupiters when she said, "If you want to know the truth about actresses, it's that we're hungry all the time. We really don't eat."

And, while Sun sign Sea Goats need to protect their bones with extra care, under the power of the planet that protects, Capricorn Jupiter, you can be less cautious. Don't be afraid to go out there and climb—but don't do anything foolish, either!

Your Lucky Foods, Element, and Spices

Capricorn Jupiter, the most important health tip is this: Keep your diet high in calcium. Use calcium supplements and drink lots of milk to keep teeth, bones, and skin in good condition.

Capricorn's cell salt is calcium phosphate, crucial for bone formation and composition of the skeleton. Lack of this mineral causes tooth disorders, problems with your knees, spinal curvature, pains in the joints, rickets, and misshapen bones. Foods that contain this mineral that you should include in your diet are spinach, lemons, kale, dandelion greens, figs, oranges, broccoli, cabbage, potatoes, oats, walnuts, and almonds. Variety is good for you, so be careful not to eat exactly the same thing every day! At least make sure you cover this every day: fresh fruits and vegetables, eggs, fish, a raw salad, yogurt, lean protein, and whole-grain breads. Oils, such as sesame and almond, are lucky for you when used topically. Your lucky herb is wintergreen; it can also be used as an oil or in extract form.

Your Best Environment

Dry, warm climates are best for you, as dampness and moisture are not beneficial for your bones. If you do live in a damp climate, please swaddle yourself in warm clothing (and especially wear good socks!). You need soothing sounds and colors—preferably of the Earth—pleasant music and people to function best. Countries ruled by Capricorn include Mexico, India, and Bulgaria; and the Capricorn-ruled cities are Chicago, Montreal, Boston, Brussels, and Oxford. You may very well find your true love in England, which has its Venus in Capricorn!

Your Best Look

For your best luck, stick with classic, strong looks and avoid the ephemeral or flashy. Your excellent figure, classically proportioned and fine, doesn't need any additional frippery, design, trend, or color to look absolutely great. Classical, mature, conservative dark colors make you the most beautiful (remember your luckiest are gray and dark green). Stick to tailored, fine-quality fabrics. And stick by the rules: Crazy, mixed-up fashions don't suit you at all. Use fashions and colors that are understated, not out there—your incredible beauty shines best in these.

Your Best Times

The luckiest times for you are when the Moon is in Capricorn, which happens every month for two to three days. Find these special days in your astrological calendar. During these days, schedule crucial meetings, tell that special person you love them, ask for a promotion or raise, push the envelope, and plan to shine. In your everyday, the best time for you to exercise is late afternoon to early evening. The luckiest times of the year for you are December 22–January 19 (when the Sun is in Capricorn); April 20–May 20 (when the Sun is in Taurus); and August 24–September 22 (when the Sun is in Virgo.)

The very luckiest time of your life—as in Lotto lucky—is when Jupiter reenters the sign it was in when you were born (Capricorn), which happens every twelve years and lasts anywhere from a few months to just over a year. This is called your Jupiter Return, it launches your luckiest streak, and is the year to top all years. You can count on the *extraordinary* to happen the very day your Jupiter Return begins. Amazing events can start happening up to six weeks beforehand, so plan ahead!

How to Be Lucky Financially

Games of chance are not for you, but long-term success strategies are, for your luck comes to you through hard work and perseverance. For your investing strategies, Capricorn Jupiter, you don't need to look for quick money; instead, invest your savings in slow-growth opportunities and stick with them. You will have great luck investing in real estate. Find apartments or houses that need fixing up, work on them—slowly and surely—and you will eventually find great, lucky profits.

Learning financial skills and investing money is lucky for you, so don't be afraid to invest in yourself—attend seminars, buy books, learn all you can about money. Once you do, develop your own investment strategy based on the principles of the tried and true. If you find a good stock to invest in, don't trade too quickly; instead, look for opportunities you can hold for years as your companies will grow and grow. In mutual funds, you should look for growth over value, and don't be afraid to pick companies that make amazing promises to change the world. If you believe in them and stick with them, they will pan out. But also make sure you use your lucky Jupiter talents of looking outside the box and being a dogged truth seeker. Use your research and investigative skills to make sure you choose your investments wisely.

Success in Family and Friendship

The Greeks called Saturn the Sun of the Night. This may be why you are nocturnal; you're at your best late at night, while the rest of the world is sleeping. Use this to your advantage. Throw late dinner parties, take evening classes, have your brothers and sisters meet you for evenings out. Take them out and show them what the night can offer. Remember, night owl Capricorn Jupiter, what fellow Jupiter George Foreman said, "Sleeping was my problem in school. If school had started at four in the afternoon, I'd be a college graduate today."

Above all, be the one who offers *The Right Words at the Right Time*. This is the title of one of Marlo Thomas's books, which offers just that. That Girl's Jupiter sign? Capricorn, of course! Both your friends and family depend on you to support them, emotionally if not in other ways as well. This brings you luck, remember, so try not to take it as a burden.

Capricorn Jupiter Excesses ♑♃	*When you go too far!*

The perfect cautionary tale for you is Bertolt Brecht's famous play *Mother Courage*. It is the story of a woman who struggles to survive with her two children against the backdrop of the Hundred Years' War, selling supplies to the army. The recurring image of Mother Courage pulling her wagon behind her, constantly trudging forward, is a good one for you to beware of in your own life. In the process of trying to get ahead in this grim existence, she loses both her children and any personal life she may have had. It is a tale of endless endurance and suffering, with no end in sight.

You must beware of making life unnecessarily difficult for yourself. Sometimes the power of this Capricorn Jupiter is so much that you can think things are more difficult than they really are. Stay aware, dear Capricorn Jupiter—it does get easier the older you become.

There can be so much pleasure in self-deprivation at times, however, that it can overwhelm your main goal—which is accomplishment. As much as you need to let yourself enjoy the journey and not sacrifice yourself for (solely) the destination, so you also need to not enjoy the pain and suffering of the journey so much that you forget your destination.

Excess #1 — *You Can Become Divorced from Feeling*

There's a haunting song by k.d. lang that goes "I've been outside of myself for so long." Lang is a Capricorn Jupiter and you know she really knows what she's talking (or singing) about. She even addresses herself in third person—"Oh, Katherine," she sighs in the song—so you know how distant and cut off she feels from herself. It's very moving (and beautiful) and painful to listen to, and describes more strongly than any other song *I've* ever heard what it is like to be "outside of yourself." A great book that does the same is *L'Etranger*, by Albert Camus, which begins indifferently: "My mother died today." It became one of the great works of the existentialist movement.

Capricorn Jupiter William James said, "The deepest longing in the human breast is the desire for appreciation," and he was certainly speaking for his and your kind.

Excess #2 — *You Can Become Cut Off from Life and Its Joys*

Another cautionary Capricorn Jupiter tale: A beggar is sitting on the side of the road when a stranger walks by. Eyeing the stranger he decides to ask the stranger for some change. The stranger looks at him and says no. Then he asks the beggar what he is sitting on. The beggar says it is nothing but an old box that he has had for years. The stranger then asks the beggar if he has looked inside and the beggar says no. With the stranger's urging the beggar looks inside the box and discovers gold.

DON'TS

✶ Involve your ego, indulge in inflated ambition, or take risks to increase your prestige.

✶ Expect anything to be done right away.

✶ Feel that life is against you, become self-pitying, self-righteous, rigid, or paranoid.

✶ Make heavy-handed judgments and become rigidly controlling or condemning.

✶ Make enemies or invite rancor, lawsuits, or ego-directed battles of any kind.

✶ Stay out long in the damp or if it's raining!

✶ Skip play breaks, listening to music, or meals to complete all your tasks.

The gold represents those treasures inside of us which have been forgotten due to life's circumstances in the closets of our minds. It's time to look in the closet and see what's in there. It might be in there because of some voice saying no or because we were willing to settle for less. As you clean out your closet, see if you discover a dream that is vibrant with challenge and excitement, a dream that is outside the box.

A suggestion would be to start with just one goal or dream and commit to exploring this dream, your unique dream for at least a year. Peek ahead a year from now and imagine how your life would be if you were living this dream. Going for your dreams is always a journey full of ups and downs, joys and sorrows, an exploration rich with meaning and purpose. Enjoy the journey!

Excess #3   *You Can Become Power Hungry*

Beware especially this particular excess, Capricorn Jupiter. Sometimes your ambition and love for power can, in unusual circumstances, lead to paranoia, self-righteousness and attempts to control everything. Richard Nixon earned his presidency through his tenacity and determination—good Capricorn Jupiter qualities. He fell from power when he tried to control everything and when he gave in to paranoia. Other examples of this excess are dark indeed: Capricorn Jupiters Saddam Hussein and Adolf Hitler are two examples of the extremes this excess can lead to.

Sun Sign-Jupiter Sign Combinations

Capricorn Jupiter is your lucky recipe for getting exactly what you want—maybe not as quickly as you'd like, but getting to the head of the line every time.

Setting impossible tasks and goals for yourself comes naturally to both Aries and Capricorn. Use your Jovian self-discipline to make them winning achievements, not haphazard hurdles. Investing your fiery Aries energy toward long-term, planned goals may seem like a restriction on your freedom, but if you use your earthy Capricorn Jupiter, your efforts—and accumulated dollars—will eventually put you in a position of greater freedom, able to do as you want, when you want. Accept your Jovian challenge and win! Aries-Capricorn astrologer Linda Goodman drew upon her Aries to be a pioneer and her Capricorn luck to make sure her trailblazing lasted—which it has! Concerned with permanence, she could proudly say, "Astrology has spanned the centuries and made the journey intact."

The Taurus Bull and the Capricorn Goat together accomplish anything with persistence, perseverance, and endurance. Both are earthy, practical, and value material attainment, yet Capricorn Jupiter asks that you forgo some comforts of the present for more future—and permanent—gains. For Jovian luck, dream *big* dreams and move with the Goat's agile tenacity rather than the slow determination of the Bull, who may be content lolling in grassy pastures while the goat climbs the heights.

In the long run—which is what Capricorn Jupiter luck is all about—you'll be rewarded with all that makes life good for you: a luxurious environment and the good things of life, security, solid achievement, a position of unassailable authority, and a love that grows richer and more sensual with the years. Allow yourself a chuckle or two along the way; you will certainly laugh all the way to the bank! Strong, silent-film star Gary Cooper, who never sought a confrontation but always won, epitomizes the Taurus-Capricorn stellar force.

IF YOUR SUN SIGN IS:

Gemini

(May 21–June 20)

You get lucky by acting in ways contrary to your airy Gemini nature: by persevering rather than experimenting, by becoming master of one or two trades, rather than Jack of all. I know, dear Gemini—sounds boring. But think of the rewards! Your Jovian placement asks you to dream impossible dreams and to challenge yourself; to engage your quicksilver mind to learn more, and do more, by setting up daunting obstacles and figuring out how to leap over or dismantle them; to continually hone your talents to perfection; and to have the persistence to woo, win, and keep that fascinating but seemingly unobtainable ideal mate. By following your Capricorn Jupiter, you have it all.

Beloved actress Hattie McDaniel used her Gemini expression in film acting and her Capricorn Jupiter to endure the long wait to become the first African American to win an Oscar.

4Ⱏ♑4Ⱏ♑4Ⱏ♑4Ⱏ♑4Ⱏ♑4Ⱏ♑4Ⱏ♑4Ⱏ♑4Ⱏ♑4Ⱏ♑4Ⱏ♑4Ⱏ♑4Ⱏ♑4Ⱏ♑4Ⱏ♑4Ⱏ♑4Ⱏ♑4Ⱏ♑4Ⱏ♑4Ⱏ♑4Ⱏ♑4Ⱏ♑4Ⱏ♑4Ⱏ♑4Ⱏ♑4Ⱏ♑4Ⱏ♑4Ⱏ

IF YOUR SUN SIGN IS:

Cancer

(June 21–July 22)

You're attuned to the Moon and tides, and this Jupiter placement gives you your heart's desire, leading you away from melancholy moods and toward your goals. Hard work and tenacity are natural to you; now, let Capricorn Jupiter expand your vision, bring you out of your cautious shell-ter and into the world of possibilities. You *will* achieve the impossible dream when you set a life aim that requires diligence, dedication to work and achievement, and the satisfaction of overcoming all obstacles. Capricorn Jupiter has a suggestion for dealing with those: Laugh at the boulders life drops in your path and the ploys you use to cope—works every time.

You carefully increase your wealth not by hiding it away, but by investment in slow-but-steady growth companies. You get the simpatico love of your life by quietly persistent wooing. And long term, dear Cancer-Capricorn, you'll reach any career goal, from science to business to the arts, if you use your Jovian lucky map. Capricorn Jupiter director Jean Cocteau created beautiful films with archetypal evocative images—including those of goats—which made him enduringly successful!

Capricorn Jupiter offers Leo some king-sized challenges, but you handle them royally and succeed brilliantly with this Jupiter placement! Envision your grandest ambitions, your highest achievements, and then set out on the long but glorious road to success. Know that it will take constant hard work, that there will be obstacles challenging all your Leo strength and cunning, that you must exercise more patience than is natural to you, but that you'll be steadily progressing toward your best self, and your greatest goals. Avoid any Cowardly Lion self-doubt by bravely facing—even looking for—daunting obstacles to conquer.

For financial success, temper your natural generosity with prudent planning and investment; you'll have funds for benevolence and self-indulgence in the end. Whatever you can dream, you can do. In love, persist and accept nothing less than your perfect royal consort. Capricorn Jupiter Emily Brontë endured not only nineteenth-century prejudice against female writers but a horribly restrictive home life to bring *Wuthering Heights*, her novel of love beyond death, to the public.

4VঙৄVঙৄVঙৄVঙৄVঙৄVঙৄVঙৄVঙৄVঙৄVঙৄVঙৄVঙৄVঙৄVঙৄVঙৄVঙৄVঙৄVঙৄVঙৄVঙৄVঙৄVঙৄVঙৄVঙৄVঙৄVঙৄVঙৄVঙ

Your earthy Virgo strengths—practicality, discernment, and desire for perfection—are reinforced by Capricorn Jupiter. Expand your possibilities by picturing in detail the brilliant murals you will paint, the political or charitable plan you'll carry to completion, the epic novel you'll write. These blueprints for success take time to realize, but with Capricorn Jupiter, you have patience and time on your side. Dreams steadily become reality as you persevere, and even obstacles lead to greater achievement!

In romance, the person worth having is worth waiting for—rein in your critical nature with patience, and avoid giving too much of yourself away. Capricorn financial advice suits your careful nature; you also achieve wealth by investing in companies with idealistically motivated growth. Heed your Jove's advice and you'll have the world your mind has created. Virgo-Capricorn H. G. Wells brought his detailed and extremely well thought out visions of the future before the public in many books (later to become films), including *The Time Machine*.

IF YOUR SUN SIGN IS:

Libra

(September 23–October 22)

Libra's Venusian beauty combines with Capricorn Jupiter to create a "steel magnolia": easy charm on the outside, with the tensile strength and never-quit attitude of the Goat on the inside. Like Libra-Capricorn Gwyneth Paltrow said, "Beauty, to me, is about being comfortable in your own skin."

Leave Libra indecision behind and get lucky—and successful—with Capricorn Jupiter's determination. With your Libra appeal, you can attract a following, whether of voters, an army, or an audience. Remember that a good leader asks more of him- or herself than of anyone else. You must have a single-minded plan and the grit to work for the sustained time to accomplish your goals. Hoard your energy; know what's worth pursuing and what's extraneous. Don't make snap judgments about people or situations; *do make considered decisions* and stick with them, including choice of a significant other. If this all sounds, dear Libra-Capricorn, too Earth-bound and utilitarian for your beauty-loving Libra soul, remember that you'll achieve lasting success and romance—and be able to afford that genuine impressionist painting to hang on your wall!

IF YOUR SUN SIGN IS:

Scorpio

(October 23–November 21)

Scorpio with Capricorn Jupiter gives you a fascinating quality, more subtle than simple charm: hidden depths, a hint (or more) of sexuality, intensity. You have incredible strength of purpose and an agenda; you know where you're going. If you follow your Capricorn Jupiter, you create your luck by linking your determination to great dreams and causes, whether in theater, as a teacher, a researcher—even a psychic. You accept hard work, persistence, and surmounting obstacles without complaining. Robert Kennedy was a prime example of the Scorpio-Capricorn crusader and fighter for causes.

For success in love, beware of being inflexible and seeing only your own viewpoint; be objective and don't forget that you must always require the most from yourself. Wait with patience till the bond (not only romantic, but psychic—Jung's "silver cord") is secured. Money will come through investing in your most secure growth potential, your own talents and ideas, so you need to continually refine and improve your best investment—*you*. Finally, Scorpio-Capricorn, don't forget to add a little levity to your long trek toward success: See the humor in life's ironies and contradictions.

Capricorn Jupiter applauds when you draw your bow and aim for the Moon, for seemingly impossible goals in work and love. It asks you to follow those arrows steadily and unceasingly—no quick, short dashes, please—to commit yourself to hard work, to be patient and persevering, seeing progress in each step and keeping in mind that the Goat will eventually lead you to the top of the highest peak.

Success in love comes from channeling your romantic longings into a worthy ideal, pursuing her or him with patience and persistence. Luck for Sadge-Capricorns comes when it knows where to find you; in other words, choose a career, cause, partner, or path and stay put. Like a fiery horse controlled by a firm hand on the reins, your Sadge energy and enthusiasm, disciplined by Capricorn determination, will make you a winner.

Sadge-Capricorn Sammy Davis Jr. had both the Sagittarius warmth and the Capricorn endurance and constant honing of his talents to become a lasting icon of success as an African American entertainer.

᷄ᚥ᪻

With both Sun and Jove in Capricorn, you get a double dip of determination; you *won't* be turned aside by difficult tasks, hard work, long hours, or years—or by anyone who might be foolish enough to try. To avoid making enemies, let "respect" be your watchword: respect for others, and for yourself, keeping ambition in check with integrity and fortitude.

You can succeed at almost anything you choose, from writing to the military, acting to philosophy to comedy, boxing ring to psychic exploration. Whatever you choose to do, you must dive deep, take it to the limit while still exercising self-control. With this intensity, you have the power to be the ultimate, a legend in your field and in yourself, so be sure that your actions are worthy of history. Invest both your time and your money in long-term—preferably idealistic—ventures, and the returns will be great, both for you and for the world. Love, for you, must be eternal and include mutual respect; it is your very private deep well, sustaining and replenishing you in all your endeavors. Double Capricorn Elaine May has used her wit and humor to move through a long and successful career, as actor, comic, and director, respected in all fields.

IF YOUR SUN SIGN IS:
Aquarius
(January 20–February 19)

Aquarius innovative intellect and humanitarianism coupled with the drive and practicality of Capricorn Jupiter gives you astounding success in a diversity of fields—literature, politics, social change, philosophy, science, music—always on the cutting edge. Use your Capricorn Jupiter to translate your vision of the future into dedicated work, always with the aim of improving your skills and reaching your goals, whatever brick walls confront you. You have the perception to find the key to the locked door and the ability to endure and to work harder than seems humanly possible. The question of how long is irrelevant; "as long as it takes" is your answer. You enter into partnerships—love and others—with the same approach: Once you see your vision clearly, you never consider turning away from it.

Capricorn Jupiter urges you to focus totally in each area of your life—career, love, finances—concentrating your Aquarius temptation toward constant exploration into an aim-oriented whole. And Aquarius' off-the-wall approach meshes nicely with Capricorn's sense of humor, lightening the arduous but so-rewarding journey. Aquarius-Capricorn Paul Newman has not only continually increased his skills and scope as film actor and director but also established his own medical summer camps (with a *Butch Cassidy* theme) to help children with life-threatening diseases; those whose luck, he says, has "been simply brutal." All the funding comes from his Newman's Own food company.

IF YOUR SUN SIGN IS:
Pisces
(February 20–March 20)

Dream your most impossible dream, Pisces, then with Capricorn Jupiter's help, give it substance. "Dreams are today's answers to tomorrow's questions," said Pisces-Capricorn famed psychic Edgar Cayce, known as the Sleeping Prophet. A vivid imagination is both your greatest asset and greatest danger; if the world seems awry, you tend to escape into wishful visions. You have so much to give the world, particularly in the arts (performing, painting, writing) or humanitarian fields (teaching, nursing, social work). *See,* in your mind's eye, what is possible—almost anything, with Capricorn Jupiter. Set the ideal as your goal; use Capricorn's determination, persistence, and ability to work as hard and as long as necessary to reach it. Imagine building the perfect castle—not a sandcastle, but one of stone-by-carefully selected-stone, arduous work, taking years—but you'll be creating something beautiful and lasting, and worth all the effort. Recognize your talents; constantly develop the skills to express them. And don't live with only a storybook romance, far away and unobtainable, but use your Capricorn Jupiter to woo and win and live happily ever after.

Relationships with Capricorn Jupiters

Finally, for those of you who aren't Capricorn Jupiter yourselves, but who have a spouse or significant other, boss, child, or parent who are, here are some great tips to help you learn how to deal! This is how the luck of someone else's Jupiter touches *your* life.

Your Significant Other

"And was it his destined part only one moment in his life to be close to your heart?" This was written by Capricorn Jupiter writer Ivan Turgenev and applies to most people's experience of a Capricorn Jupiter. Their feelings run very, very deep, but do they give freely of themselves? Not initially. If you want to keep one close to your heart, you may remember that Capricorn Jupiters like best what they have to work hardest for. They do. That's probably how they won you—didn't you feel won? Give in to it. You'll make them mad for you if you tease them or play games; they need to know there's a long-term commitment in there somewhere. Sex with a Capricorn Jupiter should always go slow, build to a smoldering fire, and then go crazy. You may have to alter your sexual rhythm for a Capricorn Jupiter mate, but I've never heard anyone complain. Not one.

The physical part of your relationship may be the easiest because they are an Earth sign. Capricorn Jupiters are surprisingly refreshing in their down-to-earth approach. They don't like pretension or games; they *love* honesty and directness. As Capricorn Jupiter Jenny McCarthy, born November 1, 1972, said, "My philosophy on dating is to just fart right away." They like everything out in the open, you see, and they do get fearful of potential problems up ahead. Just reassure them there are no monsters in your closet.

Their worst qualities are judgment, criticism, and a fondness to lecture or give teacherly type of advice. You can depend on them to be on time for all your dates, and if not, there really might be a problem. If they're with you, they care for you deeply and also respect you; and don't worry, no matter how "on top" they are (and they usually are, unlike other Jupiter signs), they do not do charity cases when it comes to their mates. You have to be top quality, too. They just might not say it all the time. You may have to train them in compliment giving.

Ah, but they do have a magnetic quality that keeps you hooked. As long as this is not an *abusive* relationship, this is fine. Never lose your will with a Capricorn Jupiter. They will fall out of love with you. And, sadly, when Capricorn Jupiters separate, it's for life. They never come back. If you've broken up with one, and you want them back, realize this, or you will waste—literally waste—years and years of your life. If you are married to one, don't mess it up. They don't give second chances. Stay married. Through the horrible and the worse. It always gets better. And you may never have such good sex with anyone else!

Your Boss

You know you've got a Capricorn Jupiter boss if they have the furnace blazing. They can't take the cold, you see. It makes them fuzzy minded. *And the most important thing is that the mind is clear*—not only their own, but that of everyone who works for them. If you want a job that allows you to have fun, to socialize, to slack off and relax on certain days, this is not the boss for you. They won't make a big deal out of it, but they expect the very best, in excruciating detail, without any emotion or excuses thrown in. It can be unnerving as there are really no social politics at all. Either you do it well, or you don't do it at all. It helps if you are as devoted to the work as they are, but the whys and wherefores don't matter; they just want you to do it—brilliantly and seamlessly. The highest quality, the best work, in the least time, and with a minimum of fuss.

A secret is that no matter how successful they are, they're always ready for the next failing or problem. As Capricorn Jupiter Dustin Hoffman said—about his flawless performances!—"A good review from the critics is just another stay of execution." So try to reassure them. They regard you as family (again, secretly) because their work really is home to them, and you'll be surprised by moments of generosity and remembrance, particularly at your birthday, your anniversary, or other appropriate times and holidays.

Your Child

In traditional astrology, a Capricorn Jupiter child is supposed to gain luck through, particularly, his or her father! Also through employers and commercial affairs. Of course, for a Capricorn Jupiter, this does not mean being handed things on a silver platter. It can mean, as happened with my Capricorn Jupiter grandfather, that his father shot him in the leg (accidentally, while hunting a fox), and then proceeded to leave the family for my grandfather to support! Apparently, this built up great strength and forbearance (so I heard many times), as my grandfather finished school, then college, then got his master's degree all while supporting his mother and five siblings (some of whom did not have as good a fate).

Capricorn Jupiter children are very conscientious, very reserved, but secretly, of course, *very* in need of love and affection (like all children), but they probably show it the least of any children. This makes it harder for them to get to, which just adds to the vicious cycle of being reserved, restrained, even cut off. The one thing this child is *not* is flighty or superficial, however they behave, and their main purpose really is not to (only) have a good time. Saturn is too much of a severe teacher—in their very *bones,* the bones being ruled by Capricorn!—for them to even have this good-time-Sally desire, let alone be swayed by it. The parents and caretakers will definitely need to be on hand to teach this kid how to have a good time. And relax. This child is born old and gets younger as he or she gets older. It is astonishing how visibly accurate this is, as well. They really will *look* older when younger but wait until they're past sixty and they'll be acting and looking like a teenager (which they probably never got to be).

Your Parent

An evolved Capricorn Jupiter is really a master, effortlessly accomplishing (as the Sea-Goat eventually did) on the material plane. An unevolved Capricorn Jupiter is thought to be excessively serious, rigid, miserly, controlling, and restricted.

The greatest thing about a Capricorn Jupiter parent is that you can *depend* on them, always, to be there, to get done what they said they were going to do, and to do it excellently. You will not need to worry at all—except about your own performance, and boy, are they strict about that! They expect the world, and they will lecture you in excessively deliberate fashion (deliberating every point, spelling it out, making it clear, teaching, teaching, teaching, or lecturing, lecturing, lecturing) until you get it—with the ethics, the principles, the history of it, the whys and the wherefores. You may feel as if you're attending a history lecture, or one on morality and ethics, many days of your childhood and even as a grownup. They can be fond of sermonizing. And it's not to hear the sound of their own voice; it's to *help you in your life.* This is the classic "Parent Lecture" Jupiter placement. Of course, the most infuriating part of it is that it *is* to help you—so how can you be mad at them? Many children do rebel and can spend years screwing up their lives just to feel a taste of freedom, independence, and to get that sound of the lecture out of their ears.

This is what astrology is all about, and, I'd venture, a civilized life.

Appreciating how different we all are, and loving these differences! Nature gives an antidote for every poison. Isn't that the miracle of life? Problems appear, but there's always a solution, and if anyone can see it, determine it, and manifest it, Capricorn Jupiter can. *Remember, Capricorn Jupiter: You are the sagacious accomplisher of the zodiac.*

Madalyn's Movie Choice for Capricorn Jupiter

Ran
This stunning epic by Akira Kurosawa portrays with terrifying dignity a warlord and his family marching to meet their grim and bloody fates (similar to *King Lear*). Set in pre-Tokugawa Japan, with Nōh motifs, *Ran* is almost chilling in its view of humanity, yet so powerful as to be universal, so profound and deeply felt, that it is wise as only a Capricorn Jupiter can be.

MARIAN ANDERSON	ROCK HUDSON
SAINT THOMAS AQUINAS	LANGSTON HUGHES
LOUIS ARMSTRONG	MOTHER JONES
PETER ARNETT	JOHN KEATS
JEAN-MICHEL BASQUIAT	B. B. KING
WARREN BEATTY	JACQUES LACAN
LUDWIG VAN BEETHOVEN	K. D. LANG
YOGI BERRA	JERRY LEWIS
LENNY BRUCE	CHARLES LINDBERGH
MARTIN BUBER	MALCOLM X
SITTING BULL	CAMRYN MANHEIM
RICHARD BURTON	KARL MARX
JOHNNY CARSON	JULIANNE MOORE
RUBIN "HURRICANE" CARTER	PAUL NEWMAN
CARLOS CASTANEDA	JACK NICHOLSON
EDGAR CAYCE	SHAQUILLE O'NEAL
CHARLIE CHAPLIN	JESSE OWENS
JOHNNIE COCHRAN	GWYNETH PALTROW
TONI COLLETTE	ROSA PARKS
SAMMY DAVIS JR.	THE PILL
MARLENE DIETRICH	BEATRIX POTTER
WALT DISNEY	COLIN POWELL
MIGUEL ESTRADA	VANESSA REDGRAVE
MEDGAR EVERS	*ROE v. WADE*
MORGAN FREEMAN	RU PAUL
RICHARD GERE	CARL SANDBURG
LINDA GOODMAN	PETER SELLERS
WAYNE GRETZKY	LEE STRASBERG
TODD HAYNES	MARY TODD LINCOLN
SEYMOUR HERSH	OSCAR WILDE
DAVID HOCKNEY	WILLIAM THE CONQUEROR
DUSTIN HOFFMAN	MARY WOLLSTONECRAFT
EDWIN HUBBLE	

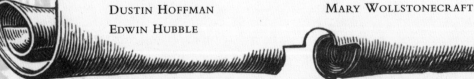

"We men of the Steppes know how to open up the womb of truth."

—Ivan Turgenev, Sagittarius Sun-Capricorn Jupiter. November 28, 1818

"There used to be a real me, but I had it surgically removed."

—Peter Sellers, Virgo Sun-Capricorn Jupiter. September 8, 1925

"Movies are fun, but they're no cure for cancer."

—Warren Beatty, Aries Sun-Capricorn Jupiter. March 30, 1937

"To live is so startling. It leaves but little room for other occupations."

—Emily Dickinson, Sagittarius Sun-Capricorn Jupiter. December 1, 1830

"Don't wait for the Last Judgment. It takes place every day."

—Albert Camus, Scorpio Sun-Capricorn Jupiter. November 7, 1913

"Never let pride be your guiding principle;
let your accomplishments speak for you."

—Morgan Freeman, Gemini Sun-Capricorn Jupiter. June 1, 1937

"How do I love thee? / Let me count the ways. / I love thee to the depth and
breadth and height / My soul can reach."

—Elizabeth Barrett Browning, Pisces Sun-Capricorn Jupiter. March 6, 1806

"People think: guy who shaves with a blowtorch.
Actually I'm bookish and worrisome."

—Burt Lancaster, Scorpio Sun-Capricorn Jupiter. November 2, 1913

"The future belongs to those who prepare for it today."

—Malcolm X, Taurus Sun-Capricorn Jupiter. May 19, 1925

"Integrity is so perishable in the summer months of success."

—Vanessa Redgrave, Aquarius Sun-Capricorn Jupiter. January 30, 1937

♒♃

The Birth of Cool

From Ovid, born March 20, 43 B.C., who advised, "Let your hook always be cast. In the pool where you least expect it, there will be a fish," to Lewis Carroll, born 1,877 years later, who said, "Everything has got a moral if you can only find it," these Aquarius Jupiters believed, and so should you, in the power of the unexpected, the new, the surprising, in every single one of us, and in life itself.

Love makes your soul crawl out from its hiding place.

— AQUARIUS JUPITER ZORA NEALE HURSTON

"Once I leave this Earth, I know I've done something that will continue to help others," said Aquarius Jupiter Jackie Joyner-Kersee, born March 3, 1962. Aquarius Jupiter luck is all about altruism and humanity. Many great Aquarius Jupiters have shocked the world into new orders by being unconventional and unusual, but at the same time they have been raging, pure idealists. Just think of Aquarius Jupiter Jean-Jacques Rousseau, who wrote, "Man is born free, and everywhere he is in chains." Believe like him in the natural goodness of humanity, use your creativity and challenge everything that stands in the way of beauty, freedom, happiness, and truth, and Jupiter will reward you mightily. You know what your ideal vision is; consider Aquarius Jupiter Michelangelo, who said of one of his sculptures, "I saw the angel in the marble and I carved until I set him free."

Aquarius is ruled by the planet Uranus, which governs electricity and therefore must *shock*—if your Jupiter is in Aquarius, you gain luck and success through spreading ideas in shockingly original and different ways. Your symbol, the Water Bearer, brings us truth and happiness in a stream of delicious, pure H_2O.

Aquarius is also known as the sign of the future, and it is the sign ruling astrology. Famous astrologers Sidney Omar and Grant Lewi are both Aquarius Jupiter. Your kind are known to be progressive, cutting edge, avant garde, and cool! Blow our minds with new ideas of space and time like Albert Einstein, shock a rigid corporation as Julia Butterfly Hill did, be a revolutionary and shake up the system like Ted Turner; whatever you do, reject conformity and embrace your unique individualism. And in your own way, and your own time. Don't embrace work for its own sake, as it can turn into drudgery. It won't matter; your fellow Jupiter John Ruskin, born Aquarius Jupiter February 8, 1819, told us "No great intellectual thing was ever done by great effort." Just relax, give birth to your own coolness, be anti-establishment, weird, unprejudiced, untraditional, innovative, radical, a rebel, new, novel, unique, quirky, unconventional . . . all these fun, cool words. Yours is the Jupiter placement of flower power!

Take a moment, look back long and deep over your life, recount as many events as you can, and ask yourself what has made you lucky or unlucky in your life. You'll notice that whenever you act like an Aquarius, *you create miracles in your life.* No matter what your Sun sign is, *you get lucky when you live out the Aquarius legacy gifted to you by the benevolent planet Jupiter.*

To take advantage of your luck, emphasize your . . .

Be inventive, original, shocking, visionary, revolutionary, progressively mind expanding, a breaker of tradition, iconoclastic, and your own thinker.	♒♃ *Aquarius Jupiter Strengths*

But because Jupiter rules excess, watch out for . . .

You can become out of touch (an ivory tower altruist), arrogant, or violently insensitive.	♒♃ *Aquarius Jupiter Excesses*

Understanding the
Aquarius Energy and Influence

Aquarius governs that time, from mid-January to mid-February, when the world outside is bleak and featureless, but the spirit inside is soaring. This is a time when literal-minded people see only bare trees and snow; imagination and unconventional wisdom is required to enjoy the season of Aquarius. In olden times, this period of the winter was one of the few times of year when no work needed to be done on the farm and in the village. It was a time when farmers and their families could step back from their daily work and enjoy art, conversation, and thinking, a season when challenging, exciting people could think of new ways of living and new ways to shake everything up. In short, this is the time for energy, creative thinking, and innovation; a time for Aquarius to experiment, to create, to share revolutionary ideas and to shine.

AQUARIUS SYMBOL
The Water Bearer

The Aquarius Water Bearer has always been the communicator of ideas, inspiration, and clarity, as represented by one pouring water into the world. There is one image that exemplifies the Aquarius motif of Water Bearer beautifully: It is the historical moment, now made famous in both book and film, of Aquarius Jupiter Annie Sullivan pouring the water from the water pump onto Helen Keller's hand. Annie then signed the word for water into Helen's hand. She had been signing words for weeks without any understanding on the child's part. In that moment, communication and clarity occurred. The meaning of the word—and the purpose of words—was understood by this deaf, blind, and mute child who had never known what language meant. The communication of the Water Bearer became a miracle as Helen Keller woke from her baffling dark world and was able to communicate with others and herself for the first time. Aquarius Jupiter, let your ideas flow like crystal-clear water into our world, bringing us lightness, spirit, and creation!

Aquarius rules the circulatory system, the ankles, and the shins. As a Bearer of Water (truth) and a thinking Air sign, Aquarius concerns itself with that which it can literally circulate: ideas and truths. The Aquarius spirit is sociable and friendly, always involved in groups and organizations—this means a great deal of circulation! So look at your own life and see how much you can expand your own circles—of friends, of mentors, of work colleagues, of those who can inspire you and vice versa—and get into communication and circulation mode. See what needs to be "circulated," whether it's news, ideals, or controversy—and keep in mind that it must be for the good of everyone. If poisons get into our circulation that can mean death. Aquarius also rules shins and ankles. Although Aquarius Suns need to be extra careful about protecting shins and ankles, you, Aquarius Jupiter, can run extra fast and be light of foot, under the power of the planet that protects!

As Aquarius rules electricity, shocking and vibrant colors are positive and bring you luck. Your colors should pop out at you, energize, and get your circulation going. Not everyone can carry off such zany, unconventional shades and have fun with them at the same time—but just think of an exception, Aquarius Jupiter Dennis Rodman—so, if you're nervous, add them as accents or buy brightly colored, electric underwear and absorb its Jupiter radiation. Sheer blue and tangy yellow-orange, your lucky hues, bring a smile to your face and fill you with happiness and enthusiasm. The cool or icy tones keep your mind clear and sharp. Add sheets, towels, china, and wall accents in your lucky colors, and observe how much more energy and clarity you have.

AQUARIUS COLORS
Electric Blue and Yellow-Orange (the Colors of Electricity)

And now your Aquarius Jupiter strengths:

〰4 *Aquarius Jupiter Strengths*

Be Inventive

In the early part of the twentieth century, an unknown scientist published a paper that overturned conventional wisdom in a spectacular way. This great

Wild and Weird Uranus

Electric blue Uranus is definitely the rebellious oddball of the solar system, the only planet that lies on its side instead of standing straight up (most likely to get a revolutionary worldview!), and its magnetic field is also different from every other planet's. It goes through forty-two years of daylight, then forty-two years of darkness during its eighty-four-year orbit around the Sun. It was discovered in 1781, during a period of many revolutions on Earth: the American, the French, and the Industrial Revolutions.

YOUR HISTORY

Uranus was the ancient Greek sky god and the first ruler of the universe. He was gifted with foresight and great vision. Having Uranus as your guiding star means you, too, have this gift.

thinker had been declared slow and backward in school because he took so long to answer a question; he famously failed all his school exams. His genius had to be expressed in an unorthodox form, one the world recognized quickly when his "Special Theory of Relativity" overturned the truth about the physical world that had been unchallenged since Newton. As a true Aquarius Jupiter, he became involved with pacifism, and together with other scientists he spoke out against the use of atomic weapons. Do you recognize Aquarius Jupiter Albert Einstein? Another Water Bearer giving it to the world.

Aquarius Jupiter Marie Curie is another genius who was recognized with two Nobel Prizes for her discoveries. Aquarius Jupiter Jonas Salk discovered penicillin. They used their inventive Aquarius ideas to improve the world for everyone. An essential part of embracing your luck as an Aquarius Jupiter is being part of, and communicating with, the world. Those who do not—and you can see how such originality could isolate one—end up throwing their luck away and being miserable.

It must be mentioned that two typically Aquarius Jupiter inventions that have become huge parts of our lives today—neon light and the polygraph machine—were also born during an Aquarius Jupiter passage, in 1902! Flood the light into your life and detect any "lies," (your kind is shockingly honest) and clear them up, dear lucky Water Bearer.

Be Original

Look at these one-of-a-kind originals, acting out their true Aquarius Jupiter colors:

Isadora Duncan, born Aquarius Jupiter May 27, 1878, invented a completely new art form based on personal expression and communion with the elements, shocking much of society with her unconventional lifestyle, nudity, and love of freedom. A true original, Isadora stunned Europe with her unorthodox dances, often done in Greek dress; she became the toast of London and Paris. The mother of modern dance, she famously said, "People do not live nowadays. They get about ten percent out of life."

Gypsy Rose Lee, raised by her mother in show business and vaudeville, became a stripper, famous for dropping her shoulder strap and gloves and little else in a comic flair, and then took up painting and writing, living for a while in an artistic community that included Carson McCullers, W. H. Auden, and all those guys in Brooklyn Heights.

And should we be surprised that two of the greatest stars of the ballet were Aquarius Jupiter? Vaslav Nijinsky, the "God of Dance," was born Aquarius Jupiter in 1880. Famous for his wild leaps and his dance as the faun, a performance that enchanted people because he literally seemed to hang in the air for several seconds as if he was never coming down. He dressed in wild faunlike costumes—eventually every day—but sadly ended his life insane. Another brilliant, artistic Aquarius original: Rudolf Nureyev, perhaps the greatest ballet dancer since Nijinsky. Born Aquarius Jupiter on March 17, 1938, Nureyev did a shocking thing to gain his freedom from the Soviet Union: While on tour, as he and his dance company were preparing to leave the United States, he employed one of his miraculous dance leaps, using it to leap over the barrier at the airport! Nureyev shocked everyone with his outrageous sexuality at the same time he amazed the world with the beauty of his dance. What a brave, shocking Aquarius gesture: fearless, inventive, and truly amazing.

Be Shocking

Even if your Sun sign is in family-oriented Cancer or critical Virgo, embrace your Jupiter energy, Aquarius—shock the pants off us! Be like *South Park*, the television show born an Aquarius Jupiter on August 13, 1997, which delights in speaking the unspeakable and breaking hypocrisy in the voices of not-so-innocent children, bringing enjoyment to millions! Or consider the example of outrageous Aquarius Jupiter Dennis Rodman. Whether dressing up as a bride, dyeing his hair crazy colors, or posing nude, this most unconventional basketball star was born Aquarius Jupiter on May 13, 1961.

Jim Carrey, born Aquarius Jupiter on January 17, 1962, has always shocked us with his silly, edgy comedy that reduces fans to hysteria. Carrey first came to public view in films like *Ace Ventura: Pet Detective* and *The Mask*, a movie about a mild-mannered man who becomes a shocking, eye-popping cartoon crazy when he wears a magic mask. He delighted in setting ambitious, crazy goals: A true future-minded Aquarius Jupiter, he wrote a check to himself for 10 million dollars when he was poor and working with his parents in a factory. He was able to cash it a few years ago, as he said he knew he would. Be like Jim Carrey, Aquarius Jupiter—pick outrageous goals that you intend to make, if you stick to your peculiar, unique vision!

The power of your Aquarius Jove is such that sometimes your actions and achievements will shock without your meaning them to. When this happens, don't worry—it means you're doing the right thing! Consider Aquarius Jupiter Nadia Comaneci's wins at the 1972

Olympics, when the judges awarded her three straight 10s—ten was considered perfect—and this breaking of tradition had thousands protesting. Or what about *Don Juan* opening during an Aquarius Jupiter passage on July 29, 1926, with the most kisses—127—in a single movie—very shocking at that time, but even for now: Do you know of a movie that contains 127 kisses?

Be Visionary

Aquarius Jupiters have the ability to see things no one else can see. Embrace your vision and change the world!

These visions can be literal. Like the vision of the Virgin Mary who appeared to Saint Bernadette, born Aquarius Jupiter on January 7, 1844. Saint Bernadette told the world of her vision; the result: the creation of the famous shrine at Lourdes.

The visions can be creative and artistic. Are you surprised that perhaps the greatest artistic genius of all time, Michelangelo, was born Aquarius Jupiter on March 6, 1475? With his remarkable sculptures *David* and *The Dying Slave*, and his spectacularly painted ceiling of the Sistine Chapel, Michelangelo revolutionized the world of art by creating his unique visions of incredible beauty. Recently, art experts restored the Sistine ceiling to its original quality, shocking many by displaying electric Aquarius colors—oranges, yellows and bright blues—in this beautiful vision.

The visions can be psychic and predictive. Madame Helena Blavatsky, born Aquarius Jupiter August 11, 1831, changed the world with her spiritual teachings of theosophy. She wrote that all religions were derived from one philosophy; she popularized Eastern concepts like karma in the West. Or think of your fellow Aquarius Jupiter Grant Lewi, born June 8, 1902. Writing in the 1930s and 1940s, Lewi, who received a master's degree from Columbia and was an English teacher, was the author of two hugely popular books that introduced the public to astrology—*Astrology for the Millions* and *Heaven Knows What*. He is still remembered for a feat that few of us would have the nerve to try: He predicted his own death at age forty-nine by a cerebral hemorrhage. He died exactly on the day he predicted—July 15, 1951. It is said that he had never been insured, but that when he saw his death coming, he took out a life insurance policy, which helped to sustain his wife and three children. See, Aquarius Jupiter, it can pay to be practical, too.

Be Revolutionary

Thomas Paine, born Aquarius Jupiter on January 29, 1737, was one of the inspirations for *both* the French Revolution and the American Revolution. One of his publications was the article "African Slavery in America," in which he condemned the practice of slavery. Paine published his most famous work, the fifty-page pamphlet *Common Sense*, on January 10, 1776. In a dramatic, rhetorical style, the document asserted that the American colonies received no advantage from Great Britain, which was exploiting them, and that every consideration of common sense called for the colonies to become independent and to establish a republican government of their own. The document went on to criticize the monarchy as an institution. Published anonymously, the pamphlet sold more than five hundred thousand copies and helped encourage with comments such as "The birthday of a new world is at hand."

Remember, Aquarius Jupiter, that many of the greatest revolutionary figures have been Aquarius Jupiter! Susan B. Anthony, born Aquarius Jupiter on February 15, 1820, campaigned her whole life for women's suffrage, worker's rights, and the abolition of slavery. She inspired the women's movement with her famous mantra "Independence is happiness." Consider Julia Butterfly Hill, born Aquarius Jupiter February 18, 1974, a great campaigner for the environment. To prevent the destruction of the redwoods, Hill lived in one, a tree named Luna, for *more than two years*. She wrote: "One makes the difference. Each one of us does." Or if you prefer a classic revolutionary, think of Fidel Castro, communist leader of Cuba, born Aquarius Jupiter August 13, 1926. His revolutionary ideas of socialized medicine and education overturned the Batista regime and brought him to power in the island nation.

Be Progressively Mind Expanding

You don't have to overturn governments, Aquarius Jupiter! The bounty of the giant planet can allow you to create change at an even more profound level by communicating your unique vision to others. Aquarius Jupiter Elisabeth Kubler-Ross changed one of our most personal and profound activities—the way we think about death. Her Six Stages of the grieving process have helped people in intense pain, leading them through, giving them a structure where before there was none. This is typically Aquarius and beneficial for you. Ensure your Jupiter power by always looking for structure—breaking it down when it's too rigid, creating it anew when there is none.

Or consider Robert Bly, father of the modern men's movement. In his book *Iron John*, Bly broke down the rigid code of masculinity, giving men a new way to think and a new way to be based on myth and legend. Bly truly broke the mold, as you should, too. As has Kofi Annan, secretary general of the United Nations, born Aquarius Jupiter on April 8, 1938. Annan is the first secretary general to be elected from the ranks of the U.N. staff, and the first black person to be elected to this position. A true humanitarian, as the greatest Aquarius Jupiters are, Annan has strived to bring together all of humanity, to, as he said, bring "the United Nations closer to the people."

Be a Breaker of Tradition

Few have overturned tradition like Aquarius Jupiter Jerry Brown, the former governor of California and current mayor of Oakland. An unconventional maverick who even today lives in a communal setting in San Francisco, where everyone sleeps on the floor and shares labor of the home and child care, Brown has been nicknamed "Governor Moonbeam" for his unconventional beliefs. He is famous for challenging the status quo. When the LAPD asked permission to march en masse—and show off their strength and power— in an L.A. neighborhood, Brown famously refused unless they marched in Beverly Hills, the wealthiest neighborhood in the city.

So Aquarius Jupiter, beware joining organizations that embrace traditionalism; they may stifle and hurt you, as Aquarius Jupiter Princess Diana discovered when she joined the Royal Family. Success came to her when she bucked the system, but in the end she had chosen too powerful a tradition to confront.

Be Iconoclastic

Aquarius Jupiter gives you the power to rise above the crowd. One famous writer set himself outside society to celebrate his passions and life: Aquarius Jupiter Walt Whitman! He wrote, "I am large, I contain multitudes," celebrating originality and individuality. Allen Ginsberg was also born Aquarius Jupiter, on June 3, 1926. He broke all forms of verse and set the world on fire with his poem *HOWL*. A founder of the Beat movement and pioneer of sixties happenings of intoxication and nudity, Ginsberg broke the molds to shock the world out of tradition and complacency. Fellow writer and Beat poet breaking with convention William S. Burroughs was also Aquarius Jupiter through his free-form novel *Naked Lunch*! Burroughs, Whitman, and Ginsberg were also pioneers of sexual liberation at a time when homosexuality and anything unconventional was unacceptable to society.

Just as Ginsberg and Whitman broke the traditional structure of poetry, and Burroughs changed the novel forever, so Aquarius Jupiter Lewis Carroll did to language itself. This brilliant, eccentric genius, born the Reverend Charles Dodgson, wrote *Alice's Adventures in Wonderland* and *Through the Looking-Glass*, books that explode language with wit, fun, and paradox. Carroll invented a special type of word, the portmanteau word (which telescoped several words into a single one), to free language from the accepted rules and structures.

So Aquarius Jupiter, find the structures and rules that you can break down! Gain the bounty of Jupiter by shattering tradition, being idiosyncratic, different, unusual, and cool. Enjoy the energy created by this very act of creation and change and rebellion, but ensure you use it to free others. Be a Rebel with a Cause and gain Jupiter's luck!

Be Your Own Thinker

In whatever you do, Aquarius Jupiter, follow the beat of your own thinking and play by your own rules. Think of the jazz great Miles Davis. This totally cool cat was born Aquarius Jupiter May 26, 1926. A brilliant jazz trumpeter (his well-known albums include *The Birth of Cool* and *Sketches of Spain*), Davis lived a wildly unconventional life. He gave everything for his art. When he played on the soundtrack of Louis Malle's *L'Ascencion du Scaffold*, his upper lip tore, but he kept on playing and refused to stop, and this resulted in the amazing shuffly, muted sound of that piece—which was adored by his many fans and critics, including me!

Try to fit in and you deny yourself the luck and power of the great planet. This has been expressed so eloquently by Aquarius Jupiter Peter Shaffer in his controversial and wildly original play *Equus*. This play is all about the dangers of becoming normalized. It features a boy, Alan, who blinds five horses he worships and is sent to a psychiatrist whose job it is to help him and make him normal, fit for society again. The psychiatrist serves as Shaffer's mouthpiece and main narrator; he regrets having to "cure" Alan, believing he is taking away the boy's passion for life as well as his sexuality. The doctor ends up feeling he has none in his own life; he envies the boy and wishes he had even a bit of his "madness."

What Makes You Lucky in Love

You probably won't be surprised to hear that finding your high school sweetheart and living behind a white picket fence isn't for you. But you don't need to go completely out into left field either.

You're definitely not one to follow conventional ideas; you won't find anyone by relying on your friends' and family members' ideas of who "might be right for you." Look for someone from a different background, in a different profession, in a different part of town. Don't be afraid to date people very different from yourself, and don't be put off if your prospective partner has unusual ideas about things.

Aquarius are made for nontraditional wedding ceremonies—as well as relationships—so, carry this over into all areas of your life. Go on unusual dates (try something creative, like a dance class or pottery design); celebrate anniversaries in idiosyncratic, uniquely personal ways; shake things up every now and then by completely changing your routine. Just make sure you do this in a kind way; embrace change for the sake of your partner and with a spirit of generosity. It's a bad sign if you feel the impulse to shock or do outrageous things in an unkind way—Jupiter will desert you.

If you marry, make sure your union is an equal partnership based on sharing. Hierarchies of any kind frustrate Jupiter's lucky energy for you. And as a last resort, you can always make your lovers laugh. Aquarius Jupiters are some of the funniest people around. Eddie Murphy, Bill Murray, Martin Short, and Mel Brooks, all born Aquarius Jupiter, have delighted in shocking us into delicious waves of laughter for years. Cheer us all up, Aquarius Jupiter, with your charm.

Your Best Career Moves

You're the lucky experimenter of the zodiac. Pick a job and a lifestyle very different from your family of origin and encourage yourself to travel and to try out many things before you settle on one dream. Your true dream will be more powerful and more easily fulfilled if you have tried out many other ones first.

Your Aquarius Jupiter energy thrives on originality, and you will succeed if you take it upon yourself to be the visionary, the unorthodox person, the person who challenges rigid hierarchies. You are especially skilled—think Einstein—at careers that require creative traits, ideas, and conceptual thinking. Design, whether for homes, products, or the way organizations are structured, would suit you. But make sure your job allows you the opportunity to work to make the world a better place. Remember: Jupiter rewards and enhances idealism.

You will not find yourself fitting in at a large traditional corporation. Avoid traditional professions like the law, accounting, and medicine unless you dedicate yourself to achieving change within their ranks (you can follow your Aquarius Jupiter energy by using even rigid professions to achieve social change). The New Economy was an Aquarius economy, and like your media-oriented Gemini colleagues, you would have found happiness working for an untraditional, experimental company like eBay or Amazon.com. Look for work as well in which you can express yourself with freedom.

Because Aquarius rules electricity and electronics, you will tend to have luck and success with any field that involves computers, electronics, television, and even aviations and the space industry. So if you're a teacher, you might incorporate more of this into your classroom; similarly, if you're an artist you might explore video performance or recording possibilities.

How to Reach Optimum Health

You must take particular care of your health, Aquarius Jupiter, and in particular make sure you eat properly and maintain a proper balance. Just think of two of your fellow Jupiters who did not: Marilyn Monroe and Karen Carpenter. Circulatory health is critical for you, so of all the Jupiter signs, exercise is of almost supreme importance. Aquarius are made for aerobics and other sports that allow you to express yourself (especially anything involving dancing). Just think of Isadora Duncan, Nijinsky, Nureyev, Gypsy Rose Lee, and, yes, Jane Fonda, the godmother of aerobics. Don electric blue workout clothes, Aquarius, and get out there on the floor and dance!

Being an aqua Aquarius, swimming is also excellent for you. Also lucky! Just ask Olympic swimmer Mark Spitz—an Aquarius by Jupiter *and* by Sun, born February 10, 1950—who won an unprecedented seven gold medals at the 1972 Olympics.

Your Lucky Foods, Element, and Spices

Aquarius' cell salt is sodium chloride, which is just common table salt! Rather than going heavy with the salt shaker (which leads to bloating, hardening of the arteries, and poor circulation) eat plenty of these foods to maintain your hydration:

tuna, oysters, ocean fish, clams, lobsters, pomegranates, chicken, radishes, spinach, squash, lentils, pecans, apples, pears, peaches, lemons, romaine, corn, and celery. Because circulatory health is so crucial, you must monitor your cholesterol intake. Concentrate on avoiding the "bad" cholesterol of processed foods and eggs yolks; instead, eat lots of the "good" cholesterol, found in avocado, papaya, and nuts. Plantains are considered a particularly "magical" food for Aquarius Jupiter.

Your Best Environment

The sheer breadth of your originality, Aquarius Jupiter, works best in desert climates. Just as your sign governs the time of year with the least natural growth in the Northern Hemisphere, you do best enjoying electric colors in the desert's atmosphere of bright, sunny days and cool nights. As you need your groups of people and social energies, bring them to the desert! (No desert island isolation for you!) Conversely, any place that contains the new, the different, and enough intellectual stimulation keeps you going and happy. Aquarius-ruled cities are Moscow, Stockholm, and Buenos Aires. In the United States, San Francisco, New York, Seattle, or Ann Arbor are best bets, or create a way-out community wherever you are. If anyone can do it, you can!

Your Best Look

You guys can carry off just about anything—and frequently do. Aquarius Jupiters are famous for slacking off in the "Best Look" department and being loved for it—everyone thinks it's charming, cutting edge, so avant garde. The truth is, though, that you Aquarius Jupiters really did forget! Just think of Einstein, known for coming into his office wearing green slippers trimmed with flowers, Marilyn Monroe, who slouched around her apartment in New York either in a stained white bathrobe or nude (she was alone; this was not for anyone's benefit), Hugh Hefner, also so fond of the bathrobe, and Julia Butterfly Hill, who didn't take a bath at all for more than two years. So, to get lucky, look as though you're not *trying,* or at the very least, wear what feels comfortable to you, don't care what others think. If you must be a fashion plate, go for what's unexpected (Eminem) or outright shocking (Dennis Rodman). Somehow, when you embrace the world of high fashion, Aquarius Jupiter, it doesn't seem right. Think of Princess Diana, the world's most glamorous woman, with her pain hidden behind designer clothes.

Your Best Times

The luckiest times for you are when the Moon is in Aquarius, which happens every month for two to three days. Find these special days in your astrological calendar. During these days, schedule crucial meetings, tell that special person you love them, ask for a promotion or raise, push the envelope, and plan to shine. In your everyday, the best time for you to exercise is early evening. The luckiest times during the year for you are January 20–February 19 (when the Sun is in Aquarius); May 21–June 20 (when the Sun is in Gemini); and September 23–October 22 (when the Sun is in Libra.)

The very luckiest time of your life—as in Lotto lucky—is when Jupiter reenters the sign it was in when you were born (Aquarius), which happens every twelve years and lasts anywhere from a few months to just over a year. This is called your Jupiter Return, and it launches your luckiest streak. You can count on the *extraordinary* to happen the very day your Jupiter Return begins. Amazing events can start happening up to six weeks beforehand, so plan ahead!

How to Be Lucky Financially

You must embrace your nontraditionalist side, and your Aquarius energy may push you away from caring too much about money. If you play the stock market, you should look for unconventional new companies that allow and encourage experimentation. Look for your luck among outsiders and those who buck the curve. If you have a chance to ignore the conventional wisdom, take it! You respect intellectual development and are not a natural gambler, but you may find the time is right to take a chance, and then go with your intuition. But you will be better off concentrating on original ideas and inventions. Tinker around with computers and gadgets in your garage, Aquarius, and you may strike gold at the patent office.

Aquarius Jupiter Ted Turner said, "As long as I could think for myself, no telling what I could accomplish." He followed his original spirit by building the first national cable channel, an unparalleled success, and he has continued to reinvent himself in unconventional ways: celebrating the Atlanta Braves, marrying the very surprising Jane Fonda, donating huge sums to the UN, and becoming a gadfly and maverick at AOL-Time Warner. It's interesting to note that Aquarius Jupiter brings one luck in anything to do with television, electronics, and cable. Follow that Aquarius Jupiter electricity to fulfill your financial dreams.

Success in Family and Friendship

Just as astronomer Tycho Brahe, born Aquarius Jupiter December 14, 1546, proved that comets are distant entities and not atmospheric phenomena, so you must see everyone as distinct entities—and your luck comes in bringing them together! You must cherish your individuality and stay out of big traditional organizations, but *don't* isolate. Cherry pick your favorite people from all professions and walks of life, introduce them, and form a new, smaller group—around you!

Remember: You are also a solitary nut and need your space. Home and family life can be an incredible source of happiness to you (just think of how you can improve the world by bringing up really nice kids)—so long as that solitude is respected. Jupiter will bless your family if you live together and work together, but also respect one another's space.

When following the luck of the humanitarian visionary, your challenge is not to give away all of yourself, as Aquarius Jupiters Princess Diana and Marilyn Monroe did. In particular, do *not* give away too much of yourself to your partners. Ensure they respect your space and freedom. Marilyn quoted Goethe before she died, "Talent is best developed in privacy."

Aquarius Jupiter Excesses ♒♒♃	*When you go too far!*

Edward Albee's play *Who's Afraid of Virginia Woolf?* is the perfect cautionary tale for Aquarius Jupiter, even described after its Aquarius Jupiter birth in 1962 by *Newsweek* in terms typical of its sign: The play was said to be "brilliantly original . . . surging with shocks." Famously filmed with stars Elizabeth Taylor and Richard Burton, the play takes place over one evening at a New England college. A professor, married to the daughter of the president of the college, entertains with his wife another academic couple. What seems like a fun party becomes a screaming, wild, drunken brawl. It soon becomes clear that the ivory tower academics of the play are completely out of touch with real-life emotions, and they wound, bicker, and skewer one another as a result.

Excess #1 — *You Can Become Out of Touch (an Ivory Tower Altruist)*

The theory as opposed to the practice. The Aquarius who suffer from this excess include the humanitarian who believes in goodness and justice for all, but can't be bothered with his or her own family and friends. Like Jean-Jacques Rousseau, who ignored the facts to produce his "Noble Savage" theory, embracing the Utopian version of the world, they stick to their intellectual beliefs even if the world contradicts them. To these Aquarius Jupiters, the abstract is so much more exciting, so much more comfortable, than the intimate. In fact, they are ruled by a fear of intimacy.

Just look at Aquarius Jupiter Queen Elizabeth II. When Princess Diana died, she stuck to protocol and tradition ("theory"), ignored the obvious facts of the British public's grief and despair, and reacted in a very cool way. She stunned the British public. It was reported that hundreds of flowers and personal notes left for Diana were found stuffed in rubbish bags, unread and unopened. This set up a massive outcry in England and turned the public against the Royal Family in a way that nothing before ever had. So, in a way, Diana did break some old traditions.

Excess #2 — *You Can Become Arrogant*

Aquarius Jupiter, watch out for the tendency to become arrogant and intoxicated with power. Here's an example of another queen: Britain's Queen Victoria, the nineteenth-century ruler of the British Empire. The royal "we" is usually traced back to Victoria, who was born an Aquarius Jupiter! She was known for her regal, cold nature, and perhaps the extent of the Empire gave the Queen license to feel and behave arrogantly.

An even more striking example of modern-day arrogance was seen at the Watergate hearings, when the shadowy advisers of President Nixon decided they were above the law and organized the break-in to a room containing secret information about the Democrats' campaign for the presidency. An Aquarius Jupiter example: John Dean, one of the masterminds of the Watergate break-in and coverup. He ended up charged with obstruction of justice and spent four months in prison.

DON'TS

* Mistrust your original sight.

* Worry too much about following fashion.

* Skip meals or starve yourself—ever, for any reason.

* Allow yourself to become aloof or arrogant with others.

* Fall into the trap of ivory tower intellectualism.

* Forget to use your creative powers to help others.

* Become trapped in rigid institutions or environments.

Excess #3 · You Can Become Violently Insensitive

Sometimes your wish to shock can backfire on you, Aquarius Jupiter. The cerebral nature of your air sign can blind you to the effect your words can have on others, and you can end up violently insensitive. Aquarius Jupiter Jane Fonda became known as "Hanoi Jane" when she carried her protest against the Vietnam War to the extent of actually coming out in support of the Viet Cong. She had failed to realize the effect her actions would have on veterans and their families. Instead of shocking people to make the world a better place, her words deepened the resentment and anguish of the war.

An example of extreme insensitivity as well as arrogance from history is that of the Ludlow Massacre, which took place under an Aquarius Jupiter passage on April 20, 1914. Twenty innocent men, women, and children were killed when the Colorado Fuel and Iron Company tried to break a strike at one of its mines. An armored car filled with men with machine guns entered the miners' camp and began firing. The result: the beginning of the modern Union Movement and eventual shame and defeat for those responsible.

Sun Sign-Jupiter Sign Combinations

Aquarius Jupiter is both fun and lucky for you. Aries: Expansive and explorative, it supports your fiery enthusiasm with an inquiring and rational approach yet an off-beat one, so you're not limited to the dull old *rules*. Aquarius Jupiter gives Aries a wider vision, tells you to use your pioneering instincts to push forward iconoclastic ideas that will benefit humanity. With your go-ahead impulses, you're the one to lead campaigns and movements, political, environmental, or class-action attacks against anything you consider unfair or harmful, or to act within your community or present position. The arts are great areas for you to introduce new concepts. Acclaimed actor and Aries-Aquarius Sir Alec Guinness ignored early advice that he had no talent for acting and went on to create, onstage and in film, an incredible variety of serious and comic roles, including seven different family members in *Kind Hearts and Coronets*! See what daring and change will do for you, Aries-Aquarius.

You'll do best working for yourself or as head of an unconventional company or organization. Investing in developing either your own ideas or talents (which you should continually work to improve) or in idealistic companies will reward you financially. In love, you'll be a happy Aries-Aquarius if your passion includes a lasting commitment to, and equal respect for, the ideas and creative impulses of your partner.

Aquarius Jupiter stimulates the Bull to charge forth into new pastures, new spheres of influence and thought. You'll find rewards by using your solid sense of what works in conventional situations and companies, and then looking at what *doesn't work now,* what could be totally changed, for new products, services, or socioeconomic plans. Taurus-Aquarius Emmanuel Kant took old concepts—God, freedom, and immortality—and presented new insights on them in his book *Critique of Pure Reason*.

Financial gains will come through looking for trends that will add to the future comfort and well-being of humanity and investing in corporations moving in this direction. Don't hesitate to trust your intuition; you'll find it's a surer guide than conventional wisdom. As for your significant other: different race, religion, not the person you or your family had in mind as "right," but you share unconventional ideas, and your gut feelings is yes? Well, that's your answer, isn't it? Just be open to the changes necessary for the blossoming of any relationship, then sit back and smell the roses.

Aquarius Jupiter reinforces Gemini's lively and fertile mental activity. Just watch out that with two Air signs urging you on, you don't take off totally into the outer ether! Gemini's ability to communicate with Aquarius' innovative style makes you a natural for artistic expression. You get public attention with your unique approach.

Maintain connection, individually and collectively; don't let love of pure concept make you lose sight of either the feelings and needs of your significant other. Your Mercurial mind can seduce you from one idea or person to the next; trust yourself to stick with, and fulfill, your wildest dreams. Gemini-Aquarius Walt Whitman blazed forth in poetry that expressed not only his epic vision of America—the diversity and power of its land and people—in the latter part of the nineteenth century, but also in shocking references to sexuality, ego, reincarnation, and the oneness of self with the universe.

Aquarius Jupiter has wonderful gifts for you, Cancer, if you let it take you in new directions, but you have to "leave home" to find your luck. Trying to fit in with the family business, traditional social norms and expectations, or tried-and-true investment plans isn't your path to success. Admit it: With urgings from your Aquarian Jove, you *do* have wild dreams you'd love to realize. Aquarius Jupiter reassures you not to be shy about presenting these ideas.

If your current love seems perturbed by your breaking the mold, communicate honestly with each other! Unleash your Aquarius Jupiter sense of humor and you'll enjoy the rewards of health, emotional fulfillment, and financial gain. Cancer-Aquarius Richard Rodgers collaborated with (among others) Oscar Hammerstein on many timeless musicals. His memorable, romantic lyrics catch the listener with incisive truths, as in this from *South Pacific*: "You've got to be taught to hate and fear," from a very young age "drummed" into "little" ears.

With the Leo sense of authority and presence, and the Aquarius daring innovations, you are the one to bring revolutionary ideas to the attention of a wide public. Leo-Aquarius famous astrologer Sydney Omarr was consulted by writers, film stars, and a U.S. president, and in a three-hour radio interview, dared to debate a well-known astronomer who rejected the science of astrology. Leo-Aquarius concepts are dramatic departures from what has gone before. You are a visionary!

Don't assume that however brilliant and valid your ideas, they exempt you from relating personally to others, whether in matters of courtesy or in forming deep relationships. You lead the way to change, but remember the ancient Roman definition of a leader: "First among *equals*." Choose your mate not as a complement to your image, but as a partner, with appreciation for his or her uniqueness. Financial success comes through your talent or your role in society, though investments in unusual new ventures bring gain.

The Water Bearer and Virgo complement each other; both are intellectual and perceptive. Use Virgo discernment and Aquarius Jupiter daring to choose jobs and investments with companies or service providers that are opening new fields. The nature of both Aquarius and Virgo is to help others; try holistic health, new educational methods, and social reform. Writing is a natural way to express yourself freely.

Virgo-Aquarius Agatha Christie originated the English country-house mystery genre and created Hercule Poirot, user of "the little grey cells" for crime solving. Your Jove urges you to see the humor in situations, to enjoy the company of others who are outside your usual social sphere. You'll be delightfully surprised to find your romantic interest piqued by unconventional people and surroundings. Trust yourself and your intuition, don't obsess about detail, and let Aquarius Jupiter be your open road to the future.

IF YOUR SUN SIGN IS:

Libra

(September 23–October 22)

I f you allow Aquarius to open your eyes you will see possibilities in your dreams and clarify them into definitive blueprints. Aquarius Jupiter tells you to trust yourself to let go of conformity to tradition. Experiment, try different jobs and avenues of expression, invest in future-oriented companies. Remember: You aren't defined by family, social position, or occupation; *you* define the world for yourself, and often for others, by acting upon your visionary ideas. For Aquarius Jupiter, "different" is *good,* and your Libra charm will come through, anyway.

Libra-Aquarius President Dwight D. Eisenhower, America's victorious military leader in World War II, was courted by both Democrats and Republicans (that Libra balance); as candidate for the latter, "I like Ike" (Libra charm!) was the campaign slogan. His Aquarius Jupiter was revealed when he warned the public of the dangers of the "military-industrial complex"—quite an unorthodox statement from a general!

You find love not by looking for the "right" person but for someone who is launching a far-out project and thinks fashion sense means you sense what you'd like to wear. A bit odd? Yes, but lucky for you—and exciting!

IF YOUR SUN SIGN IS:

Scorpio

(October 23–November 21)

A quarius Jove with Sun in Scorpio is a world changer, an explorer of fields unthought of by others. Bring your deep intuitive sense of the world's needs and your tenacity of purpose to translate feelings into ideas that revolutionize society. Your complete certainty of the necessity and worth of your projects is very often proved right, but don't lose touch with the pulse of the times and the feelings of others. Scorpio-Aquarius General Charles de Gaulle had a reputation for arrogance and made enemies by his strong stance for what he saw as right for his country, though he was a tower of strength in France's time of need in World War II.

Aquarius Jupiter's biggest message to you is to retain your inner conviction but *communicate,* open up to others, especially in social situations and with love partners. Don't look for someone conventional or nonchallenging; you need an equal partnership. Listen to ideas of others, not only for personal happiness but also to spur your learning and intellectual growth.

Sagittarius fire ignites in Aquarian air, burning with crusading zeal, moving into the future, and leaping into the midst of new situations and people. Probably some of the most famous leaps ever made were by Sagittarius-Aquarius ballet legend Nijinski, who seemed to be immune to gravity. Nijinski's most famous role with the Ballet Russe was in *L'Apres Midi d'un Faur*. Its music and frank sensuality shocked Paris in the early 1900s but opened the way for new forms and more experimental expression in the hitherto rather delicate world of ballet. Aquarius Jupiter suggests innovative ideas—the more daring, idealistic, and off the wall, the better!

Select causes wisely, with clarity of insight unclouded by conventional attitudes or by your own prior assumptions. Remember that you should seek to improve not only the world but yourself. Your skills lend themselves to physical expression, including sports and performance; don't seek jobs that are sedentary. In love, look for partners who excite you intellectually as well as physically, and with whom you share visions.

The Capricorn Goat with the luck of Aquarius Jupiter reaches the peak, the mountaintop. If you allow Aquarius Jove to liberate your thinking, inspire you to great dreams and innovative ideas, and push you to communicate them to the public, you will have all the success you can handle. Being Capricorn, you'll be in for the long haul. Remain true to yourself and your vision; it will often be something that the world needs to see or hear. The children's novels of Capricorn-Aquarius Horatio Alger are perfect examples of this combination: *Strive and succeed* was his message to the world.

Accept Aquarius Jupiter's help on the long road by adding a sense of humor—it might even turn out to open a new career field—and by seeking out situations where you can enjoy the company of friends and, especially, new acquaintances. Leave behind ideas of success being a matter of achieving conventional goals, positions, and social contacts. Capricorn-Aquarius, you'll find your perfect partner in "disguise": different race, religion, viewpoint, rather than the one you might have expected. Share love, ideas—and laughs!

IF YOUR SUN SIGN IS:

Aquarius

(this makes you double)

(January 20–February 19)

These folk are like Alice's Red Queen, created by a favorite double Aquarius: "She runs so fearfully fast that I could never catch her." A double helping of iconoclastic, out-of-the-blue, dare-all, you-mean-there-are-*rules*? approach to life. You can go anywhere with double Aquarius, and lead the world, kicking and protesting, into a better future, making people aware that the old way of looking at government, society, and life in general just won't do. Double Aquarius Gypsy Rose Lee gave people a *lot* to see and also communicated true, and entertaining, observations on show biz and family relations in her autobiography, which became the play and film *Gypsy*.

You'll find success if you remember that your job—writing to performance, sociopolitical theory to active protest, sports to anything else in the world—is to *shake it up*. The only danger is letting your excitement about saving the world blind you to personal needs. Remember, double Aquarius, not to engage your mind, shift into high gear, and leave your emotions, and those of others, behind. Your significant other shouldn't echo you but excite your interest with his or her differences of origin, look, and opinion. Don't forget (you often do) to invest, always with ventures and companies that are just beginning.

IF YOUR SUN SIGN IS:

Pisces

(February 20–March 20)

Pisces-Aquarius not only explore all aspects of the world but show it to the rest of us in a way we'd never have dreamed of, and because of the power of their visions, make them a part of our way of seeing things forever after. The stunning black-and-white pictures of nature photographer Pisces-Aquarius Ansel Adams are etched in our minds, especially those of the California Sierras. And beyond that, his photos have made people conscious of the natural treasures of the American landscape and the need to preserve them. Though Pisces-Aquarius often use visual representation, others have given the world new visions in philosophy, comedy, dance, song, marketing, and economics.

Use your Aquarius Jove in any field, any job; let yourself dream, then bring those dreams out into the world. Don't become discouraged by an imperfect world. Just set out to change it! You have no trouble accepting those whose background, lifestyle, and thought differ from your own. Remember to stay totally true to your own individuality. Even in business, you succeed principally by concentrating on implementation of new commercial concepts rather than on gain, but you often do well financially, especially when investing in these new ideas. In love, Aquarius intellect balances your emotional Pisces nature; if it's a match of equals, you'll live out beautiful dreams of perfection.

Relationships with Aquarius Jupiters

Finally, for those of you who aren't Aquarius Jupiter yourselves, but have an Aquarius Jupiter spouse or significant other, boss, child, or parent, here are some tips to help you learn to deal! This is how the luck of someone else's Jupiter touches *your* life.

Your Significant Other

It is traditionally said that Aquarius Jupiters *bring luck to others,* so however infuriating this one's behavior occasionally is—and there's no way to get around that—they will, even if you separate (and they do that a lot, too) bring luck into your life. Even though you may not realize it until later.

They love to be intrigued, and they love ideas and discussion. So if your idea of a good time is to cuddle while watching TV in silence every night, this is probably not the mate for you. Confessions don't really turn them on, either; they love a mystery and they love to analyze.

One thing you can count on with an Aquarius Jupiter significant other is lots and lots of others! Friends, family, acquaintances—they'll all be brought into the home, along with stray cats, dogs, people met on the street—Aquarius Jupiter truly believes the more the merrier. They find their luck and ideas through groups and organizations, remember, so don't try to stifle their impulses completely—just at some point, explain that you might like some time alone, just the two of you, for, well, *you* know.

Aquarius Jupiters do love their freedom, so take care that you're not dating forever. If you want to take it to the next level, you'll probably have to be the one to do so. Marriage is a biggie for these guys, and most will resist as long as possible. Once married, however, you can count on them to be honest and generally stay faithful because of their high ideals.

Your Boss

There are fewer bosses who are Aquarius Jupiter than any other Jupiter sign, so if you've got one—*wow!* First, they don't believe in hierarchy, authority, or any of that other class-strata crap. They may insist on your calling the company a *co-operative,* and to refer to yourself as a partner, an equal sharer, even though you're not (and may never be). All in all, they're really not comfortable being in this kind of "unfair" position, and if they seriously try to fit the role, they may make everyone as uncomfortable as they feel. All that said, there are lots of positive benefits of having an Aquarius Jupiter boss.

They're fun. They don't believe in rules (which may make your life more difficult day to day, but we're back to fun again). They look at you as a friend, not an employee, but then they look at everyone as a friend. They really don't care—not nearly as much as other bosses anyhow—how you dress, how you act, about your idiosyncrasies. They do care that you're honest, ethical, and don't cheat. They abhor arrogance, although they can be quite arrogant themselves. They like the unexpected, the unusual, the new, the surprising. So you can tell them your ideas anytime, and you can be sure they'll listen. On the other hand, when you're just trying to get something done, it can be very distracting if they suddenly decide to change the entire office around for better Feng Shui, or launch a huge philosophical discussion with the accountant, cleaning lady, man off the street, right by your desk—and expect you to join in, of course, if you want to. What you will learn from your Aquarius Jupiter boss, however, will be invaluable: flexibility, respect for all people, and how to get on with anyone, at any time. It might even loosen you up and introduce you to new ideas and the idea of fun and innovation in your work.

Aquarius Jupiter Mel Brooks said, "I only direct in self-defense." No self-respecting Aquarius Jupiter will be pedantic or over-bearing; if they're your boss you're going to have quite a ride!

Your Child

This child is unique, there's no two ways about it—or there are many, many ways . . . I'll start again. This child will probably be a puzzle even to you, dear parent, for many years, as they change from interest to interest, refining their talents, and homing in on what makes them really happy. They are brilliant and they can sense things in the future and hear things that haven't even been said yet, but it's all a matter of manifesting their thoughts into something concrete. They love to experiment, investigate, and explore. Give them a telescope, a detective kit, a canvas and they'll be in seventh heaven. Most important, give them access to music and music lessons because the number of Aquarius Jupiter musicians is extraordinary. Please try to respect their disdain for rules. Give them progressive schooling if you can, or at least something that will challenge their minds. At the same time, teach them structure and discipline because that's what they need. They're free spirits who need their space and sometimes that means being insensitive to their siblings and not taking care of them as they should (or as you want them to).

Explain why they should in terms of *ethics,* and they will get it immediately and become the best big brother or sister who ever existed. As a matter of fact, you can save yourself years of agonizing heartbreak and time this way: Anything you want them to do (within reason—they're very rational, remember), if you put it to them in terms of how it will help society or it's a matter of ethics, or discovering a new form of house-cleaning, or you just want to try out this new gadget (to clean their room) to see how it works . . . get my drift? That's the way to discipline your child and make them cooperative. ("Cooperative" is a golden word to them; your household and family should be a cooperative and then they will participate much more willingly.) Even if they don't have the vocabulary yet, this is the way to go. Don't give orders—it just won't work. Yelling, punishments, and arguments are even worse. Their Uranian antennae are very fine, and the tiniest thing can set them off. Plus they think about and analyze everything. They most likely will argue for "kids' rights." Don't pry into their privacy and do allow them plenty of friends. Unlike other children or teens who may need to be protected from negative influences, they're just friends with the whole world and need to stretch their wings. Their friends may be kind of weird (definitely varied) and that's okay.

Your Parent

Your Aquarius Jupiter parent will help you decide—really help you decide for yourself, as opposed to telling you—where you want to go in life and what you want to do with yourself. They are excellent life counselors: As Aquarius, their sign governs the eleventh house of hopes, greatest ideals, and wishes, and where the hell you're going in life.

Are you sure your parent isn't really a hippie or flower child in disguise? This is a parent of the old anti-establishment stock. They think the world of you—and think in an abstract, intellectual way that children have all the answers, anyway (before being corrupted by society)—but you may wish they would just tell you what to do like other parents do sometimes. It's easier than having to think it all out, ponder all the human ramifications, and make a fully informed decision yourself. Your Aquarius Jupiter parent is way cool, but they do sometimes forget the basic parent stuff. They might tell you to stand up for yourself, argue for what you believe in (when you don't even know yet), and there's bound to never be a dull moment. They may have been a bit lacking in the affection and hugs department—and that can be your job, to teach them this. They and you will be glad you did. They think you're amazing even if they don't say it or show it all the time. They're as proud of and tickled by you as anything.

The one problem can be that, sometimes, your Aquarian parent, caught up in changing the world, will forget to give basic love and affection. You may feel more than a little bit lonely—don't forget to tell them (once you bring them back to Earth they will reward you with renewed attention). And that's what astrology is all about! Appreciating how different we all are, and loving both similarities and differences.

Nature gives an antidote for every poison; isn't this the miracle of life? Problems happen, but always, somewhere, there's a solution, and Aquarius Jupiters will think of a solution so original no one's ever thought of it before. *Remember, Aquarius Jupiter: You are the humanitarian Water Bearer of the zodiac.*

Madalyn's Movie Choice for Aquarius Jupiter

Magnolia
Ten jaded Los Angeles individuals experience life revolutions through *Water Bearer* rain that washes away their lies and loneliness. Its shocker of an end—frogs raining from the sky and an ensemble singing of Aimee Mann's exquisite "Save Me"—serves as a moving catalyst to bring the divided group together, with beauty, redemption, and love.

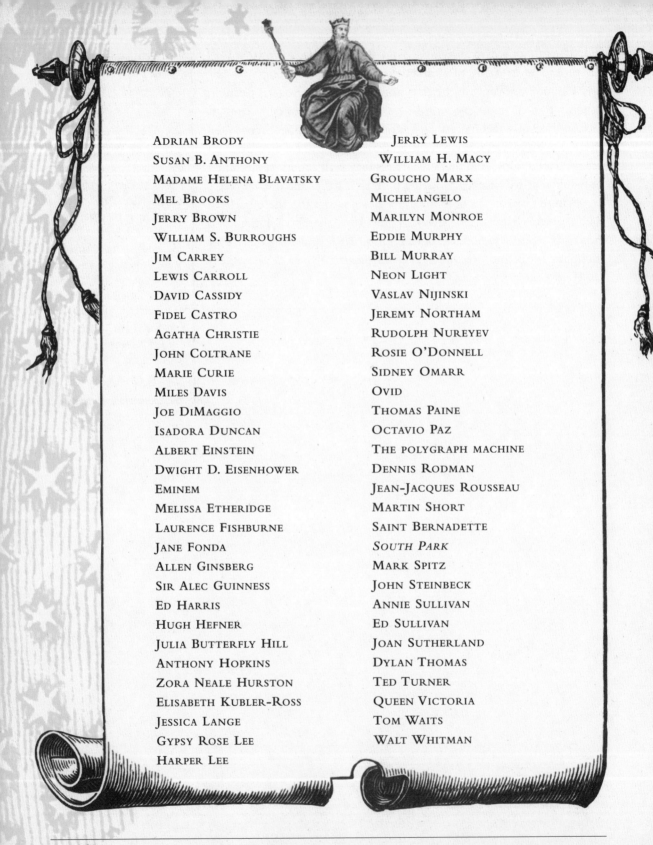

ADRIAN BRODY

SUSAN B. ANTHONY

MADAME HELENA BLAVATSKY

MEL BROOKS

JERRY BROWN

WILLIAM S. BURROUGHS

JIM CARREY

LEWIS CARROLL

DAVID CASSIDY

FIDEL CASTRO

AGATHA CHRISTIE

JOHN COLTRANE

MARIE CURIE

MILES DAVIS

JOE DiMAGGIO

ISADORA DUNCAN

ALBERT EINSTEIN

DWIGHT D. EISENHOWER

EMINEM

MELISSA ETHERIDGE

LAURENCE FISHBURNE

JANE FONDA

ALLEN GINSBERG

SIR ALEC GUINNESS

ED HARRIS

HUGH HEFNER

JULIA BUTTERFLY HILL

ANTHONY HOPKINS

ZORA NEALE HURSTON

ELISABETH KUBLER-ROSS

JESSICA LANGE

GYPSY ROSE LEE

HARPER LEE

JERRY LEWIS

WILLIAM H. MACY

GROUCHO MARX

MICHELANGELO

MARILYN MONROE

EDDIE MURPHY

BILL MURRAY

NEON LIGHT

VASLAV NIJINSKI

JEREMY NORTHAM

RUDOLPH NUREYEV

ROSIE O'DONNELL

SIDNEY OMARR

OVID

THOMAS PAINE

OCTAVIO PAZ

THE POLYGRAPH MACHINE

DENNIS RODMAN

JEAN-JACQUES ROUSSEAU

MARTIN SHORT

SAINT BERNADETTE

SOUTH PARK

MARK SPITZ

JOHN STEINBECK

ANNIE SULLIVAN

ED SULLIVAN

JOAN SUTHERLAND

DYLAN THOMAS

TED TURNER

QUEEN VICTORIA

TOM WAITS

WALT WHITMAN

"The one thing that doesn't abide by majority rule is a person's conscience."
—Harper Lee, Taurus Sun-Aquarius Jupiter. April 28, 1926

"I don't care to belong to a club that accepts people like me as members."
—Groucho Marx, Libra Sun-Aquarius Jupiter. October 2, 1890

"Rage, rage against the dying of the light / Do not go gentle into that good night."
—Dylan Thomas, Scorpio Sun-Aquarius Jupiter. October 27, 1914

*"I am able to play monsters well. I understand monsters.
I understand madmen."*
—Anthony Hopkins, Capricorn Sun-Aquarius Jupiter. December 31, 1937

"Don't do drugs, don't have unprotected sex, don't be violent. Leave that to me."
—Eminem, Libra Sun-Aquarius Jupiter. October 17, 1973

"Spoon-feeding in the long run teaches us nothing but the shape of the spoon."
—E. M. Forster, Capricorn Sun-Aquarius Jupiter. January 1, 1879

"Comedy is a man in trouble."
—Jerry Lewis, Pisces Sun-Aquarius Jupiter. March 16, 1926

"It is never too late to be what you might have been."
—George Eliot, Scorpio Sun-Aquarius Jupiter. November 22, 1819

"All of us are stars and deserve the right to twinkle."
—Marilyn Monroe, Gemini Sun-Aquarius Jupiter. June 1, 1926

*"After I die I shall return to the earth as the doorkeeper of a bordello
and I won't let a one of you in."*
—Arturo Toscanini, Aries Sun-Aquarius Jupiter. March 25, 1867

♓♃

We Reach Divinity

From Leonardo da Vinci, born April 15, 1452, who said, "There shall be wings! Once you have tasted flight you will long to return," to Jodie Foster, born 510 years later on November 19, 1962, who said "Normal is *not* something to aspire to, it's something to get away from," these Pisces Jupiters believed, and so should you, in the power and ecstasy of "the other"—that magic world where everything is possible and nothing is forbidden.

When any person suffers for someone in greater need, that person is a human.

— PISCES JUPITER CESAR CHAVEZ

Because Jupiter in Pisces is considered in astrology to be "dignified," Jovian extra luck magnifies Pisces Jupiter's already incredible stash. Pisces was thought to be ruled by Jupiter, until 1846, when its proper ruling planet, Neptune, was discovered. "Reason is our Soul's left hand, Faith her right, / By these we reach divinity." The greatest of all the metaphysical poets, Pisces Jupiter John Donne, wrote this— perfect to express the metaphysical *and* Pisces Jupiter! You are both.

Your luck, Pisces Jupiter, comes when you look to your intuitions and your psychic self, when you indulge flashes of genius and create magic. Fantastic creative energy comes to you once you free yourself of practicality and the "normal" and let yourself go to those hidden places in your unconscious—like fellow Pisces Jupiters Leonardo da Vinci, Sir Isaac Newton, and Charles Darwin. All saw beyond their own time, and, indeed, are hailed as genius. Psychic Evangeline Adams drew on her Jove for the sixth sense and used it to advise J. P. Morgan and others successfully on the stock market. Sigmund Freud gained fame and influence exploring the unconscious (a word he invented) and freeing himself to think of the "forbidden." Wilbur Wright dreamed of the sky and allowed himself to imagine flight; Michael Jordan literally freed himself to physically fly on the basketball court and became a legend. To tap into Pisces Jupiter's energy, you must explore creativity as much as you can.

You will get lucky when you explore the origins of life as Pisces Jupiter Charles Darwin did. As a botanist, a gardener, an artist, or a healer, you in essence are exploring water as our first primal experience—in the womb—as our origin of life, and also water as representing the unconscious, that below the surface, prebirth. Think on this: The first test-tube baby, in 1986, and the first American child to be born from her mother's frozen egg, in 1998, are Pisces Jupiter. This sounds like Pisces magic! "Creation" is literally the key word here.

Take a moment, look back long and deep over your life, recount as many events as you can, and ask yourself what has made you lucky or unlucky in your life. You'll notice that whenever you act like a Pisces, *you create miracles in your life.* No matter what your Sun sign is, *you get lucky when you live out the Pisces legacy gifted to you by the benevolent planet Jupiter.*

To take advantage of your luck, emphasize your . . .

Be internal, imaginative, an artist of the emotions, a healer, humane, childlike, a vast thinker, spiritual, and compassionate to the underdog.	♓♃ *Pisces Jupiter Strengths*

But because Jupiter rules excess, watch out for . . .

You can become holier than thou, a martyr, or in love with illusion.	♓♃ *Pisces Jupiter Excesses*

Understanding the Pisces Energy and Influence

Pisces rules the time so close to spring, to light and warmth, that you can almost taste it. No other sign is more tempted to daydream about the future. It is the very end of winter, the time of the year when the cold lingers on and on. Because of the cold, Pisces likes to be in a small place to feel warm and safe, to take care of all those out in the cold, all those weak and suffering—you, Pisces, are the kindest sign of the zodiac. Your sign represents the last season of rest, on the ancient calendar, before the busy Aries season of planting, tilling, and work begins with the spring. Pisces is everyone's last chance to think and to dream.

Your sign is also the twelfth and last sign. The conclusion, as it were, that cuts away the crystallizations or fallacies of the past, telling what should be forgotten and what should be pruned away before the rebirth of spring. Pisces, your energy is about anticipated regeneration, the time when everything and anything seems possible and we retain the capacity for surprise. Your sign is also about forgiveness, about transcending the superficial for the profound, as well as compassion to all who suffer and are weaker.

<hr/>

PISCES SYMBOL

The Two Fish

Pisces Fish swim in the watery depths of feeling and knowledge. Like the dream world where all their secret thoughts and desires reside, the water is dreamy and fluid, shadow and light, from warm to cold, rippling and constantly moving. The two Fish swim in opposite directions, signifying extremes and choices. Yet the Fish *are one with the water as they swim, and so must you be, dear Pisces Jupiter.* Remember this carefully for your luck if you *are* Pisces Jupiter, for water signifies the unconscious, emotion, ESP, and the imagination—where you should spend most of your time for greatest success. Finally, Pisces compassion, spirituality, and self-sacrifice—pure love, in other words—is reflected in the first symbol used for Christ, which was of two fish.

Believe it or not, true Pisces don't even have feet—they have fins! *Pisces is the only sign that exists purely in water;* thus, their difficulty with keeping their feet on the ground (or feet touching the Earth at all). Like Hans Christian Andersen's Little Mermaid, who hurt so terribly when she traded her beautiful fins for clumsy human feet, Sun sign Pisces always have an awful time with their feet. They get more aches, corns, and bunions than any other sign, and get very concerned with wearing the purest socks and the best shoes. This you don't need to worry about if your Jove, the planet that protects, is in Pisces; instead, emphasize the feet perhaps with an ankle chain or a tattoo. Fly rather than walk—at least a few feet above the ground, anyway—or, at least with your dreams.

4)X4)X4)X4)X4)X4)X4)X4)X4)X4)X4)X4)X4)X4)X4)X4)X4)X4)X4)X4)X4)X4)X4)X4)X

These are the colors of the magical sea and its fairytale underworld. Think of shiny, blue-green turquoise, the shade of a pebble plucked from the bottom of the sea, or a fish on the coral reef. Or celadon green, color of the waves and the foam. For your best luck, wear these colors but also think about them (you are so internal, Pisces!). Allow your mind's eye to bathe in the pale green and washed-out blues of the sea. Feel the deep waters rushing over you. When you've fully relaxed, wear turquoise accents, like Southwestern jewelry, and flowing scarves and ties, and experiment with painting your room sea green. Jupiter's bounty will come to you if you wear light jade jewelry and if your plates and pottery are celadon.

And now your Pisces Jupiter strengths: ♓♃ *Pisces Jupiter Strengths*

Be Internal Just as Pisces is the last sign of the zodiac and traditionally thought to be the end of the year's cycle, your Jupiter luck comes if you concentrate on endings and their sisters, beginnings. You must be willing to prune away all that is dying around you and within you. And you must nurture the new growing within, like the tendrils and roots growing beneath the surface just before spring. Remember the outer bleakness of late winter conceals the fertility and growth beginning underground, the prelude to spring.

Pisces Jupiter is all about the primal waters of fertility. Is it a surprise that the first test-tube baby, Louise Brown, was born Pisces Jupiter on April 16, 1986? Or that the first child in the U.S. born from her mother's frozen egg was born Pisces Jupiter on May 8, 1998? The most influential modern expert on child care, Dr. Spock, born May 2, 1903, was a Pisces Jupiter. This influential man introduced new ideas on childrearing and encouraged a softer, more compassionate approach, emphasizing positive, kind parenting. His *Baby and Child Care* continues to be a best-seller.

Dr. Spock was accused at the time of encouraging America to raise a generation of self-indulgent children (many said spoiled brats). Yet concentration on the self and the mind is lucky for Pisces Jupiter. You are nothing if not internal; as Pisces Jupiter Charles Darwin—again, an expert on heredity and, in a sense, fertility—said, "The highest possible stage in moral culture is when we recognize that we ought to control our thoughts." Allow yourself to embrace your internal journey, Pisces Jupiter, dig deep into who you are!

Be Imaginative

Pisces Jupiter encourages creative genius. Yet your Jupiter energy rewards a specific type of creativity: the ability to create new worlds hewn only from the cloth of the imagination. Professor-author Pisces Jupiter J. R. R. Tolkien, an expert in ancient languages of the Vikings, created an entire new world when he wrote *The Hobbit* and *The Lord of the Rings*. Not only did Tolkien create centuries of history and myth for his Middle Earth, he also invented entire languages, like Elvish, with fully realized grammar, syntax, and vocabulary. His creation continues to delight us in the *Lord of the Rings* movies.

Since Pisces embraces both light and dark, your fellow Jupiters are as capable of being dystopian as utopian. Pisces Jupiter George Orwell, born Eric Blair, created a future world in which totalitarian government was taken to its logical conclusion. His nightmare future, *1984*, is one of the most influential books ever written. And Pisces Jupiter Margaret Atwood, prizewinning Canadian author, has made her name creating dark and disturbing visions of the future. Her book *The Handmaid's Tale* sketches a future ruled by religious and reproductive fundamentalism. Her novel *Oryx and Crake* describes a hideous world destroyed by genetic engineering (a betrayal of natural fertility, so precious to Pisces).

Perhaps the greatest filmmaker of all time, another genius who has given us new worlds to explore, was also Pisces Jupiter: Orson Welles. In *Citizen Kane*, he created the life of a fictional magnate out of whole cloth and made it completely believable, both light and dark. And in his famous radio broadcast *War of the Worlds*, he convinced Americans that Martians were invading, whipping up a national hysteria.

Be an Artist of the Emotions

Pisces Jupiter, you must be concerned with the emotions, particularly those universal feelings associated with human predicament and suffering on Earth. Remember: Your mission is to explore light and dark but for a noble purpose: to bring compassion to those who are suffering.

The most deeply human and humane artists have been Pisces Jupiters. Looking deeply into the human condition and revealing it in all its beauty is a hallmark of the art of Pisces Jupiter Rembrandt van Rijn. Rembrandt didn't paint art for the sake of it, art that was "just" experimental, cerebral, or witty—he painted works that are deeply human, universal, humane. The faces in Rembrandt paintings are pure genius; you can see all their experience, hundreds of years in each, all their sorrows, joys, disappointments, hopes, and desires—in short, the entire human condition. Even Rembrandt's self-portraits never shied away from presenting truth in all its depths; for instance, they show him growing bitter as an old man.

Rembrandt was not the only Pisces Jupiter to embrace truth in his art. Fellow Pisces Jupiterian Leonardo da Vinci balanced scientific genius with luminous, lovely paintings. His Mona Lisa, in all her mystery, seems to embrace the human condition in a way that cannot be put into words. Similar genius is the hallmark of Pisces Jupiter composers Vivaldi, Schubert, and Mendelssohn. All composed in a way that speaks deeply to the human condition.

Pisces Jupiter Francis Ford Coppola is known for movies that capture the deepest emotions and human values. Many of his films, like *The Godfather* and *Apocalypse Now*, are concerned with the creation of self and person. They are like dark essays on human nature. We see in *The Godfather* how a man is created, what makes him fail and succeed, as Michael Corleone struggles to fill his father's footsteps. Tom Hanks's character said in *You've Got Mail* that everything a man needs to know about life is in *The Godfather*—so very Pisces Jupiter!

DOS

✳ Invent new worlds of the imagination with your artistic creativity.

✳ Embrace fertility and change, your unconscious and your dreams.

✳ Work actively to help those in need.

✳ Think about the big questions of life.

✳ Swim and bathe as much as you can.

✳ Try yoga and dance, and enjoy your body.

✳ Reacquaint yourself with the child within.

✳ Hang out with more Pisces, Scorpios, and Cancers.

Be a Healer

Your natural role is as a healer, Pisces Jupiter. You have the gift. The most important nurse in history, Florence Nightingale, was born Pisces Jupiter. She was the famous Lady with a Lamp who saved thousands of lives in the Crimean War, creating an entirely new profession of modern nursing and revolutionizing medical care.

Another great pioneer, the psychiatrist and writer R. D. Laing, was born Pisces Jupiter on October 7, 1927. Laing's most famous book, *The Divided Self*, explored the world of schizophrenia, attempting to understand what it is really like to experience the disease. Laing shocked the world of psychiatry by proclaiming an unusual degree of sympathy and empathy for the mentally ill. He famously said, "We are effectively destroying ourselves by violence masquerading as love."

A third famous Pisces Jupiter hero was born January 4, 1809, in Coupvray, France. An injury to his eyes at age three left him permanently blind. Louis Braille, at the age of fifteen, developed his ingenious system of reading and writing by raised dots—and two years later adapted it for musical notation! He said of the blind, "We do not need pity, nor do we need to be reminded that we are vulnerable. We must be treated as equals, and communication is the way we can bring this about." This kind, compassionate man gave his life to service—he opened the eyes of the blind. Be so ambitious, Pisces Jupiter!

Be Humane

Perhaps the most important legal document ever written was "born" Pisces Jupiter on June 15, 1215! For the first time in history, a legal document limited the powers of absolute government and gave the people some powers set in stone. The Magna Carta was the first charter of rights for the English common people; in this document, King John bound not only himself but his "heirs, for ever" to grant "to all freemen of our kingdom" the rights and liberties the great charter described. As Winston Churchill said, "Here is a law which is above the King and which even he must not break." Without the Magna Carta, we would not have a United States Constitution or our democratic society. It is typical that such an advance happened in a Pisces Jupiter passage.

Should it surprise you that the U.S. president who did the most for freedom and compassion was also Pisces Jupiter? Abraham Lincoln, born the same day as Charles Darwin, February 12, 1809, led the United States through the Civil War and issued the Emancipation Proclamation, freeing the slaves. He still stands for freedom and compassion; in a letter written in 1864, he wrote: "I see in the near future a crisis approaching that unnerves me and causes me to tremble for the safety of my country . . . corporations have been enthroned and an era of corruption in high places will follow, and the money power of the country will endeavor to prolong its reign by working upon the prejudices of the people until all wealth is aggregated in a few hands and the Republic is destroyed." Pisces Jupiter, is our present akin to what Lincoln feared? Help us find our way.

Be Childlike

As Pisces Jupiter Margaret Atwood said, "Another belief of mine: that everyone else my age is an adult, whereas I am merely in disguise." Yes, Pisces Jupiter, you must embrace the child within and never give up on childish things. Is it a coincidence that the invention of crayons, our favorite as children—whose smell brings us right back to kindergarten—happened in a Pisces Jupiter passage in 1902?

One of the most famous authors of children's books, Louisa May Alcott, was born Pisces Jupiter. A Sagittarius Sun, Alcott embraced the sunny, smiling face of childhood. She began publishing fairy stories for children but became famous with her novel *Little Women*, a work that described a family of girls brought up by an eccentric father much like her own. A pioneering "career woman" who held deep beliefs about compassion and the rights of the less fortunate, Louisa May Alcott embraced her Pisces energy and received the luck and success of Jupiter.

Pisces Jupiters, you must stay young at heart. As Pisces Jupiter Tina Turner, born November 26, 1938, said, "I will not give in to old age until I become old. And I'm not old yet!"

Be a Vast Thinker

Pisces Jupiter, your skill is not in detailed analysis of life. Leave such intense critical thinking to Virgos; instead, stick to the big picture. Thinking of the grandest, most profound ideas and concepts is lucky for you.

Many of the greatest scientific thinkers who considered the "big picture" have been Pisces Jupiter. Your fellow Jupiter Sir Isaac Newton, born December 25, 1642, was a genius at mathematics, physics, and astronomy. He said, "If I have seen farther than others, it is because I have stood on the shoulders of giants." He discovered new ideas about movement, which he called his three laws of motion, and ideas about gravity, the diffraction of light and force. Typically for a psychic Pisces Jupiter, he held an equally strong interest in alchemy, mysticism, and theology. His fellow physicist Johannes Kepler, born December 27, 1571, was also concerned with motion. His three laws of planetary motion revolutionized astronomy.

Other Pisces Jupiters who have considered the most profound, big ideas explored the dark side as well as the light. Friedrich Nietzsche wrote some of the most revolutionary philosophy ever written in books, like *Beyond Good and Evil* and *The Genealogy of Morals*. Nietzsche rejected Christian morality and the distinction beyond good and evil—unlucky for a Pisces Jupiter, who must always ensure their exploration of the dark side is done for the good of all. He went mad and his work was unfairly used to justify the Nazi regime. Another deep thinker and Pisces Jupiter, Sigmund Freud, who explored the dark side of the human unconscious, made a more positive contribution. Freud saw his revolutionary work as medical research intended to heal the sick and suffering. Because he used his knowledge of the dark side to help those in need, Freud has left us a more positive, healing legacy.

Be Spiritual

Embracing faith and spirituality is superimportant for Pisces Jupiter—this really brings you luck and benefits! Not only do you have a natural knack for it yourself but you can use your spirituality to help others with unparalleled success.

Deeply spiritual singer-songwriter Stevie Wonder, born Pisces Jupiter May 3, 1950, has achieved astonishing success by exploring his faith in his many classic songs. Unusually for a music artist outside the gospel or Christian genres, many of his most popular songs are about faith in God and God's compassion. His fellow Motown artist (and Pisces Jupiter) Marvin Gaye was also known for the spiritual power and healing of his music. Listen to a Stevie Wonder record, Pisces Jupiter, and hear the faith and compassion in his voice.

Jeane Dixon, born Pisces Jupiter, was one of America's most famous psychics. In the 1950s, she gained the national spotlight with her amazing and convincing political revelations. In the May 13, 1956, issue of *Parade* magazine, Dixon said that the 1960 presidential election would be won by a Democrat who would "be assassinated or die in office." According to a Catholic newspaper, Dixon faithfully attended Mass each morning. *A Gift of Prophecy: The Phenomenal Jeane Dixon*, written by syndicated columnist Ruth Montgomery, published in 1965, has sold more than 3 million copies. Dixon and her predictions, during the past several decades, have been a staple for the supermarket tabloid industry.

Be Compassionate to the Underdog

Pisces Jupiter, be tolerant, give up criticism, be a bleeding heart! Jupiter rewards your sympathy for the underdog, the victim, and the outcast—this brings you luck! As Pisces Jupiter Felix Mendelssohn said, "I dislike nothing more than finding fault with a man's nature or talent; it only depresses and worries and does no good; one cannot add a cubit to one's stature, all striving and struggling are useless there."

One of the greatest activists for the silent and suffering was born Pisces Jupiter on March 31, 1927. At a time when no one even considered their plight, he brought public attention to the suffering of migrant workers. He worked hard to change the terrible abusive conditions under which they worked, changed much for the better, and worked as a union organizer, and so much more. Do you recognize compassionate Pisces Jupiter Cesar Chavez?

Many other social workers and activists have been Pisces Jupiter—this path brings you success. Harry Belafonte used his fame as a singer to campaign for civil rights in the 1963 March on Washington. Russell Means dedicated himself to working for Indian rights and obtained the Lakota name Oyate Waconia ("Works for the People"). He became the first national director of the American Indian Movement. Means, who also founded the American Indian Anti-Defamation Council in 1990, led the seventy-one-day siege at Wounded Knee in 1973. He said, typical Pisces Jupiter: "Life for each person is a journey to personal vision. Each person starts at some point on the rim of the Great Circle with their own perspectives, and if they follow the path of the spirit, they begin to move toward the Center."

What Makes You Lucky in Love

O Pisces Jupiter, without the stimulus of emotional exchange, your life is very empty. The intellectual type of person has very little appeal for you. Unless you can reach empathetic union with your companion, you become bored, restless, and unhappy. It is so important for you to have a friend or lover who understands your feelings of otherworldliness and to whom you can confide your dreams and secret wishes. Before you allow intimate relationships of this nature, the person must demonstrate deep love, loyalty, and understanding. You are more discriminating as a rule, in your choice of confidants than you are in ordinary love affairs.

You go straight to people's hearts. You suffer personally when you see those you care about sorrowful or in pain. You can often touch and stir in others emotions they didn't realize lay within them.

Even if you are a practical, hard-headed Taurus by Sun, if your Jupiter is in Pisces, indulge your dreams. Be whimsical on dates and encourage your prospective partners to fantasize and dream with you. Do not concentrate on practical things or Jupiter's luck will desert you. Try out creative activities, like painting or storytelling, on dates, or take your date to a fantastical movie. Express your artistic self in every way you can; the partner who responds will be perfect for you.

Last but not least, don't forget "Sexual Healing" by fellow Pisces Jupiter Marvin Gaye. This delicious, life-affirming, and yummy love song should be listened to by everyone—and especially by lovers.

Your Best Career Moves

You need work that is not overly intellectual. It's not uncommon for Pisces Jupiter to go from job to job until they find what they like. You simply can't stay somewhere where you are unhappy, and you cannot endure work that involves pressure, deadlines, or too much repetition. You need to express your altruistic urges—you want to be of use to humanity in some way—so don't work or live in a highly competitive or disharmonious environment. You charm and get along with your employers and coworkers, but you must have a function you consider worthwhile. If you understand this need, you should be able to save yourself a lot of time and aggravation the next time you have to change positions or go looking for satisfaction elsewhere.

Artistic work is great for you, especially acting or dancing. You love to communicate directly to an audience and feel their emotional response. Without the stimulus of emotional exchange, your life is very empty. You must recognize your luck comes through your acutely sensitive, psychic, and spiritual nature. It's sometimes hard for you to distinguish between reality and daydreams, mainly because so many of them come true!

As a Pisces Jupiter, stretch the boundaries of work and explore where your mind will go and won't go. Use your imagination and creativity, always, always, no matter what type of work you do. Let yourself think and make time for it—but also make time to just sit and not think at all! Trust yourself and your own judgment and make sure your work allows you to use your imagination and even your fantasies. You can apply your unique, magical energy by talking to everyone you work with and coming up with new ideas for your company. Don't stick to one job; move around a lot and sample different roles at the workplace. Your Pisces energy allows you to undertake multiple roles and to slip seamlessly from one to the other. You could probably work as a psychic if you had the stamina for it. You must, however, avoid self-indulgence; you will find success working for others, as it will be difficult to be your own boss!

How to Reach Optimum Health

Your everyday prayer is a seaweed-and-saltwater bath—as you are a fish, after all! These are especially good for the feet, which Pisces rules. Due to your affinity with water, it is good to drink as much as possible and also to swim and bathe—rather than shower—on a regular basis. This has the effect of lessening the possibility of water retention and bloat—water will find its way to Pisces Jupiter any way it can!

The best way to stay in good health is to make sure you get *at least eight hours of sleep a night.* You cannot miss your sleep! Due to the glamorous late-night aspect of Neptune living, sleep and rest can often be neglected. Add on the effects of overstimulated responses and heightened sensitivity and you can feel utterly exhausted just by the end of a workday. Drinking alcohol and comfort eating should not be the way to relax for you; it overburdens your sensitive nervous system and ages you faster than any other sign. Moderation is key! Remember that Neptune rules the thalamus, which transmits stimuli in the brain, from and to the sensory organs, which is one reason you are so sensitive to the environment you are in. You have the most extraordinary antennae! Take good care of them and they will take good care of you—and your good looks.

Your Lucky Foods, Element, and Spices

Seaweed is just as fantastic for you applied to the tongue (in yummy seaweed salad and sushi) as it is applied externally. Add to this a daily intake of kelp (derived from seaweed), which helps correct the thyroid and reduces bloat, and you're sitting pretty! The thyroid is only something to watch if you're over-exhausting yourself or particularly stressed. Go easy on the coffee (overstimulation), the soda, and the salt (bloat).

Pisces' cell salt is iron (ferrum phosphate), which is necessary for hemoglobin in the bloodstream to carry oxygen from the lungs to all the other cells. (See, we're still dealing with fish having to breathe out of water!) So eat foods that are rich in iron every day and take iron supplements as well. Every single day: spinach, liver, kidneys, egg yolks, barley, lean beef, lamb, dried beans, raisins, oysters, beets, onions, prunes, apricots, apples, grapes. Generally, you function best with lots of protein, yogurt, and nuts—and water, water, water!

Your Best Environment

Pisces Jupiters are known for creating their *own* perfect environment, more than any other Jupiter placement. Frank Lloyd Wright was a Pisces Jupiter, as are many of our greatest landscapers. Not only do you have a magic touch growing plants and trees, but you know exactly the kind of environment you need and can manifest it. For your finest luck and health, create what is soothing and beautiful, nurturing and joyous, for your senses. You need nature and peace to do what you do best; and the sound of running water is like a magic key opening your energy. If you do not live near water, even a stream or pond, make sure you have a fountain of water you can physically hear near you at all times. Even a few arranged stones, with a trickle of water, as in a Japanese Zen garden, will do the trick. Keep one by your bed and in one other room of your house for best luck.

Your Best Look

Elegant, like a fish. Smooth, feminine, deftly moving, slippery, lithe. *The clothes really don't matter*—it's the movements of the body. Anything that shows your whole form, your movements, and particularly your feet is going to be a hit. Perhaps it's that mermaid thing again, but *the way you walk* is particularly important for your attraction and appeal. Try to dress streamlined, in one straight line of color, and only one flowing accessory, say a scarf.

Think Pisces Jupiters Natalie Wood, Vanessa Williams, or Fairuza Balk for the women; Cary Grant, Wesley Snipes, or Ryan Phillippe for the men. Pisces Jupiter men contain that great feminine elegance that is so sexy and should be emphasized. Pisces Jupiter women also look extremely attractive as men—consider Pisces Jupiter Hilary Swank in *Boys Don't Cry*. The more feminine and curvy, in both sexes, the better. Pisces Jupiters just do not look as good underweight—Earth signs like Capricorn or Virgo Jupiter can carry the stick look off well, but not Pisces! Just think of stunning Pisces Jupiters Harry Belafonte, Ingrid Bergman, and Drew Barrymore. (Pisces Jupiters who get thinner will look immediately older and not as healthy; I won't name any names, but you can check your list! This is simply an amazing phenomenon with Pisces Jupiter.) Go for black for streamlining, turquoise and light green, and anything shimmering and sparkling like moving currents of water.

Your Best Times

The luckiest times for you are when the Moon is in Pisces, which happens every month for two to three days. Find these special days in your astrological calendar. During these days, schedule crucial meetings, tell that special person you love them, ask for a promotion or raise, push the envelope, and plan to shine. In your everyday, the best time for you to exercise is in the evening. The luckiest times of the year for you are February 20–March 20 (when the Sun is in Pisces); June 21–July 22 (when the Sun is in Cancer); and October 23–November 21 (when the Sun is in Scorpio).

The very luckiest time of your life—as in Lotto lucky—is when Jupiter reenters the sign it was in when you were born (Pisces), which happens every twelve years and lasts anywhere from a few months to just over a year. This is called your Jupiter Return, and it launches your luckiest streak. You can count on the *extraordinary* to happen the very day your Jupiter Return begins. Amazing events can start happening up to six weeks beforehand, so plan ahead!

How to Be Lucky Financially

Your luck, Pisces Jupiter, is magical and deep. Avoid gambling games, as you have a compulsive nature. But take chances on yourself and you will win big. Your Pisces energy finds success through change, and you will be able to survive any upheaval that results with aplomb and success. In fact, you will do best financially in periods of great upheaval and uncertainty. Take risks, Pisces Jupiter, and act on your psychic hunches. You will find Jupiter rewards you.

Even if you are a terribly practical Virgo Sun or a Capricorn Sun working hard to gradually improve your lot, if your Jupiter is in Pisces, you must keep close to a psychic connection. Don't make rigid plans or rely on logical thinking alone.

Dreams are particularly lucky for you. Jupiter makes the saying "Follow your dreams and the money will follow" come true for Pisces Jupiters. An example: Pisces Jupiter filmmaker Nia Vardalos, born September 24, 1962. Despite all the naysayers, she worked for years on her personal dream, a one-woman show about growing up in a crazy Greek family. One lucky night Hollywood producer Rita Wilson attended the show in Chicago. The result: the movie *My Big Fat Greek Wedding*, based on Vardalos's show and screenplay, the most profitable romantic comedy of all time.

Success in Family and Friendship

As much as you enjoy lively company, you need regular periods of seclusion. Without these, you can become irritable, depressed, and fiercely self-critical. This can make you tough to live with!

More than any other Jupiter sign, you must choose your friends and partners carefully. You need to be with someone who is your true soul mate, nothing else will do! Make sure you spend a long time dating before you make a major commitment. And balance the courting period with times when you completely withdraw into solitude. You must then examine your dreams and the innermost part of your being to ask if this person is truly right for you. The same is true with friendships. You should not make friends easily (even if you're a fun-loving, gregarious Sun sign Sadge!). Have few friends but make your connection with them deep and psychic, on the dream level.

You can choose your friends but not your family. Exercise your Pisces Jupiter compassion here: Take care of the family members that need you and help them heal their wounds. Perhaps your strongest relationship, however, will not be with friends or family but instead with your personal vision of God. You have an affinity for religious and spiritual communities. Explore connections between the imagination, spirituality, and the emotions, like your fellow Jupiter poet John Donne.

Pisces Jupiter Excesses ♓♃	When you go too far!

The great Pisces Jupiter cautionary tale is *Don Quixote* by Miguel de Cervantes, born Pisces Jupiter on September 29, 1547. *Don Quixote* tells the story of a hapless knight, born well after the end of the age of chivalry, who rides around accompanied by his faithful servant, Sancho Panza. The don has spent years reading chivalrous romances and dreaming about knightly escapades, and he believes his life mission is to travel around Spain fighting dragons to secure damsels in distress. An unfortunate victim of his own dreams and illusions, Don Quixote is actually quite mad: The monsters he fights are really windmills. Yet he remains a gentle figure of fun, protected by Sancho Panza, who does everything to prevent Quixote realizing the ludicrousness of his beliefs. Why is this a cautionary tale? Well, Pisces Jupiter, you gain access to the world of dreams. Just make sure you take a reality check once in a while or you may discover you have deceived yourself through your own illusions.

Excess #1 *You Can Become Holier Than Thou*

As you dispense your endlessly helpful advice, it can turn to criticism, and thence to outright insults. The desire to help coupled with your need to see the world as perfect can bring severe disappointment, childlike crankiness, a tendency to blame others, and the need to protect yourself at all costs. This is where the holier than thou comes in, infuriating to others and often completely unseen by you. This position can also protect you from seeing your own mistakes. Because you can be so outside of everyone else's reality, you must be particularly careful about addiction to alcohol and/or drugs, which has taken the lives of some of our finest, including Pisces Jupiters Billie "Lady Day" Holiday, dead at the age of forty-four, "the little sparrow" Edith Piaf, and indirectly Marvin Gaye, who was murdered by his own addicted father.

Excess #2 *You Can Become a Martyr*

So much sacrifice, so much offering of your services and your help can become a nightmare, which will have your helpee(s) running for the woods. You can become overly fond of reminding people what a giver you are, and how much you are there constantly for your family and friends, and how underappreciated you are. Some healthy acknowledgment of a human ego—anyone's right—would not be amiss here. The problem with that is that it makes you like everyone else, and that's the last thing you want to be.

Pisces Jupiter King of England Henry II felt so bad about locking up his wife for thirty years (but what could he do?); felt so bad about

---

DON'TS

❋ Get trapped on the dark side.

❋ Allow your success as a healer to go to your head.

❋ Use alcohol or drugs to excess.

❋ Forget that you are human like everyone else.

❋ Lounge forever in a wet bathing suit or keep your feet cramped up in sweaty shoes.

❋ Criticize others who fail to live up to your idea of perfection.

❋ Fall into dishonesty or self-deception.

having his best friend, Thomas Beckett, murdered (but he couldn't help it)—when people around you start being hurt, and you're still feeling *so* bad yourself, it's time to look at your actions, not only your feelings.

Excess #3 *You Can Become in Love with Illusion*

The worst form of deception is self-deception, and Pisces Jupiter can go in this direction when reality becomes too much to bear, or when moodiness, disappointment, or laziness takes over. A certain amount of nonacceptance, even self-hatred, comes in here, as the need to distort reality means you're not very fond of accepting it in the first place. You can have certain myths about yourself, which do not hold up under a harsh light. This is what I call the Blanche du Bois "paper lantern" approach to life, where the light must be soft, hiding much of reality, and your need to be in a fantasy world is so strong that you will begin to lie even to yourself.

Pisces Jupiter Karl Rove has created an entire picture of an environmentally friendly, pro-education, democracy-abiding administration for President Bush. Another great PR picture: "the Cruses"—Tom Cruise and Penelope Cruz (both Pisces Jupiter) spin the fantasy of their romance also for PR purposes.

Sun Sign-Jupiter Sign Combinations

✦)✦(✦

The zodiac year begins with Aries and ends with Pisces, so with this combination you have your bases covered. You have a fiery impulse for beginnings and the visionary intuition and ripe wisdom of Pisces Jupiter to bring projects to maturity. Whatever you do must be creative; make an art of anything, from painting to farming, music to mechanical invention, sports to merchandising. Draw on your Pisces Jupiter and use imagination to visualize something fine and beautiful arising from your work; hone your skills and use your Aries energy to create that picture.

Work *with,* but not *for,* others. Aries-Pisces' second challenge is to *listen* to the voices of others, the voice of history, and to your own inner voice. For gain, invest in art (literally) and in companies in transition; back new inventions or presentations. In love and sex, let Pisces Jupiter enrich passion with magical overtones. You'll revel in being the costar of a storybook romance.

✦)✦(✦

Taurus earth grows fertile when watered by Pisces Jupiter. Yours is a fine, practical nature, Taurus; Pisces Jove gives you the imaginative insight and compassionate empathy to apply this practicality to the comfort and well-being of others. You have Freud, Dr. Spock, and Florence Nightingale to follow. You have much to offer in physical and mental health areas as practitioner or administrator, in child care, and as a worker or teacher in environmental studies and stewardship. With your appreciation of the good things in life, try creative cuisine.

Combine your Taurus grasp of reality and ability to follow a line of study with your Jovian intuition and you'll offer revelations to the world. Be open to questioning and changes in order to grow. This is true for investing, too. With your partner, ignore practicality and dream, fantasize, and share sensuality and love—the greatest art of all.

IF YOUR SUN SIGN IS:

Gemini

(May 21–June 20)

Pisces Jupiter offers Gemini the experience of a lifetime, for a lifetime. You'll get lucky if you take advantage of Pisces emotional creativity and receptivity, dive in, and allow the currents to carry you to new shores! You'll increase the scope of your quick intellect by getting in touch with your emotions and imagination, opening yourself to dreams, and returning, refreshed, to the world of movement and action. This approach works for investment growth, too. And try the Pisces Jupiter recipe for a good love life: Take imagination, compassion, and emotion, add dashes of fantasy and whimsy, and sip lazily by a tropical sea. Ummmm!

Or let Cole Porter say it for you. A debonair Gemini-Pisces, American-born but considering himself Parisian (he lived for years in Paris), Porter wrote hundreds of romantic, witty, but oddly haunting songs for musicals and films. Among the most unforgettable are "I've Got You Under My Skin" and (how very Gemini) "Don't Fence Me In." As one says, *"You're the bubbles in my champagne."* That's you, Gemini-Pisces.

IF YOUR SUN SIGN IS:

Cancer

(June 21–July 22)

The Cancer Crab and the Pisces Fish—sounds like the start of a whimsical children's story. That's Pisces Jupiter, and with Cancer's emotional nature, it goes right to the heart. Cancer-Pisces author George Orwell had no less an effect, though a much darker one, on readers. His novel *1984* was not only eerily predictive of today's world, but became synonymous in people's minds with the evils possibly inherent in technology and conformity. Trust yourself; don't dam up your flowing talent with doubt or let it dissipate into moods, nor allow others to impose their influence. Bring out the rich colors you *feel* through painting and design, landscape architecture, or in any field connected to creating living environments. You can speak to our hearts with acting or writing; start jotting down those ideas for a magical tale, perhaps for children. For gain, don't hold on tightly but dabble in investments here and there.

With dates or your significant other, share dreams and creative fantasies but maintain individuality, and let whimsy add a lighter touch. Light candles, float them on lilypads, and read stories to each other.

isces Jove brings you luck and opens a new world to you. The Lion finds himself strolling, not on the sun-baked veldt but in gardens of huge flowers, up marble staircases, in underwater grottos; he talks with a bird, hears a fish singing—it's the interior world of your Pisces Jupiter, the place where you can explore possibilities and find answers, even to questions you'd never thought of asking.

As a Leo, you're sure of your outward presence. Now you find who you are inside. Leo-Pisces actor Hilary Swank drew upon talents she didn't know she had in the acclaimed film *Boys Don't Cry*. For her portrayal of Brandon Teena / Teena Brandon, a woman who discovers a different self inside, Hilary won an Academy Award. You can bring this infinitely varied creativity to your work, whether in acting (Leo is the natural performer of the zodiac), stage or film directing, art or design or fashion. And with your Pisces Jove intuition combined with Leo command, you could do well as an investment adviser. Not a bad reward for letting yourself dream now and then. Your love life will also be richer. Let Pisces Jupiter encourage you to fantasize, play, and dream together.

isces Jupiter brings to you, Virgo, a certain magic, delicate nuances of feeling that give you pleasure but also entice you toward unknown, vaguely unsettling realms. You sense there's "something more"; with your honest and critical approach, you strive to discern what it is. Accept it, Virgo-Pisces—it's your inner voice, the echo of your dreams. Go with it to find your love, your creativity, your success.

Oscar-winning Virgo-Pisces actor Ingrid Bergman captured this magic in her radiant screen performances. She brought to her acting both the fine sense of truth, of concern with right and wrong, and an elusive emotional vulnerability typical of Virgo-Pisces. One felt that she represented each of us caught between irrational feeling and rational thought. That is both the challenge and the luck of your Pisces Jupiter. In all areas of life, including financial, take advantage of changing patterns but retain your critical judgment and individual viewpoint. You find the perfect partner in one who respects your private emotional space but shares dreams and a strong psychic connection.

Libra

(September 23–October 22)

The legendary French actor Sarah Bernhardt epitomizes Libra charisma blended with Pisces Jupiter's evocation of deep, fluid emotions. She communicated to nineteenth-century audiences the passions and nuances of the great roles, from Saint Joan to Salome (even Hamlet!—an example of the mutable, male-female aspect of Pisces Jupiter), moving them to tears and etching herself in their hearts as "the Divine Sarah."

Your creativity and psychic intuition will find expression also in teaching, diplomacy, or arbitration (getting to the heart of the matter), or in anything concerned with children. Pisces Jupiter understands both the whimsical fantasies and the fears of childhood. But be careful, Libra-Pisces, that while listening to all sides, all voices, you're not too swayed by their emotional influence; know who *you* are at the deepest level and stay true to yourself. This is particularly important in romantic relationships and situations where you might be tempted to give away yourself, your goods, and your autonomy. After dreamy delving into the watery ebb and flow of emotions, balance the scales with physical activity, only a walk in the garden.

Scorpio

(October 23–November 21)

A Scorpio Sun with Pisces Jupiter creates an intensity that gives Scorpio-Pisces the power and persona to leave an indelible impression on the world, to change history, as did Scorpio-Pisces revolutionary Leon Trotsky. In an entirely different way, talented but tragic Fanny Brice—singer, entertainer, creator of "Baby Snooks"—became an American legend in the 1920s and '30s. This Sun and Jupiter combination draws on Scorpio's primal motivation and relentless pursuit of goals and on Pisces dreams of a perfect world. How does this influence both a revolutionary and a singer? Trotsky was a dedicated crusader for socioeconomic justice; his unrelenting stand on issues conflicted with Stalin (who was also a Pisces Jupiter) and led to his assassination. Brice squandered her talent and life in passionate devotion to her faithless gangster-lover, Nicky Arnstein. Both refused to give up their image of perfection. But you can use your Pisces Jupiter for success and happiness, as have many others, by avoiding extremes.

Find your inspiration in the depths of your being but refuse the temptation to totally live there. Invest in your own talents, choose a significant other who lightens you.

Sagittarius energy and abandon with Pisces emotion and immersion in dreams—*La Vie en Rose!* So sang Sagittarius-Pisces Edith Piaf, crying out with exuberant gusto and heart-rending pathos for a life of beauty and love. This French cabaret singer who became an international icon of her kind blended an aura of street waif with sexual innuendo; her haunting voice was rich with emotion. Though admired and loved, she gave in to the Pisces temptation to escape in drink and drugs; each morning she woke with her voice gone and would drink scalding coffee to bring it back.

Sadge-Pisces are intensely emotional, combining enthusiasm and action with belief in dreams; they want to paint the town bright red— and deep purple! Take time to listen to voices, from your psyche's to that of your granny and the neighborhood grocer. Scan the market for investment opportunities in sports or underwater salvage—diversity and whimsy. Then back to dreams and playful fantasies, especially with your significant other—and *La Vie en Rose!*

4)(4)(4)(4)(4)(4)(4)(4)(4)(4)(4)(4)(4)(4)(4)(4)(4)(4)(4)(4)(4)(4)(4)(4)(4)(4)(4)(4)(4)(4)(4)(4)(4)(4)(

Edgar Allan Poe—and a number of other writers who touch our deepest fantasies, fears, and funny bones—was a Capricorn-Pisces. It's all that Pisces imagination and the tenacity of the Goat, who just won't veer from the trail of a good idea or story. Poe did succumb to the Pisces tendency to stay in the murky depths, but his literary explorations of the darkest unconscious have never been equaled.

To acquire the luck of your Pisces Jove, to tap into your own creativity and your psychic abilities, you do need to dare the depths. Find success, too, as a cartographer, a hydrologist—or a psychic! Where but in the Capricorn-Pisces combination do you find such diverse geniuses as Isaac Newton and Johannes Kepler, Arthur Miller and Jeane Dixon? Or just get inspiration for a new look you'd like to try or where to vacation this year. Despite Poe et al., you can *enjoy* your Pisces Jove. Let it take you into the social swim, from street party to posh; eavesdrop to catch the tone of the world. Financial rewards come from inventions, yours or those of others.

Aquarius-Pisces can be *anything:* composers, winning players, presidents, scientists, film stars, generals, psychics, social activists, dancers. That's free-spirited, dare-all Aquarius with Pisces Jupiter dreams, empathy, and intuitions. Aquarius-Pisces actor Jeanne Moreau, a quirky mix of liberated intelligence and femme-fatale sensuality, was called by Orson Welles "the greatest actress in the world." "We're here right now. Tomorrow, we'll be someplace else. So why nostalgia?" asks Moreau.

Aquarius-Pisces is a law unto itself; a rich interior with far-ranging intellect to carry your visions to the world. Your luck comes when you direct these talents into creative channels, set high standards for yourself, and remain open to new insights through change. You, Aquarius-Pisces, are the dolphin, a creature that lives in both water and air. Swim below the surface to feed your soul; come up to breathe. Never neglect either aspect of your being. Your partner must be someone who is, like yourself, deeply emotional and incessantly curious. Together you play, diving and leaping, sharing the joy of being, perhaps forever, perhaps . . . but *why nostalgia?*

IF YOUR SUN SIGN IS:
Pisces
(this makes you double)
(February 20–March 20)

If music be the food of love, play on. So many of you are musicians! Antonio Vivaldi, classical composer of melodic Baroque music, and Lawrence Welk, the "polka king," who reached the hearts of middle America with his "champagne music," are two wild opposite examples! Pisces Jupiter combines with Pisces Sun to *lift* the mood. Your Jove gives you a chance to take your dark-blue dreams, your romantic nostalgia, and put a lilt to them—in music, art (including cartoons), acting, or radio astronomy (those sound waves from deep space!). Your ability to glide through any scene, from pier to palace, and converse easily with all, could lead to work with the physically or mentally challenged, or as entertainment director on a ship.

With a double helping of Pisces, you're happiest living in your imagination, and you still need to beware of alcohol or other substances for escape, but Pisces Jove helps you to turn your dreamy ideal into something for the world. Financial rewards come from unexpected sources and intuitive investments. You will find your perfect partner at a concert, a folk-dance club, or singing in the rain. *Play on!*

Relationships with Pisces Jupiters

Finally, for those of you who aren't Pisces Jupiter yourselves, but have a Pisces Jupiter spouse or significant other, boss, child, or parent, here are some tips to help you learn to deal! This is how the luck of someone else's Jupiter touches *your* life.

Your Significant Other

"Come live with me, and be my love, / And we will some new pleasures prove," wrote Pisces Jupiter John Donne, and this can be exactly how you are invited to play with your Pisces Jupiter. It will be delightful, like recovering paradise lost, like playing in a paradise that never disappeared—at first. Two babes in a wood, naked, exclaiming over each other—how fantastic! But allow some details of the day to get in, or problems with scheduling, an absolutely urgent matter of theirs that cannot be put off for one solitary single second, and you may feel suddenly left out in the cold. They are known to be able to turn right around and look at you as if you're a complete stranger in the worst of times. It's a self-protective device, but it can hurt like hell.

Love is a strange world for them. They'll willingly give it to everyone, to humankind in general (and any stray animals get it automatically), but if it becomes difficult with their lover or mate, they can withdraw in two seconds flat. They love the act of seduction even though they're often quite shy. Enchantment is where they're at.

They are wonderful when you first start dating. Things will proceed smoothly if you give them space and allow them occasional solitude. They will construct imaginative, wonderful events for you. Explore each other's dreams and be prepared to spend time letting the relationship grow. The thing is, they are looking for a soul mate (nothing else will do). If it's not a deep relationship, it won't work out. But if it is true, deep love, you are guaranteed a rich, fulfilling connection like no other. A perfect love can be yours with your Pisces Jupiter.

Your Boss

Remember that Pisces Jupiter bosses find their luck and success in the artistic and creative, so that means that you might have to do more of the practical work and thinking. They are most often found in any area that involves illusion and glamour—from television to magazines, from modeling to advertising.

You could be lucky enough to work for one involved in their real calling, which is healing, charity, or a spiritual movement of some kind, but then you really would have to cover all the practical details, as your boss could be in a deep meditation, trance, or hypnotic state most of the day. Schedules will be difficult to maintain. These fish are also given to daydreaming, hurt feelings, and some ups and downs in the emotional department—they can tend to hog the feeling part of the environment around them—so learn to be sensitive to them and give them slack when they need it. Their ideas are apt to be brilliant. They will sympathize with you, be your best friend, and be the best shoulder to cry on as well. Then, suddenly, they may just disappear. Do not take this personally; it's only the Fish's disappearing act, which they need for their psychic survival. Pisces need for rest and recovery is legendary, as they psychically pick up everything and can be over-whelmed by what's around them. Do not push. If they could just be the thought tank (as in fish tank), and not have to deal with the mun-dane every day, they'd be happy.

Your Child

Pisces Jupiter children are pure love and need pure love to help them grow. Filled with endless imagination, their play hours are a delight to behold as they will liter-ally invent and see anything they want. Often they see things no one else can see in real life, too! They are born with perhaps the highest psychic quotient of any of the Jupiter placements and will be picking up the invisible along with some spirits from the day they are born.

To discipline them will take a great deal of patience because they live so in their own world. What works instantly, however, is if you play and imagine right along with them. If you want them to do something (particularly if they don't want to do it), cast yourself in the role of a magic character and they will go right along. Pretend you're in the Hogwarts School of Witchcraft and Wizardry and you're giving them a test or an order to be invisible, and they'll do whatever you say! Just be gentle, don't ever abuse your power, and respect their need for quiet and harmony. Make sure colors around them are soothing, espe-cially lucky turquoise and sea green. They will be devoted to their sib-lings but may also feel left out and demand more attention.

Your Parent

From your Pisces Jupiter parent you will learn the value of keeping your imagination, crea-tivity, and dreams all your life. Hopefully you haven't rebelled so much that you've given up on all that inner stuff and gone off and joined the army or something like that!

Theater or film is where you've really been groomed to go if you had a Pisces Jupiter parent. Your Pisces Jupiter parent is not exactly what you'd call a big disciplinarian—great for playing with, for telling stories, for cuddling, and if you need a good cry. He or she is very gentle and doesn't like to fight. They offer to do anything for their kids. But sometimes you need to be the grownup and take care of *them*. There is a part of them that stays like a child all their lives, which is fun if you're playing with them but can be hell in a crisis situation or when you really need them. They can be the ones having a nervous breakdown over *your* breakup or divorce or your losing your job. Sometimes it's a little much to have to take care of them on top of losing your own mind. They can get so involved in their own agenda that you may have to nudge them a little for your own.

This is what astrology is all about, and, I'd venture, a civilized life. Appreciating how different we all are, and loving these differences! Nature gives an antidote for every poison. Isn't that the miracle of life? Problems appear, but there's always a solution, and if anyone can see it, prophecy it, and heal it, Pisces Jupiter can. *Remember, Pisces Jupiter: You are the creative genius of the zodiac.*

Madalyn's Movie Choice for Pisces Jupiter

Breaking the Waves
This odd love story set on a remote island in the Hebrides contains more emotion than I've ever seen in one film! A handheld camera zooms right up to the characters' *skin*, uncomfortably close to their inner storms. A horrific twist proves that Pisces self-sacrifice indeed saves and heals. Its end magic is impossible to believe— yet you believe it.

DREW BARRYMORE

SARAH BERNHARDT

HARRY BELAFONTE

INGRID BERGMAN

LOUIS BRAILLE

MIGUEL DE CERVANTES

CESAR CHAVEZ

CHARLOTTE CHURCH

FRANCIS FORD COPPOLA

TOM CRUISE

PENELOPE CRUZ

LEONARDO DA VINCI

CHARLES DARWIN

LEONARDO DICAPRIO

JEANNE DICKSON

JOHN DONNE

WALKER EVANS

PATRICK EWING

FIRST TEST-TUBE BABY BORN

JODIE FOSTER

SIGMUND FREUD

CARY GRANT

BILLIE HOLIDAY

BOB HOPE

INVENTION OF CRAYONS

PETER JENNINGS

MICHAEL JORDAN

JOHANNES KEPLER

R. D. LAING

JAY LENO

ABRAHAM LINCOLN

THE MAGNA CARTA

RUSSELL MEANS

HENRY MILLER

ARTHUR MILLER

DEMI MOORE

JEANNE MOREAU

SIR ISAAC NEWTON

FRIEDRICH NIETZSCHE

FLORENCE NIGHTINGALE

JOYCE CAROL OATES

GEORGE ORWELL

ESTELLE PARSONS

S. J. PERELMAN

EDITH PIAF

HENRY PLANTAGENET

EDGAR ALLAN POE

COLE PORTER

REMBRANDT VAN RIJN

CHLOE SEVIGNY

FRANK SINATRA

WESLEY SNIPES

DR. BENJAMIN SPOCK

HILARY SWANK

QUENTIN TARANTINO

DAVID THEWLIS

J. R. R. TOLKIEN

LEON TROTSKY

TINA TURNER

ANTONIO VIVALDI

JOHANN VON GOETHE

ORSON WELLES

VANESSA WILLIAMS

STEVIE WONDER

NATALIE WOOD

FRANK LLOYD WRIGHT

WILBUR WRIGHT

"Everyone wants to be Cary Grant, even I want to be Cary Grant."

—Cary Grant, Capricorn Sun-Pisces Jupiter. January 18, 1904

"When you don't have this dying and becoming, you are only a sad guest on this earth."

—Johann Von Goethe, Virgo Sun-Pisces Jupiter. August 28, 1749

"Everyone is like a butterfly, they start out ugly and awkward and then morph into beautiful, graceful butterflies that everyone loves."

—Drew Barrymore, Pisces Sun-Pisces Jupiter. February 22, 1975

"Isn't she lovely? Isn't she wonderful? Isn't she precious? Made from our love?"

—Stevie Wonder, Taurus Sun-Pisces Jupiter. May 3, 1950

"I knew that, like Mike Mulligan and Mary Ann, I would always do better if people were watching."

—Jay Leno, Taurus Sun-Pisces Jupiter. April 28, 1950

"The world is not to be put in order. It is for us to put ourselves in unison with this order."

—Henry Miller, Capricorn Sun-Pisces Jupiter. December 26, 1891

"You can cage the singer but not the song."

—Harry Belafonte, Pisces Sun-Pisces Jupiter. March 1, 1927

"One must still have chaos in one to give birth to a dancing star."

—Friedrich Nietzsche, Libra Sun-Pisces Jupiter. October 15, 1844

"One has no right to love or hate anything if one has not acquired a thorough knowledge of its nature."

—Leonardo da Vinci, Taurus Sun-Pisces Jupiter. April 23, 1452

"Sure they love my singing. But me? I'm tired. I'm human. I want to be loved."

—Billie Holiday, Aries Sun-Pisces Jupiter. April 7, 1915

ACKNOWLEDGMENTS

First, I must praise Esmond Harmsworth, without whom Jupiter would have remained forever in the sky. His vision, humor, and dedication kept me going—in true Leo Jupiter style.

A debt of gratitude to the excellent folk at Penguin: Julie Saltman, Nancy Sheppard, Trena Keating, Emily Haynes, Gretchen Koss, and Acadia Wallace.

For help with research, copyediting, and helping this book go forward, my special thanks to Julie Mars, Whitney Woodward, Rema Badwan, Sandra Shagat, Jhennah St. Clair, Adrienne Zicht, Jennifer Gates, and Felissimo NY.

For invaluable help with ideas and editing, the brilliant George Shea.

For caring inspiration, Guy Barzilay, Sally Christian, Kristin Fontana, Linda Frasier, Jill Jarnow, Cassidy Phelps, and Mark Prezorski.

For emotional support throughout, and for those invaluable "Jupiter Interviews," my loving thanks to: Sara Bliss, Kathy and Liz Conrad, Deede Dickson, Sharon Eisenhauer, Cindy Gallop, Jeffrey Geiger, Marietta Gubuan, Wheeler Jackson, Pamela Keogh, Kay MacCaulay, Steven Manuel, Sarah Pollitt, Serge Raoul, Leta Rath, Elizabeth Redfield, Eva Saleh, Gabrielle Selz, Jennifer Simpson, Katherine Taylor, Quia, Juno and Jules, and last, but never least, Robert Haufrecht.

Above all, to Heidi Howell, Sheela Patra, Rebecca Bruce, and Deidre de Franceaux, the best girlfriends one could ever hope for.

Finally, to my mother, Donna Todd, without whom I could never have finished this book. For her love, belief, beauty, fun, irreverence, and double Aries drama, I am forever amazed and grateful.